The collection is distinctive in providing commentary on the role of commissions from 'outsiders', who approach the topic from an academic and critical perspective, as well as unique insights from the first-hand experience of members of the commissions themselves. The contributors are leading voices in debates on religious and cultural diversity, and they provide expert commentary on the four diversity commissions, as well as offering lively and at times competing accounts of how to achieve greater social cohesion. The result is a fascinating account and critique of the role of expert commissions in developing societal responses to diversity.

Professor Lucy Vickers, Oxford Brookes University

Public Commissions on Cultural and Religious Diversity

In several Western countries, expert commissions composed of academics, public figures, politicians and community organisers have been established by governments or civil society to reflect on the changes and challenges of an increasingly plural society. Commission recommendations on how to 'manage' diversities successfully have shaped national narratives and affected law and public policies, yet research on the workings of such commissions remains rare.

This book focuses on the experiences of expert commissions in the UK, France, Quebec and Belgium. Furthering the debate on commissions' potential and limitations it draws on the first-hand experiences and introspection of former commission members and close observers, along with outside perspectives and critique from independent scholars.

Building on its companion volume (*Public Commissions on Cultural and Religious Diversity: Analysis, Reception and Challenges*), this book engages with core concepts of identity, nationality, citizenship, freedom, equality and accommodation. It will appeal to researchers and students of public policy, sociology, anthropology, law, religion, politics, history and migration studies, as well as policymakers and anyone with a general interest in current debates on ethnic, cultural and religious diversity.

Katayoun Alidadi is Assistant Professor of Legal Studies at Bryant University in Smithfield, Rhode Island, and a Research Associate at the Max Planck Institute for Social Anthropology in Halle, Germany.

Marie-Claire Foblets is Director of the Department of Law and Anthropology at the Max Planck Institute for Social Anthropology in Halle/Saale, Germany and Professor of Law at the Catholic University of Leuven, Belgium.

Public Commissions on Cultural and Religious Diversity

National Narratives, Multiple Identities and Minorities

Edited by
Katayoun Alididi and
Marie-Claire Foblets

Routledge
Taylor & Francis Group
LONDON AND NEW YORK

First published 2018
by Routledge
2 Park Square, Milton Park, Abingdon, Oxon OX14 4RN

and by Routledge
711 Third Avenue, New York, NY 10017

Routledge is an imprint of the Taylor & Francis Group, an informa business

British Library Cataloguing in Publication Data
A catalogue record for this book is available from the British Library

Library of Congress Cataloging in Publication Data
Names: Foblets, Marie-Claire, 1959- editor. | Alidadi, Katayoun, editor.
Title: Public commissions on cultural and religious diversity. National narratives, multiple identities and minorities / edited by Marie-Claire Foblets and Katayoun Alidadi.
Description: Abingdon, Oxon ; New York, NY : Routledge, 2018. | Includes bibliographical references and index.
Identifiers: LCCN 2018000134| ISBN 9781472471758 (hardback) | ISBN 9781315602851 (e-book)
Subjects: LCSH: Cultural pluralism—Government policy. | Cultural pluralism—Government policy—Case studies. | Minorities—Government policy. | Minorities—Government policy—Case studies. | Religious minorities—Government policy. | Religious minorities—Government policy—Case studies. | Governmental investigations. | Governmental investigations—Case studies.
Classification: LCC HM1271 .P868 2018 | DDC 305.8—dc23
LC record available at https://lccn.loc.gov/2018000134

ISBN: 978-1-472-47175-8 (hbk)
ISBN: 978-1-315-60285-1 (ebk)

Typeset in Times New Roman
by Swales & Willis Ltd, Exeter, Devon, UK

Expert commissions in discussion – drawing by Fahimeh Elyas

Contents

List of figures and tables

Notes on contributors

Katayoun Alidadi is Assistant Professor of Legal Studies at Bryant University in Smithfield, Rhode Island, and a Research Associate at the Max Planck Institute for Social Anthropology in Halle, Germany. Her research focuses on comparative law, human rights and the intersections of law and religion. She is the author of *Religion, Equality and Employment in Europe: The Case for Reasonable Accommodation* (Hart, 2017) and co-editor of *A Test of Faith: Religious Diversity and Accommodation in the European Workplace* (Ashgate, 2012) and *Belief, Law and Politics: What Future for a Secular Europe?* (Ashgate, 2014).

Jean Baubérot is Honorary President at the Ecole Pratique des Hautes Etudes (EPHE) in Paris where he successively occupied the chair 'History and Sociology of Protestantism' and the chair 'History and Sociology of Secularism'. He is the author of over 20 books in the area of sociology of religions and secularism, including *Les sept laïcités françaises. Le modèle français de laïcité n'existe pas* (Maison des Sciences de l'Homme, 2015) and *La laïcité falsifiée* (La Découverte, 2012). He was a member of the French Stasi Commission (2003).

John R. Bowen is the Dunbar-Van Cleve Professor in Arts and Sciences at Washington University in St. Louis, and recurrent Visiting Professor at the London School of Economics. He has been studying Islam and society in Indonesia since the late 1970s, and since 2001 has worked in France, England and North America on problems of pluralism, law and religion, particularly contemporary efforts to rethink Islamic norms and civil law. His books include *Why the French Don't Like Headscarves* (Princeton University Press, 2007), *Can Islam be French?* (Princeton University Press, 2009), *A New Anthropology of Islam* (Cambridge University Press, 2012) and *Blaming Islam* (The MIT Press, 2012).

Patrick Charlier is Co-Director of the Belgian equality body Unia (the Interfederal Centre for Equal Opportunities, formerly called the Centre for Equal Opportunities and Opposition to Racism). From 2010 to 2015 he was a member of the Management Board of the Fundamental Rights Agency (FRA) in Vienna, and since 2007 he has been an alternate member of the European

Commission against Racism and Intolerance (ECRI) in Strasbourg. Unia's predecessor was tasked with following up the recommendations in the Belgian Round Tables on Interculturalism report.

Nathalie Denies is a jurist with Unia and the co-author, with Dounia Bouzar, of *Diversité convictionnelle: Comment l'appréhender? Comment la gérer?* (L'Harmattan, 2014).

Nadia Fadil is Professor of Anthropology at the Interculturalism, Migration and Minorities Research Centre (IMMRC) at the Catholic University of Leuven, Belgium. Her primary research interest pertains to the presence of Islam as a lived and embodied reality in Europe. She has published extensively on these issues in academic journals (such as *Social Anthropology*, *Feminist Review* and *Ethnicities*) and has contributed numerous book chapters to edited volumes. She has co-authored a book on this topic (*Leeuw in een Kooi. De Multiculturele verbeelding in Vlaanderen*, 2009). She was involved in the Belgian Round Tables on Interculturalism as a participant and moderator of the final commission discussions.

Marie-Claire Foblets is Director of the Department of Law and Anthropology at the Max Planck Institute for Social Anthropology in Halle/Saale, Germany and Professor of Law at the Catholic University of Leuven, Belgium. She has held various visiting professorships both within and outside Europe. She has published widely on issues of migration law, including the elaboration of European migration law after the Treaty of Amsterdam, citizenship/nationality laws, compulsory integration, anti-racism and non-discrimination, etc. In the field of anthropology of law, her research focuses on cultural diversity and legal practice, with a particular interest in the application of Islamic family law in Europe, and more recently in the accommodation of cultural and religious diversity under state law. From September 2009 to November 2010 she was the co-chair of the Round Tables on Interculturalism in Belgium.

Jean-François Gaudreault-DesBiens is Dean of the Faculty of Law at the Université de Montréal and Canada Research Chair in North American and Comparative Juridical and Cultural Identities. His research interests are constitutional law (domestic and comparative), civil liberties, legal theory and epistemology, and the sociology of legal cultures. His work currently focuses on the legal treatment of religious claims in multicultural liberal societies, on the relations between the civil law and common law traditions in a globalised economy, and on the legal theory of federalism.

Ralph Grillo is Emeritus Professor of Social Anthropology at the University of Sussex, where he was formerly Dean of the School of African and Asian Studies and founding director of the Research Centre for Culture, Development and the Environment. He has had a long-standing concern with transnational migration and ethnicity in Africa and Europe, but since the mid-1990s he has focused on cultural diversity and its governance in France, Italy and the UK. More recently he has been working with anthropologists, lawyers and political scientists on cultural and religious diversity and the law in Europe and North America. He has a particular interest in the 'legal industry' that has grown up around Islam, which is the subject of his book *Muslim Families, Politics and the Law: A Legal Industry in Multicultural Britain* (Ashgate, 2015). Other publications include *Pluralism and the Politics of Difference* (Oxford University Press, 1998) and, as (co-)editor, *The Politics of Recognizing Difference: Multiculturalism Italian Style* (Ashgate, 2002), *The Family in Question: Immigrant and Ethnic Minorities in Multicultural Europe* (Amsterdam University Press, 2008), *Legal Practice and Cultural Diversity* (Ashgate, 2009) and *Regulation of Speech in Multicultural Societies* (Routledge, 2015).

Christian Joppke holds a chair in sociology at the University of Bern, Switzerland. He is also a Visiting Professor in the Nationalism Studies Program at Central European University, Budapest, and Honorary Professor of Political Science, University of Aarhus, Denmark. He is a leading authority in the field of citizenship and immigration, lately with an interest in religion. His most recent books are *The Secular State Under Siege: Religion and Politics in Europe and America* (Polity Press, 2015) and *Is Multiculturalism Dead? Crisis and Persistence in the Constitutional State* (Polity Press, 2017).

Sébastien Lacroix is a PhD candidate in Philosophy (Université Laval, Canada). His doctoral research project is interdisciplinary in nature, encompassing reflections from legal philosophy, political philosophy and Canadian constitutional law. Specialising in the numerous questions surrounding the Hart-Dworkin debate in philosophy of law, Sébastien has interests in issues regarding multiculturalism, nationalism, constitutionalism and judicial review, vagueness in the law, and metaethics.

Geoffrey Brahm Levey is Associate Professor and an Australian Research Council Future Fellow in Political Science at the University of New South Wales, Sydney. The author of many journal and book contributions, his recent publications include, as editor, *Authenticity, Autonomy and Multiculturalism* (Routledge, 2015) and *The Politics of Citizenship in Immigrant Democracies: The Experience of the United States, Canada and Australia* (Routledge, 2015). He is also the editor of *Political Theory and Australian Multiculturalism* (Berghahn, 2012, 2008) and a co-editor of *Secularism, Religion and Multicultural Citizenship* (Cambridge University Press, 2008).

Karel J. Leyva is a PhD candidate under the joint supervision of the University of Montréal and the École pratique de hautes études (Paris-Sorbonne). He is a researcher for the international project 'Cultural and Religious Diversity in Four National Contexts: A Comparative Study of the Dynamics of Identity and the Regulation of Religion', directed by the Chair in Management of Cultural and Religious Diversity at the University of Montréal. His recent publications include 'Multiculturalisme et laïcité en France: les trois républicanismes du rapport Stasi' in *Dialogue: Canadian Philosophical Review* (2015) and 'Debating Intercultural Integration: From the Commission for Intercultural Dialogue to the Round Tables on Interculturalism' in *Public Commissions on Cultural and*

Religious Diversity: Analysis, Reception, and Challenges (2017). His research focuses on liberal and republican responses to the challenges posed by cultural and religious diversity. He is currently an advisor to the Canadian federal public service on policies regarding diversity and inclusion.

Jocelyn Maclure is Professor of Philosophy and co-holder of the research chair La philosophie dans le monde actuel at Université Laval (Quebec, Canada). His publications include *Secularism and Freedom of Conscience* (Harvard University Press, 2011, co-authored with Charles Taylor), and *Quebec Identity: The Challenge of Pluralism* (McGill-Queen's University Press, 2003). He was an expert analyst for the Bouchard–Taylor Commission in 2007–2008, and is currently advising the Quebec Department of Immigration, Diversity and Inclusion on a new policy statement on interculturalism.

Maleiha Malik is Professor of Law at King's College, London. Maleiha Malik's research focuses on the theory and practice of discrimination law. She has written extensively on discrimination law, minority protection and feminist theory. She is the co-author of a leading text titled *Discrimination Law: Theory and Context* (Sweet & Maxwell, 2008). She is, along with Dr Jon Wilson from the Department of History at KCL, the co-ordinator of the AHRC project on 'Traditions in the Present', which explores the relevance of 'tradition' in contemporary societies. Her current research focuses on the intersection between sexual and cultural equality, and it explores the adjustments that may need to be made to feminist theory to accommodate increasing cultural pluralism. She was a member of the Commission on Religion and Belief in British Public Life (CORB, 2015).

Tariq Modood is Professor of Sociology, Politics and Public Policy, the founding Director of the Centre for the Study of Ethnicity and Citizenship at the University of Bristol and the co-founder of the international journal *Ethnicities*. He was awarded a MBE for services to social sciences and ethnic relations in 2001, was made a Fellow of the Academy of Social Sciences (UK) in 2004 and elected a Fellow of the British Academy in 2017. He served on the Commission on the Future of Multi-Ethnic Britain, the National Equality Panel and the Commission on Religion and Belief in British Public Life. His latest books include *Multiculturalism: A Civic Idea* (2nd edn; 2013); and, as co-editor, *Multiculturalism Rethought* (2015), *Multiculturalism and Interculturalism: Debating the Dividing Lines* (2016) and *The Problem of Religious Diversity: European Problems, Asian Challenges* (2017).

Bhikhu Parekh is Emeritus Professor of Political Philosophy at the University of Westminster, a fellow of the Royal Society of Arts and of the Academy of the Learned Societies for Social Sciences. He has written about political thinkers as diverse as Jeremy Bentham, Karl Marx, Hannah Arendt, Michael Okeshott and Isaiah Berlin. Since the 1970s he has written and published extensively on multiculturalism and post-colonialism, including two highly influential books, *Rethinking Multiculturalism* and *A New Politics of Identity*. Parekh served on the Rampton and Swann Commissions, and was Deputy Chair and later acting Chair of the Commission for Racial Equality. He was Chair of the

Runnymede Commission for Multi-Ethnic Britain (1998–2000), and oversaw the production of the Commission's report, *The Future of Multi-Ethnic Britain* (2000). He was also involved, as a patron, with the work of the Woolf Institute Commission on Religion and Belief in British Public Life (CORB, 2015).

Eugenia Relaño Pastor is Senior Research Fellow in the Law & Anthropology Department at the Max Planck Institute of Social Anthropology in Halle (Germany) and has worked as Legal Adviser in the Migration and Equal Treatment Area at the Spanish Ombudsman. She holds a doctorate in Law and bachelor degrees in Political Science and Sociology. Her research expertise includes international religious freedom, comparative law, religious minorities, equality, multiculturalism and immigration. Her recent publications include 'Religious Pluralism in Liberal Democracies' (*Islam, Human Rights and Secular Values*, Liberte, 2012); 'Etudes de cas de discriminations religieuses en Espagne' (*Les discriminations religieuses en Europe: droit et pratiques*, L'Harmattan, 2012); 'Participation of Muslim Minorities in the Spanish Mainstream Society' (*The Interrelation between the Right to Identity of Minorities and Their Socio-Economic Participation*, Martinus Nijhoff, 2013).

Varun Uberoi is Senior Lecturer in Political Theory and Public Policy at Brunel University. His research focuses on political theories of multiculturalism, nations and nationalism. He has published widely on how to conceptualise multiculturalism, national identity, unity in culturally diverse societies, government policies of multiculturalism and nation-building policies. His most recent book is an edited collection with Tariq Modood entitled *Multiculturalism Rethought* (Edinburgh University Press, 2015). His recent articles include 'The Parekh Report: National Identities without Nations and Nationalism' (*Ethnicities*, 2015) and 'Legislating Multiculturalism and Nationhood' (*Canadian Journal of Political Science*, 2016).

Patrick Weil is Senior Research Fellow at the French National Research Center in the University of Paris 1, Pantheon–Sorbonne and was a Visiting Professor of Law and Peter and Patricia Gruber Fellow in Global Justice at Yale Law School. His work focuses on comparative immigration, citizenship, and church–state law and policy. His most recent book is *The Sovereign Citizen: Denaturalization and the Origins of the American Republic* (University of Pennsylvania Press, 2013). Among his other recent publications are 'Citizenship, Passports, and the Legal Identity of Americans: Edward Snowden and Others Have a Case in the Courts' (*Yale Law Journal Forum*, 2014); 'Headscarf versus Burqa: Two French Bans with Different Meanings' (*Constitutional Secularism in an Age of Religious Revival*, Susanna Mancini and Michel Rosenfeld ed., Oxford, 2014) and 'From Conditional to Secured and Sovereign: The New Strategic Link between the Citizen and the Nation–State in a Globalized World' (*International Journal of Constitutional Law*, 2011). In 1997, he completed a mission and a report on immigration and nationality policy reform for Prime Minister Lionel Jospin, which led to the implementation of new immigration and citizenship laws adopted the following year. Professor Weil was a member of the French Stasi Commission (2003) established by Jacques Chirac.

Foreword

Commissions set up to investigate and propose ways of dealing with issues of public significance are a modern phenomenon going back no further than the nineteenth century. They require a particular kind of political culture which did not exist earlier. They presuppose that the government policy should be based on the best available knowledge, that the prevailing diversity of views on a subject should be narrowed and hopefully reconciled, that the citizens have a stake in and views on how an issue is decided and are not prepared to leave it in the care of the government, and so on. Made up of experts in the relevant area, commissions were expected to produce reports making recommendations to guide the government, educate the public, and raise the level of public debate. As more and more of their preconditions became historically available, commissions began to proliferate. Initially set up and funded by the government, they later came to be set up by non-governmental bodies or groups of private individuals anxious to highlight a particular injustice or investigate a particular situation, as with the Jean-Paul Sartre commission to investigate the American actions in Vietnam or the citizens'commission to investigate the extent and causes of massive anti-Muslim violence in the Indian state of Gujarat.

Commissions vary greatly in their nature and objectives. Some are triggered by a particular crisis and intended to suggest ways of dealing with it. Others are proactive and deal with large issues long before they become urgent. Both types are reflective and make policy recommendations, but differ in respect to which of these is more dominant and forms its focus. Some commissions again begin with a broad agreement on their goal and are concerned to interpret and realise it in different situations. Others are expected to arbitrate between different goals, introduce greater clarity about the intended objective, and are more discursive in nature. All commissions function within the broad limits set by the prevailing public opinion as otherwise their recommendation would lack popular support. However, some may stretch these limits in various degrees and become controversial, whereas others are far less ambitious. Unless it is poorly led or dealing with a highly contentious issue, no commission is so deeply divided as to break up without an agreement.

Sometimes commissions have overrepresentation of minorities, which might even equal that of the majority. Such commissions are used by their minority representatives either to push their view and get the majority to agree to or to reach a consensus with the majority. In the latter case there is a danger of the minority community in the society at large feeling dissatisfied with the final report because it does not go as far as the minority would like it. In the former case the report would be out of step with the dominant public opinion and please the minority but alienate the majority. All commissions need to guard against these dangers.

The present volume is a fascinating and long overdue study of the ways in which commissions get set up, function, run into difficulties, produce their reports and have short-term and long-term impacts. It selects four commissions from as many countries, namely the Runnymede Commission in the UK (also called the Parekh Commission), the Stasi Commission in France, the Bouchard-Taylor Commission in Quebec and the Foblets-Kulakowski Commission in Belgium, all addressing common issues thrown up by our culturally and religiously diverse societies.[1] Although all but one of these were set up by the government, they all maintained their distance from it. They addressed both the government and their fellow-citizens, discussed general principles as well as practical issues, and contributed to a broad national consensus at each level. Their recommendations initially provoked negative reactions among the media and the political class, some more than others, but in each case these died down after a while and were sometimes replaced by a sympathetic response. This is as true of the Runnymede Commission, the target of the greatest hostility, as of the Stasi Commission, which enjoyed the greatest public support.

All four commissions, which produced their reports during the first decade of the present century, addressed closely related issues and came up with broadly similar but differently accented and contextualised answers. The Runnymede Commission had the widest remit and made the largest set of recommendations covering almost all areas of British public life, including what it means to be British and why 'Britishness' involves not just a common body of values but also a shared national narrative. The Stasi Commission had the limited objective of articulating the French concept of *laïcité* and what it entailed. The Bouchard-Taylor Inquiry's remit was wider than the Stasi but narrower than Parekh; it explored the role of religion in public life, implications of equal treatment of different religions, and more generally the basis and content of reasonable accommodation. The Foblets-Kulakowski inquiry had a broadly similar remit to Bouchard-Taylor, though a little narrower, and was concerned with the religious neutrality of public authorities in Belgium but also more generally. Each of these commissions is distinguished by certain core ideas which it articulates in its own national context, but whose implications are wider and, when judiciously used, can be sources of guidance in other countries.

In their own different ways, the four commissions have several assumptions in common. They all accept cultural and religious diversity as a stubborn and inescapable fact of modern society, though with different degrees of enthusiasm. Although their reasons vary, they all agree, some such as Parekh more readily than others, that minority differences should be respected. They also agree that respect for differences should go hand in hand with the fostering of national unity so that neither unity nor diversity is severely compromised. For all of them there are limits to the differences a diverse society can and should tolerate, and these are set by the body of common values arising out of the society's own historically evolving moral identity. This gives some pre-eminence to the voice of the majority and the prevailing moral structure. Beyond a certain point this cannot be helped because no society can jump out of its skin or bracket out its identity and face its minorities with a clean slate. And nor is it wrong, because a society has a right to its identity, to its past and its history, just as much as and even perhaps more than the minorities have to theirs. It is, of course, crucial that the majority's values should be inclusive, rationally defensible and amenable to revision when shown to put the minorities at an unfair disadvantage. The four commissions show in their own different ways how to broaden the society's values without rendering them incoherent and how in general to integrate its past with its inevitably more plural future.

Contributors to this excellent volume discuss the four commissions and the core insights of their reports with considerable skill and insight, and enrich the ongoing global conversation on how to meet the demands of unity and diversity in a fair and collectively beneficial manner. Sometimes directly but more often through the way they structure their essays, the authors also indicate how such commissions should function in future and arrive at and present their recommendations, valuable advice in an age in which the role of such commissions is bound to increase in response to the widely felt need to deepen and regenerate our democratic culture. I know of no other collection that offers the kind of comparative analysis this volume does, let alone one that does it so well.

Bhikhu Parekh
House of Lords

Note

1 They were not all called commissions. I use this term broadly to refer to all public inquiries of the kind described above.

Acknowledgements

This volume presents a mix of retrospection and critique on the work of four diversity expert commissions in Britain, France, Quebec and Belgium. Most of the contributions in this collective volume were presented at an international conference titled *Religious and Cultural Diversity in Four National Contexts*, convened by the Department of Law and Anthropology of the Max Planck Institute for Social Anthropology on 26 and 27 June 2015 in Halle/Saale, Germany. A few chapters were solicited afterwards to establish a balancing attention across the four commissions and experiences. The goal from the start was to document insights of both formal 'insiders' and 'outsiders' to these commissions, and we want to thank all contributors who shared their experiences and analyses at the conference. It should be noted that 'insiders' who share in their chapters the dynamics of negotiation and compromise that inevitably take place behind the scenes all speak in their own name, and their views may not necessarily reflect official statements coming from the commission or be attributed to other commission members. We thank Jean Baubérot, John R. Bowen, Patrick Charlier, Nathalie Denies, Nadia Fadil, Jean-François Gaudreault-DesBiens, Ralph Grillo, Christian Joppke, Sébastien Lacroix, Karel J. Leyva, Geoffrey Brahm Levey, Jocelyn Maclure, Maleiha Malik, Tariq Modood, Eugenia Relaño Pastor, Varun Uberoi and Patrick Weil for contributing to this collective volume. It was a pleasure working with all of our contributors, and we hope we can build on these collaborations in the near future. Lord Bhikhu Parekh was so gracious as to write a highly insightful foreword to this volume. We are very grateful to him for his time and engagement.

The conference also included various discussants and conference participants who, though they may not appear as individual chapter contributors, made the discussions rich and passionate and rendered the resulting chapters and our introduction to the volume more sharp, nuanced, and complete. These include Peter Beyer, Joseph David, Alison Dundes Renteln, Anver Emon, Mark Goodale, Michele Graziadei, Marie-Eve Larivière, Solange Lefebvre, Werner Menski, Vladimiro Zagrebelsky and Olaf Zenker.

We would like to thank all members and service staff of the MPI institute, particularly Ute Gradmann and Kateřina Marencakova, for making sure the conference, with many participants taking international flights and during a hectic time at the Institute, ran as smoothly as could be. We also thank the indefatigable Dr Brian Donahoe, the Senior Editor at the Max Planck Institute for Social Anthropology, for his meticulous work in preparing the book proposal and coordinating the author contributions. Gita Rajan reviewed and language-edited a number of chapters in this volume. Iranian-Belgian artist Fahimeh Elyas generously created a commissioned visual representation of the challenges of 'commission work'.

This book, together with its companion volume *Public Commissions on Cultural and Religious Diversity: Analysis, Reception, and Challenges*, edited by Solange Lefebvre and Patrice Brodeur (Routledge 2017), is part of a joint research project funded by the Canadian Social Sciences and Humanities Research Council (SSHRC). Thanks and appreciation for the ongoing collaboration go to the investigators (Solange Lefebvre, Jean-François Gaudreault-DesBiens) and members, namely, Lori Beaman, Peter Beyer, Céline Béraud, Tariq Modood and James Beckford.

At Routledge, we want to thank Claire Maloney, Editorial Assistant for Politics and International Relations, and Robert Sorsby, Editor for Politics and International Relations, for their patience, invaluable advice and support while we were preparing the manuscript for production. We also thank production editors Zoe Everitt, at Taylor and Francis, and Julian Webb, at Swales & Willis, and copy-editing freelancer Jonathan Hoare for their meticulous work and for making sure the book production process stayed on schedule and progressed without hiccups.

Last but not least, we thank our families for the love, support and patience that we often take for granted but without which this book could not have been the same.

Katayoun Alidadi and Marie-Claire Foblets
January 2018

List of abbreviations

Bouchard-Taylor Commission: The Consultation Commission on Accommodation Practices Related to Cultural Differences (Quebec, 2008)

CJEU: Court of Justice of the European Union

CMEB: Commission for Multi-Ethnic Britain (UK, 2000)

CORB/CORAB: Commission on Religion and Belief in British Public Life (UK)

ECHR: European Convention on Human Rights

ECtHR: European Court of Human Rights

Foblets-Kulakowski Commission: Belgian/Federal Round Tables on Interculturalism (Belgium, 2010)

Gerin Commission (Parliamentary Commission to Study the Wearing of the Full Veil in France): *La Mission d'information sur la pratique du port du voile intégral sur le territoire national* (2010, France)

Parekh Commission: Commission for Multi-Ethnic Britain (UK, 2000)

Stasi Commission: *La Commission de reflexion sur l'application du principe de laïcité dans la Republique* (France, 2003)

Unia (Belgium): Interfederal Centre for Equal Opportunities (formerly the Centre for Equal Opportunities and Opposition to Racism)

List of abbreviations

[illegible]

1 Introduction

A retrospective on diversity commissions: from set-up and working methods to impacting policy

Katayoun Alidadi and Marie-Claire Foblets

The perceived need to *manage* cultural and religious diversity has been felt particularly in several Western countries, where relations between the historically entrenched and culturally dominant population and more recent inhabitants from increasingly diverse ethnic, cultural and religious backgrounds have created or exacerbated tensions and conflict. How can peaceful co-existence, tolerance and social cohesion be advanced in such contexts? Various Western countries have turned to a particular tool of public policy and set up commissions, often composed of academics, public figures, politicians and community organisers (in short: 'expert commissions'[1]), to reflect on the changes and challenges of an increasingly plural constituency in terms of culture, ethnicity/race and/or religion and to offer recommendations to governments and institutions on how to 'govern' or 'manage' such diversities successfully. The main motives behind establishing such commissions can thus be to promote public reflection and consultation and to collect evidence-based policy recommendations, but they can also be aimed at appeasing tensions or legitimating certain laws or policies that can be either accommodating or rather restrictive[2] towards minorities.

This volume focuses on four recent and prominent experiences of expert commissions, namely those of the UK, France, Quebec and Belgium. Listed in chronological order, and named after the respective chair(s) of the commissions, these include the Parekh Commission (Commission for Multi-Ethnic Britain, report issued 2000), the Stasi Commission (France, report issued 2003), the Bouchard-Taylor Commission (Quebec, report issued 2008) and the Foblets-Kulakowski Commission (Belgian Round Tables on Interculturalism, report issued 2010). The experiences with expert commissions in these four Western democracies form the distinct focus of this volume. To be sure, commissions set up by governments or civil society organisations to investigate and take stock of challenges surrounding religious, cultural and/or ethnic diversity form a distinct phenomenon of our time.[3] In order to present a mix of retrospection and critique, with the benefit of hindsight, and in keeping with the aims of the conference at which preliminary versions of the essays were presented,[4] this volume documents insights of both formal 'insiders'[5] and 'outsiders' to these commissions. The narratives, insights and lessons included will be of interest to students, researchers, and policy-makers, as well as to the wider public interested in learning about the nature and impact of expert commissions or knowing more about the dynamics of negotiation and compromise that inevitably take place behind the scenes.

Together with its companion volume, *Public Commissions on Cultural and Religious Diversity: Analysis, Reception, and Challenges* (Routledge, 2017), edited by Solange Lefebvre and Patrice Brodeur,[6] this volume presents the findings of a comparative research project, entitled *Religious and Cultural Diversity in Four National Contexts: Comparative Study of the Dynamics of Identity and the Regulation of Religion.*[7] Both volumes share their topical inquiry into the phenomenon of expert commissions and their relevance and value for the *management* of multiple identities and minority relations in a post-multicultural context.[8] Yet the objective and approach of each is distinct. The first volume includes in-depth studies of the issues and controversies examined by the four said commissions as well as similar commissions in Morocco, Singapore, Norway and Australia, in order to analyse, *inter alia*, how issues were perceived, what results and impact the commissions had and how the media covered the reports. In contrast, the current volume zeroes in on the distinct British, French, Quebec and Belgian experiences, comparing these in time and space, with the emphasis being less on an exegesis of the reports than on their core contributions to shaping or shifting national narratives, (critical) insider insights, as well as some hardships and pitfalls that were key for lessons to be learned for the future. Together, these volumes raise and explore key issues surrounding the use and value of the oft-used but under-analysed and under-theorised method of deploying commissions to address the challenges of diversity today.

Towards a better understanding of the role of expert commissions

The work and the final reports of expert commissions often provide detailed policy recommendations, grounded in elaborated theoretical bases, and offer much-needed insights and inspiration to other democracies faced with some of the greatest challenges to equality, fairness, and justice today. Indeed, important lessons can be drawn from the experiences of various countries where expert commissions have assessed the situation and the underlying problems, listened to the public's opinions and perspectives on the ground and, on the basis of this valuable work, provide recommendations on how to move forward and ideally create opportunities out of the challenges presented. Nevertheless, expert commissions should not be regarded as the *deus ex machina* of modern nations, a guaranteed technique that produces perfect and sustainable blueprints which, if followed, will solve all cultural and socio-economic issues in society. Thus, in analysing such commission experiences, which involve inevitable limitations, there may be ways to move forward in a more promising fashion. Members of commissions are well placed to consider what could have been done differently. Perhaps, listening across borders will enable new commissions to emulate promising aspects and prevent similar problems and pitfalls, for instance how to prevent misconstruction of general approaches or specific recommendations by either politicians or the media, and facilitate the 'success' (however defined) of the important work generally done in these commissions. That was our aim in pursuing this project.

What were the experiences of members who were asked to chair or join these commissions? How did they handle the attention, critique and/or praise, and what lessons did they draw for themselves? How were these reports received and perceived, both at the time of their issuance as well as years later? What happened to the policy recommendations, carefully considered and crafted but nonetheless sometimes misunderstood and/or misrepresented? Do such expert commissions, both individually and taken as a whole, fulfil the promise of addressing the challenges of multicultural diversities, however defined or framed in the particular time frame and national context, and offer guidelines from which other countries going through transition, transformation and crisis could benefit? Such questions inform our analysis of expert commissions in the four national contexts, about which misconceptions abound. Some contributions in this volume are written by former members (or close insiders) of one of the four said expert commissions. In other chapters, we hear from experts, who – with the benefit of hindsight – comment on, analyse and critique the approaches taken and the proposals advanced in the respective socio-political and institutional contexts. It is the combination of these insights that produces a fruitful debate.

The UK, France, Quebec and Belgium as case studies

There are many similarities but also differences between commissions. For one, expert commissions can be set-up by governments or state institutions. They can also be rooted in civil society. While the French, Quebec and Belgian commissions were initiated by the respective governments, the Parekh Commission, also known as the Commission for Multi-Ethnic Britain (CMEB, 2000), was set up by the Runnymede Trust, a non-departmental public body in the UK (where there is a long tradition of using Royal Commissions to address particular public concerns). The Parekh Commission was a reflective rather than a reactive commission; in other words, it was not set up in response to a particular crisis or controversy, but rather meant to move the discussion on national identity and minority relations forward. This forms a point of contrast with the Stasi Commission and the Bouchard-Taylor Commission, both of which were triggered by particular precipitating events.[9] The Belgian Round Tables on Interculturalism (and CORB, another example addressed in this volume) were also established to address various challenges in a broad and long-term, less immediate fashion.[10]

Differences as well as commonalities are also reflected in the key topics addressed in the respective reports. Overall, expert commissions place emphasis on those questions that appear the most relevant, urgent or delicate in the context of their particular country. These topics differ considerably: in France, it was the question of upholding the principle of *laïcité* that was the focus of attention, while in Québec, it was the question of reasonable accommodation. In the UK, the

Parekh Report was interested chiefly in the question of identity ('Britishness') and, lastly, in Belgium, the recommendations gave priority to the thorny question of the neutrality of public authorities.

The composition of the commissions varies as well, and the contributions brought together in this volume make it clear that this can make a crucial difference. In some cases members from across ethnic, cultural and religious lines were included in order to increase representativeness and/or legitimacy. We will return to this issue. Not only are the identity and affiliation of members significant, but also the size of the commission, since it can only be expected that not all members will agree on all issues. Even a commission of two will necessitate compromise. For instance, the Bouchard-Taylor Commission consisted of the two prominent co-chairs, Gerard Bouchard, a historian, and Charles Taylor, a philosopher, who co-signed the report. While there was an advisory committee of 15 experts, these experts are not signatories to the report. And even here there are *post facto* signs that some commission recommendations were a compromise between the co-chairs (see Chapter 9 by Maclure and Lacroix). In larger commissions, the level of disagreement and the way it is handled and allowed to sift through to the final report will differ. For example, the Belgian Round Tables Report shows signs of a rupture between different ideological stances that could not be bridged on core issues, and two dissenting opinions are annexed to the report. One could read the Stasi Report as representing the consensus reached by all members, but upon closer look, there, too, was dissent. The Stasi Commission's 20 members all agreed on the importance of the foundational French concept of *laïcité,* but held divergent views concerning its interpretation and application.

The French Stasi Commission was the only one of the four commissions headed by (and named after) a politician rather than an academic, i.e. the late Bernard Stasi, a former ombudsman of the (French) Republic (*médiateur de la République*). The Parekh Commission was chaired by Lord Bhikhu Parekh and comprised 22 members, with Tariq Modood as advisor. It should be clear that chairs (and members) can provide credibility and legitimacy, as well as access to resources. Budgets and duration of the commission proceedings (turnover) are other points to consider. Each commission deliberated for varying lengths of time: two years for the Parekh Commission, six months for the Stasi Commission, and one year each for the Bouchard-Taylor and the Belgian Round Tables. While the Parekh Commission Report is over 400 pages, and the Bouchard-Taylor Report comes in second at over 300 pages, the Stasi Report is 78 pages and the Belgian Round Tables report is 68 pages.

The work of the commissions in the four countries has suggested areas of convergence that could prove relevant for the aims of other countries that are facing new manifestations of cultural and religious diversity that will need to be accommodated in one way or another. The project's comparison of the findings of the commissions follows two main lines of investigation: first, the identification of the controversies concerning collective identity and ethno-religious diversity, and second, the examination of the management and regulation of cultural and religious diversity on a country-by-country basis. All commissions derived their mandate from the given constitutional and legal frameworks, which include fundamental rights to equal treatment, freedom of expression, and freedom of religion or belief, but they take into account the socio-political context and the challenges presented, in order to (re)interpret norms and values and to give them a more concrete meaning.

Overview of chapters

Part I. The importance of Britishness: The Parekh Commission Report and beyond

The Parekh Commission, whatever its shortcomings, is widely considered to have moved forward the discussion on national identity and the contested idea of Britishness, as Tariq Modood, Varun Uberoi and Ralph Grillo show in their respective essays. In the UK, where there has been no shortage of commissions, the Parekh Commission constitutes a significant juncture. It advanced understandings or recommendations that later, against some odds, became commonplace in law, policy or practice. The Parekh Commission can be separated from the three other commissions because of its timing. Having issued its report before 9/11, it predates the 'turn towards religion' in the public sphere and minority narratives, which strongly feature in the reports of the Stasi, Bouchard-Taylor and Foblets-Kulakowski commissions. Indeed, taken as a whole and in that timeline, the main concern has shifted from ethnic and cultural diversity to religious diversity, as Maleiha Malik notes in her essay on the more recent Commission on Religion and Belief in British Public Life (CORB), of which she was a member.

Tariq Modood's essay considers the place of religion in relation to post-immigration diversity in Britain and British national identity, as reflected in three national commission reports stretching over 30 years, namely the Swann Report (1985) and the reports of the Commission on Multi-Ethnic Britain (CMEB, 2000) and the Commission on Religion and Belief in British Public Life (CORB, 2015). While the Parekh Commission explicitly included religious identification within its conceptualisation of 'multi-ethnic' (up to that point unprecedented for a report framed under racial equality), it remained ambivalent about whether Britain was a plural nation or a post-nation. Reactions to the CMEB report were fierce, with it being called unpatriotic and out of touch. Yet years later, many of the recommendations in the report were adopted, in particular those relating to religion in the public sphere (with the establishment of CORB), and subsequent social changes brought about shifts in discussions of British national identity, which became important for both majorities and minorities.

Varun Uberoi argues that the Parekh Commission Report reflects a valuable way to think about national identities, and illustrates in particular an evolution in Bhikhu Parekh's thinking about the nature and worth of national identities, which has been documented in some lesser-known publications from the 1970s onwards. The concept of national identity need not imply the existence of a nation in the sense of a morally, culturally and ethnically uniform polity, as few policies today are culturally uniform (hence the idea of Britain as a post-nation). As Uberoi writes, the CMEB's Parekhian approach to national identity, which can be captured in the idea of national identity without a nation, offers a way to envisage the future of a Britain to which minorities can fully belong. Ralph Grillo, in his wide-ranging essay, explores the shift from 'the relatively benign framework for the accommodation of minorities . . . giving way to "muscular liberalism"'. While his well-documented observations point to a pessimistic outcome for the accommodation of religious and cultural diversity in Britain, British multiculturalism has historically taken shape through what has at times been 'a painful process of dialogue and negotiation'. There are reasons for optimism, as Grillo emphasises that 'pragmatic quotidien accommodation is still apparent on the ground and in legal practice'. The question is whether tradition can be sustained in an era which lacks sympathy with 'otherness' and looks to criminalise it. The latest British commission, CORB, is the subject of Maleiha Malik's essay. As Malik writes, CORB's terms of reference can be distinguished from Parekh's, as the aim was to address 'how to re-imagine religion and belief in public life rather than issues of migration or minorities'. Malik registers another important shift (from the Parekh Commission) that involves regulating religious diversity, including through soft law mechanisms, as she aims her focus on minorities and gender relations.

Part II. **Laïcité** *facing religious diversity: France's Stasi Commission and its legacy*

Turning to France, Patrick Weil, a member of this presidential commission that was tasked with reflecting on the significance of the French concept of *laïcité* in the midst of integration struggles, recalls the public reception of the Stasi Commission Report and notes a striking difference between the political and the academic reactions. Whereas, for many academics, the Stasi Commission became a symbol of illiberal French policies, for others the proposal to adopt a ban on religious symbols in public schools did not go far enough. Weil offers insights into the recommendations of the Stasi Commission, including an idea he launched on pluralising the public holiday schedule. Further insight can be derived on the Stasi Commission from the fruitful dialogue included in this volume between Jean Baubérot, a member of the Stasi Commission and the only one to abstain from signing the final report, and John R. Bowen, an American anthropologist who happens to be a 'distant' outsider yet astute observer of the French debates on religion, law and politics. Baubérot compares the working method of the commission to that of a funnel ('*entonnoir*'): starting with larger societal challenges and ultimately arriving at the proposition of a law on the wearing of religious symbols in public schools, which culminated in the ban with which the commission became synonymous. Bowen, however, notes that from the outside it appeared that the commission's agenda did not really revolve around a general reflection on French secularity (in which case, other matters such as state subsidies or the place of religion in national curricula would be discussed), but rather with 'problems in and around poor schools that are linked to the rise of Islam'. Yet for their parts, Weil and Baubérot both emphasise not only the depth but also the breadth of the Stasi proposals – 26 in all – and lament to a certain extent that this fact remains unrecognised and unappreciated in the common treatment of the Stasi Commission. The essay by Karel J. Leyva, a researcher of a new generation, offers much-needed commentary on the composite character of the Stasi Commission Report. Leyva compares the Stasi Commission Report and the Belgian *Commission du dialogue interculturel* (Commission for Intercultural Dialogue (2005)), which was the predecessor to the Round Tables on Interculturalism (2010). In doing so, Leyva adopts an analytical framework of two ideal-typical republican positions. Leyva notes that the term 'communitarian' was used to discredit the reports produced by both Belgian commissions, yet he argues this assertion is tenuous at best.

Part III. Reasonable accommodations revisited: Quebec's Bouchard-Taylor Commission

A third commission – Quebec's Consultation Commission on Accommodation Practices Related to Cultural Differences – is the subject of essays by Jocelyn Maclure, who was an adviser to the commission, Sébastien Lacroix, and Jean-François Gaudreault-DesBiens. The Bouchard-Taylor Commission has enjoyed considerable attention abroad. Jean Baubérot, in his part, remarked that, despite piercing criticism, he considered the Bouchard-Taylor Commission to have functioned better than the Stasi Commission, referring to significant differences in budget and turnover timeline (luxuries, he notes, the Stasi Commission lacked). In 2006–2007, Quebec was the site of heated debates on the accommodation of religious diversity, triggered to a significant degree by the Canadian Supreme Court's *Multani* decision involving a Sikh schoolboy who sought to wear a kirpan at school.[11] Maclure and Lacroix note that many years later, there is near unanimous public opposition to the Supreme Court's decision to allow the boy, under certain conditions, to wear the kirpan at school. The decision is considered not only judicially activist, but also an imposition of Canadian-style multiculturalism, which risks conflicting with Quebec's interculturalism. It was against this backdrop that the Bouchard-Taylor Commission was perceived as a 'consultation commission'. Its final report, titled *Building the Future: A Time for Reconciliation*, issued in June 2008, concluded that the 'reasonable accommodation crisis' was a 'crisis of perception' rather than a crisis at the level of practice or day-to-day encounters.[12] Yet, following the formation of the commission, cultural anxiety would soon resurface and become politicised in Quebec, including in the divisive Quebec Charter of Values debate.[13] Thus, as Maclure and Lacroix note, the debate on religious accommodation and secularism continues and may once again come to the fore in the 2018 provincial elections.

In his comment on the essay by Maclure and Lacroix, Jean-François Gaudreault-DesBiens utilises the apt concept of 'small nations' to address the public anxiety in the context of religious diversity in Quebec. By 'small nations', Gaudreault-DesBiens refers to 'nations that share a sentiment of cultural anxiety or that repeatedly find themselves confronted with the same existential problems'. Such societies need not be geographically 'small' countries – indeed some large nations too have behaved as 'small nations' recently – or even nation states *per se*, which makes the notion relevant to the Quebec (as well as Flemish) context. In such contexts, where the 'Other' is seen as a threat to the nation's existence but also provides the basis for the self-definition of the nation, Gaudreault-DesBiens warns that 'fear fosters, or reinforces, social cohesion', which explains the title of his essay: 'Cohesive fear'. He places the debates over reasonable accommodation in Quebec within this framework, but it should be clear that cohesive fear is 'a phenomenon that can easily be instrumentalised by populist movements of all stripes' in the West and beyond.

Part IV. Belgium's experiment with interculturalism: The Round Tables experience

Some of the same dynamics that could be observed in Quebec were present in the Belgian context, where some opinion makers relied heavily, though selectively, on the Bouchard-Taylor Commission findings. Marie-Claire Foblets, one of the chairs of the Belgian Round Tables on Interculturalism, describes her experiences with expert commissions in a context where expectations were high and criticism was ongoing and fierce. With *la politique du compromis* as the distinct Belgian political export product, the dynamics of compromise figured prominently in the internal commission debates. Yet even here some differences in point of view seemed insurmountable and would become a feature of the commission's report, offering legitimacy on the one hand, but becoming an easy target for criticism on the other. The Round Tables' predecessor, the *Commission du Dialogue Interculturel* (2005), had avoided taking a position on a variety of core issues (choosing, in contrast, an analytical and anthropological approach) that the Round Tables would aim to address head-on.

As Nadia Fadil writes, some commentators saw the Round Tables as a 'second chance' to make a difference on key themes related to multiculturalism and minorities in the Belgian political, educational, social and economic sphere. Fadil herself was involved not as a member of the Round Tables but as a 'moderator' of the internal closing meeting and was thus privy to the dynamics that underpin the written recommendations. As an outsider witness to some intense and sensitive discussions, she offers a fascinating narrative of what happens when the role that minorities play in the social space is inverted in the specific commission setting. Minority members were fully aware that their 'multicultural' positions deviated from the dominant view and were, at best, marginal in Belgian public debates. This was an important issue, especially as religious and cultural minority members were 'overrepresented' in the Round Tables, which, like most commissions, was not (nor sought to be) 'democratically' constituted, nor reflected, for instance, the actual religious composition in a given constituency. Again, this feature can be an easy target of criticism. Yet to an important extent, this very possibility is what enables commissions to transcend existing political and ideological cleavages (which does not mean that commissions always do so) and makes commissions 'extra-political' tools distanced from election cycles, political scandals and populist movements.[14] A striking commonality between the commissions was that none of the four mandates was tied to the agenda of any particular legislature, political party, labour union or other interest group. The members of the commissions took the view that this ultimately enabled them to be more creative and, above all, to reflect on a more long-term basis, not having to fear losing favour with a critical electorate that is often conservative on matters concerning (new) minorities and their rights. Particularly in countries where the positions of the various political parties and certain interest groups are diametrically opposed on certain points, when it comes to outlining a policy of sustainable integration, which seems to be the case in most democracies today, the idea of drawing on an independent commission of experts might be worth considering. In fact, since the start of the comparative research project in 2012,[15] several other countries have drawn upon that model to create similar commissions with a view to making policy recommendations, notably in Australia, New Zealand, Norway, Spain, Singapore and South Africa, to name a few.[16] In each of these countries, people with knowledge and experience in a wide range of questions relating to the accommodation of cultural and religious diversity were given an open-ended mandate to make practical recommendations.

In the Belgian Round Tables context, 'minority views' were indeed expressed and shaped the report, but the commission was also highly concerned with 'making a difference'; thus, the viability of the proposals was of utmost importance. Taking on positions that were too 'rogue' would have run counter to that larger aim. Fadil's essay focuses on the discussions on religious symbols in public schools and public offices, showing how the 'contradictory and ambivalent dynamics' of, on the one hand, representing marginalised perspectives while, on the other hand, trying to embrace positions that bridge differences impacted (or failed to impact) the report and the subsequent political and public discussions in Belgium.

Patrick Charlier and Nathalie Denies of Unia, the Belgian equality body, monitor the ongoing Belgian debate. Charlier, the co-director of Unia, was an active participant in the Round Tables process, and as Unia's predecessor was tasked with promoting the implementation of the report's recommendations.[17] While their essay offers an important critique of the Round Tables, it also offers a promising perspective, noting that a large number of proposals, despite initial backlash, have in fact either been adopted or in some ways have made an impact on the ground. In the approach Charlier and Denies advocate under the heading 'finding the common denominator', which draws on the work of French anthropologist Dounia Bouzar, it emerges that Unia's position is not necessarily aligned with those expressed in the Round Tables Report. To 'find the common denominator' means finding a solution that can satisfy 'the greatest number of people'. As they put it, 'This implies that the answer to an individual request (change of hours, specific food, etc.) must not only satisfy the person making the request but also be of benefit to everyone'. Charlier and Denies argue that this approach avoids two pitfalls, namely, imposing one view of the world as the superior and universal standard and introducing specific treatment for one part of the population, which is seen to lead to segmentation. Some may argue that this approach, which is reminiscent of utilitarianism rather than a rights-based approach, diverges considerably from one based on reasonable accommodation for minorities. Reasonable accommodation is argued to 'lead to the exacerbation of differences and even the attribution of identities'. Certainly, some accommodations made for minorities can be productively extended to cover a greater number of persons, but one can ask what happens when no such good-for-all 'common denominator' can be found. Indeed, rejecting any and all 'special treatment' can serve a *status quo* that fails to offer a level playing field. Yet it remains a crucial perspective to include in this volume because it is a position defended by the Belgian equality body, which tracks the impact of the Round Tables closely.

Part V. Engaging retrospective insights, forging ways forward

Despite their reliance on extensive consultations, research, and evidence-based approaches, expert commission reports in the end remain rooted in the normative, particularly where controversial approaches and proposals are advanced. In this context, it is important to take stock of normative debates surrounding the evolution of multiculturalism and the dilemmas faced by liberal democracies. The essays by Christian Joppke, Geoffrey Brahm Levey and Eugenia Relaño Pastor interrogate some crucial and overarching political and legal debates that contextualise commission work. In his essay, Joppke asks what we should make of the proclaimed (but also denied) 'death' or 'retreat' of multiculturalism. Distinguishing religion from language as multicultural claims-maker, he traces the ways in which Islam has become imbricated in multiculturalism's current retreat in Europe. Nonetheless, Joppke points to 'a persistent form of legal multiculturalism underneath a changing political reality' in the British sharia councils.

Though much debate in expert commissions has revolved around minority accommodation, the institutionalised advantaging of the dominant cultural majority lacks equal attention. Levey's essay starts from a working assumption that 'if we wish to take seriously the cultural interests and attachments of minorities we must also take seriously the cultural interests and attachments of established majorities'. Yet, as he writes, 'the privileging of the majority culture is deemed to raise questions about equal citizenship and fairness and broader issues of inclusiveness and belonging'. Addressing the key question of the basis and legitimacy of the majority's cultural precedence, Levey hews closely to the examples in the four expert commissions as well as to the *Lautsi* affair to argue that the 'interpretation of symbols for the purposes of assessing majority cultural precedence should not be anchored to some form of "originalism" but rather to how they are predominantly perceived in our own time'.

Commission work has a complex relationship with jurisprudence. Some of the commissions' proposals are built on prior national jurisprudence, while others are framed to depart from existing judicial lines. Eugenia Relaño Pastor links commission work with the case law of the European Court of Human Rights, a significant legal discourse that has a strong impact not only on the particular parties involved but also on the human rights adjudication cases throughout Europe. Pastor focuses on the value of pluralism in the court's case law (e.g., pluralism was deemed 'indissociable from a democratic society' in the *Kokkinakis* opinion).[18] Based on a close reading of Article 9 jurisprudence (on freedom of religion or belief), Pastor shows that values such as tolerance, respect and pluralism are seldom employed in a 'neutral' fashion, and post-*Kokkinakis*, the so-called Islamic cases have marked significant turns, to the degree that the Court can no longer be said to focus on the minimal conditions for pluralism to flourish, but rather on its limits. Pastor provides a multiplex and fierce critique of the directions recent case law has taken, but in so doing offers an antidote when she suggests that the European Court should posit 'pluralism' as the central question for human rights adjudication. 'It is through pluralisation that there might emerge a reconstituted idea of human rights, and it is through a contextualised legal analysis of the individual's own perception of religious normativity that human dignity can be better protected'.

Because diversity commissions by their very nature are rooted in a certain socio-political context and imply a particular legal background, cross-border borrowings seem precarious and fragile. As Grillo notes in his essay, the reports may be using similar language, concepts, and terminology, but the meaning they attribute to the various terms is clearly not always the same ('integration', 'accommodation', 'intercultural' are offered as examples). Thus, at the very least, any 'comparative borrowing' necessitates a careful approach; the 'transplantation process' is fraught with difficulties. Commissions will consider, in the best case, social realities 'on the ground' as well as the state of the legal frameworks, and these will necessarily be different in other countries and in other times, despite the appealing idea of adapting concepts and theories. In a comparative essay based on a close reading of the British, French, Québécois and Belgian reports, Katayoun Alidadi takes on the topic of reasonable accommodation for religion or belief, charting and linking the respective approaches, positions and recommendations regarding accommodating *religious* needs. She argues that reasonable accommodation, as an application of the 'basic principle of reciprocity', fulfils a barometer-like function in pluralist societies regarding openness to (visible) diversity in public life. 'Reasonable accommodation' means that there is an obligation, on the part not only of the government but also of employers and private service providers, to contribute to increasing opportunities for participation by people who face particular obstacles.[19] An approach that makes 'reasonable accommodation' its cornerstone in effect redistributes some of the burdens of difference to the majority society, so that people can participate in society despite physical or social limitations or obstacles. In all four expert commission reports, attention is paid to measures that can increase the effective chances of participation of target groups whose disadvantages are due, among other factors, to their religious beliefs and practices and their ethnic or cultural background. All reports also address the issue of 'reasonable accommodation' to some extent. However, the comparative exercise reveals a cross-national conversation which has not been immune to misunderstandings and (un)conscious misreadings that have already crept into conversations between expert commissions. In this regard, the essay refers to examples from the Stasi Commission (borrowing from the 'Canadian concept' of *accommodement raisonable*) as well as from the Belgian Round Tables (which borrowed, among other terms, the concept of 'harmonisation practices' from the Bouchard-Taylor report). The immediate and direct usability of reports thus may be restricted, but this does not mean that cross-cultural and temporal comparisons or assessments by future commissions have no value. Indeed, alternative experiences or traditions can be a source of self-interrogation, and the antidote may be in increased and more careful listening to prior commission experiences rather than in too readily adopting concepts, ideas and solutions.[20]

From establishment to working methods and reactions

While using commissions to address societal challenges is hardly unique to national identity and minority issues, there are distinct features that merit investigation in the case of ethnic, cultural and religious diversity. To that end, we want to focus on two overarching themes that derive from the essays in the volume. These involve (1) the establishment and the working methods of diversity commissions and (2) the reception of the reports (both the immediate reactions and the long-term impacts). Finally, we include some retrospective insights in the conclusion.

Establishment and working methods, or outer influences and inner workings

As noted above, one way to distinguish the genesis of commissions is on the reflective/responsive or proactive/reactive scale. However, this need not mean that the commission reports only strictly address the 'topics at hand'; rather, commissions routinely consider their mandate and area of reflection in a characteristically broad fashion. This allows for overarching reflections and proposals, which may have a better potential to stand the test of time.

Further, consultations can be a more or less important feature of the commission process. All four expert commissions collected and integrated comments from outside experts and the general public in their deliberations, but some were more dedicated than others to the 'voices on the ground'. The Stasi Commission was criticised for not including (more) testimonies of women who wore the hijab, while on the other end of the spectrum, the public fora created within the consultative framework of the Bouchard-Taylor Commission had the counterproductive effect of giving voice to various xenophobic public tendencies. The four commissions being analysed here went about their tasks in quite different ways; some, like the Bouchard-Taylor Commission in Quebec, relied largely – if not exclusively – on (public) hearings, while others, such as the Parekh

Commission, preferred to work behind the scenes, eschewing publicity before their work was completed. In any case, the implications of such outreach are hardly defined or measurable, and the commission remains composed of 'experts'. Experts come from a wide variety of backgrounds and have different affiliations and contributions, but they remain elite participants. Indeed, issuance of reports is often accompanied by fierce criticism of the 'unrepresentative' composition of the membership. Whether justified or not, this does make clear that commissions cannot meet (nor should be held to) strict standards of representation. Membership and working methods often go hand in hand, and some essays offer particularly striking insights on the commissions' working methods. For instance, Nadia Fadil presents a focused description of the inner workings and negotiation dynamics behind the Foblets-Kulakowski Round Tables Report. She notes that when minorities obtain a prominent position within such commissions, a position they do not have in the larger society, something happens to the power dynamics in the group that can influence the recommendations included in the report as well as the reception of the report and its public or political credibility. Varun Uberoi and Tariq Modood engage in a similar exercise with regard to the workings of the Parekh Commission. Both authors illustrate how the conceptions developed by Bhikhu Parekh (and other key advisors, including Tariq Modood) influenced the report. Regarding the French example, the dialogue between Jean Baubérot and John Bowen points to the serendipitous developments that can shape public reports with broad socio-political influence. Furthermore, it should also be clear that members of a commission can come out against certain commission proposals, as indeed compromise is part of the internal negotiation dynamics, or simply because people can change their minds or consider the context changed. Whereas the Bouchard-Taylor Report proposed a ban on visible religious symbols for high-level state officials or employees (such as the president and vice-president of the National Assembly, judges and Crown prosecutors, police officers and prison guards),[21] Charles Taylor subsequently came out as no longer in support of such ban while Gérard Bouchard remained committed to such ban as justified by state neutrality.

Reports rarely address the inner processes of negotiation and 'compromise', allowing at times contentious settlements to appear as standing commission consensus. Therefore, the glimpse behind the curtain that we gain from insiders is both important and relevant to the lessons that expert commissions hold in store. As Ralph Grillo rightfully notes, 'the politics of it all – not political theory, not political philosophy, but *politics*, the power relations revealed by the internal workings of the commissions on the one hand, and their relations with the world in which they operate',[22] is always central.

Immediate reception of the reports versus long-term effects

Various essays in the volume, including those authored by 'commission insiders', including Tariq Modood, Maleiha Malik, Marie-Claire Foblets and Patrick Weil, recall that the immediate reception of reports routinely included highly critical, sceptical and even hostile reactions. In the different political contexts, proposals, especially those aimed at accommodating minorities, have triggered critical voices and at times even *ad hominem* attacks on the chairs or members of a commission. Modood speaks of 'wilful misunderstanding of the Parekh Report so that it could be accused by the right-wing press of embracing an overly minority perspective and of being unpatriotic, and in effect dismissed as an expression of an out-of-touch liberal intelligentsia'. Even the initiators of expert commissions have distanced themselves quite strikingly from (parts of) a report, e.g., Jack Straw in the case of the Parekh Commission and the political reactions following the Bouchard-Taylor Commission recommendation to remove the crucifix from the Quebec National Assembly. Commission work is not for the faint of heart, and, particularly, guarded academics and sheltered academics in particular may encounter conflicting views in shapes or forms they have not experienced before. A report can also enable specific actions, offering legitimation for legislative initiatives, as did the Stasi Commission (and the Gerin Commission in the case of the French 'burqa ban'; see the chapter by Maleiha Malik).

Yet when taking a long-term perspective, a different image may emerge. As Tariq Modood (UK) and Patrick Charlier (Belgium) note, many recommendations are in fact realised over time (even though the Round Tables were not necessarily credited).[23]

With all their inherent limitations, can expert commissions then be viewed as viable agents of change? Are commission activities conducive to the general pursuit of 'making a difference' or of specific ambitions, whether it be appeasing public tensions (even if temporarily) or finding productive ways to tackle social divisions and struggles in particular social areas? Or are commissions simply ahead of their time, foreshadowing future developments ahead of the curve? Even if often subtle and indirect, the impact of the commission deliberations and reports on laws and policies, or just on social attitudes if nothing else, cannot be ignored. Malik notes that 'although a non-statutory commission cannot directly legislate, it acts as a form of "soft regulation" that influences the development of law and policy'. Commissions, then, after all, could be credited with moving along the (national) narrative on diversity, national identity and minorities.

Retrospective insights and a look ahead: the continued relevance of commissions

A comparative overview of the commissions under study points to an important cross-national shift towards framing challenges and tensions in terms of religion, religious differences and identity. But more and more, as the West's perennial 'Other', Islam is also the targeted subject of anti-terrorism and security measures. Minority issues have gone from being framed as accommodation to security concerns, since many recent terror acts have been linked to fundamentalist Islamist groups. Religious community leaders try to distance themselves from terror acts and their perpetrators (e.g. in the UK, refusing to conduct funerals for Islamist terrorists[24]), yet in the public opinion, terrorism and Islamism seem to be converging ever more closely with each terror event, leaving behind a threadbare social fabric. If safety is the primary concern of the political elite and majority populations, then minorities, for their part, seek to not fall prey to the collateral damage in the form of continued and aggravated discrimination, stigmatisation and other effects of Islamophobia. These respective concerns are linked, but in the end also separate, and should be disentangled. Terrorism harms us all, irrespective of our political, religious or ideological identities, and the fight against it should unite rather than divide us. At the same time, equal rights and anti-discrimination laws are invaluable on their own terms. Discrimination and stigmatisation feed into segregation and promote radicalisation, but accommodation loses much of its meaning if the core reason is to prevent radicalisation. Yet in a security-focused frame of mind, it becomes even harder to make the case for acting positively and proactively to sustain diversity instead of just tolerating it.

And still, global forces have only increased the urgency of finding successful ways to integrate newcomers. The 2015 refugee crisis saw people fleeing Afghanistan, Syria and other war-torn Middle Eastern and African countries and arriving on the territory of the member states of the European Union. This presents and will continue to present enormous political, legal and social challenges for Europe as a whole and its member states. Once the situation has more or less stabilised, hundreds of thousands of people of non-European origin will have been authorised to make a new start within Europe. Their presence will effect significant changes in the demographic, social and cultural profiles of their host countries. In the face of these developments, public authorities have been compelled to clarify the ground rules for their participation: throughout Europe, legislators have put in place anti-discrimination policies and provided for the creation of equality bodies to support victims of discrimination and to advise them in the steps they could take when experiencing discriminatory practices and exclusion. It is not easy to achieve a balance: to this day, many long-established cultural and religious minorities continue to feel inadequately recognised within the wider society in various parts of Europe. Moreover, in several European countries we can see a tendency within the majority society to revert to an emphasis on a foregone national identity. This attitude is fuelled by a feeling that the 'natives' are no longer 'at home' in their own country and reinforced by a lack of the necessary knowledge to become properly informed about the situation and the identity of the 'other' or to be able to establish contacts – in short, to be able to bridge the cultural divide. The situation in certain suburbs and regions of Europe has become so tense that some people have concluded that the project of multiculturalism has failed in Europe.

How, then, against the backdrop of this troubling situation, can we nevertheless envisage, with fresh hope and without lapsing into naïveté, the situation that European member states will have to face in the years to come? To conclude that the project of multiculturalism has failed completely is too dismissive and ultimately unhelpful since it does not do justice to the empirical reality, which is much more nuanced. A more constructive step at this point would be to revisit the experiences of recent years – as a number of countries have done – in order to see what specific lessons one could draw from the integration policies that have been pursued in various democratic states. Some measures have undoubtedly produced undesirable consequences or even had counterproductive effects, or else have simply remained ineffectual. Nevertheless, some policies have worked and managed, within their own contexts, to achieve their original goals, namely, better integration of persons of immigrant origin into the society.[25]

In the process of 'commissioning', that is, asking a commission of experts to answer a series of questions about diversity, integration and '*le vivre-ensemble*', there is some potential for a thoughtful and productive engagement in situations of crisis. However, it must be noted that commissions come with their own limitations and are not to be considered an easy 'fix' by any means. First, one may object to what Ralph Grillo calls a 'phenomenon of our time, which is the continuous detailed scrutiny (in the public sphere and in the academy) of diversity and diverse populations and what to do with them'.[26] Second, the role of power relations both within such commissions and surrounding their mandate and work often goes unrecognised, despite the fact that such power relations may at times be of key significance in understanding the reports issued and recommendations advanced. Further, the reactions to the commissions and their reports signal some limitations as well. The vast majority of the reactions to the reports of the four commissions – and this is true in particular of the responses voiced via the media – were critical, if not sceptical. With the exception of the French Stasi Commission Report, the reports were received with much consternation once they were made public, and indeed at times elicited vehement reactions. Reports and recommendations were at times (grossly) misrepresented. All that is important to bear in mind when assessing the possibility of setting up a commission of independent experts, as such commissions face formidable tasks. At times, it may not appear feasible to call for maximum openness to cultural and religious diversity given the current preoccupations with the future of European societies and the role that recent waves of migration might play in it. Against the backdrop of the failings of European migration policies referred to above, there is reason to be concerned about the feasibility of an intercultural/multicultural project.

Public opinion has in previous years been alarmed by successive crises in immigration policy. In the months prior to the publication of the Belgian report, for example, the immigration office in Brussels had been ordered by a number of judges to pay heavy penalties in individual asylum cases, with the aim of obliging the government to assume its responsibilities in relation to potential homeless refugees. Other factors preceding the report, in part, also coloured public perception of the commission and the report, such as the high number of regularisations that had in recent years been issued through the discretionary powers of the competent minister, giving rise to a great deal of anxiety regarding the lack of transparency of the regularisation policy and thus running the risk of being perceived as a policy of impunity. Finally, many people felt that there ought to be a more consistent deportation policy. This anxiety in turn had a negative impact on the willingness of the majorities to build a society together with newcomers and thus to welcome the direction taken in most of the recommendations contained in the commission reports.

Nevertheless, the successful implementation of many of the proposals of these commissions gives us grounds for cautious optimism, and we would like to draw this section to a close on a hopeful note. After having compared the experiences of the four countries – Belgium, Canada, France and the United Kingdom – and although the publication of the reports initially led to a wave of highly critical reactions, in each of the four countries the reports also drew the attention of the public authorities, including both local and/or central administrations and legislators. The fate of proposals in commission reports is linked to the continued willingness of policy-makers to keep the topic on the agenda. Fortunately, as we mentioned earlier, the work of the experts has been able to serve, in all four cases under scrutiny here, as a continued source of inspiration for practical measures intended to support and guide the integration of new minorities into the wider society. This experience, ultimately very positive, could in turn serve as inspiration, both in *tempore non suspecto* and in critical times of transition, transformation and crisis.

Notes

1 The terms 'diversity/expert commission' and 'public commission' are used interchangeably throughout the introduction.

2 The French Stasi (2003) and Gerin (2010) Commissions offer examples of commissions whose work facilitated the adoption of restrictive regulations on religious signs, respectively banning ostentatious religious symbols in public schools and banning full-face veils in the public space.

3 Ralph Grillo, who has studied the work of the many consultations and/or reports in the UK that have sought to address the situation of minorities in Britain over the last 50 years, and increasingly in the last 20, identified over 130 interventions from the mid-1960s to 2015 (100+ since 2000), initiated by various ministries (principally the Home Office), non-departmental public bodies and NGOs. His work illustrates how minorities have been subjected to official and unofficial scrutiny, consultation, report, and policy recommendations.

4 Contributions were presented at an international conference convened by the Department of Law and Anthropology of the Max Planck Institute for Social Anthropology; 'Religious and Cultural Diversity in Four National Contexts', 26–27 June 2015, Halle/Saale, Germany.

5 Commission members express their personal views in their chapters, which do not necessarily reflect official statements coming from the commission or that can be attributed to other commission members.

6 Solange Lefebvre and Patrice Brodeur (eds), *Public Commissions on Cultural and Religious Diversity: Analysis, Reception, and Challenges* (Routledge, 2017).

7 The joint research project is funded by the Canadian Social Sciences and Humanities Research Council (SSHRC). The team members are: Principal Investigator, Solange Lefebvre (Religious Studies, 'Chaire religion, culture et société', Montréal); Co-investigators: Jean-François Gaudreault-DesBiens (Law, Montréal), Lori Beaman and Peter Beyer (Sociology and Religious Studies, Religion and Diversity Project, Ottawa); Marie-Claire Foblets (Department of Law and Anthropology, MPI Halle/Germany); Céline Béraud (Sociology, Maison de la Recherche en Sciences Humaines, CNRS/Caen, Paris); Tariq Modood (Political Science, Centre for the Study of Ethnicity and Citizenship, Bristol) and James Beckford (Sociology of Religion, Warwick).

8 While we use management of diversity in a more general sense, encompassing rights-based approaches, 'diversity management' in business settings – a practice that started in the US and since has been exported around the world – can have a particular connotation 'related to the upcoming neo-liberalism, where market is promoted as a means

of regulation rather than rights and legislation'. For the dilemmas and contradictions of diversity management in this stricter sense, see Evangelina Holvino and Annette Kamp, 'Diversity management: Are we moving in the *right* direction? Reflections from both sides of the North Atlantic', *Scandinavian Journal of Management* (2009) 25, 395–403.

9 It must be noted that even when commissions are set up to address particular challenges, they tend to interpret their mandate broadly and the concerns discussed transcend the instigating factors.

10 In Belgium, the federal government tasked the commission with 'making recommendations to the government with a view to reinforcing the success of a society based on diversity, respect for cultural specificities, non-discrimination, integration and shared values' (Government agreement of 18 March 2008).

11 *Multani v. Commission scolaire Marguerite-Bourgeoys*, [2006] 1 S.C.R. 256, 2006 SCC 6.

12 G. Bouchard and C. Taylor, *Building the Future: A Time for Reconciliation*, Final Report of Consultation Commission on the Practice of Accommodations Related to Cultural Differences, Quebec, 2008, 18.

13 The Charter of Values (Bill 60) was introduced by the governing Parti Québécois (PQ) in 2013, under Premier Pauline Marois, and included provisions to ban ostentatious religious symbols in public institutions. The bill died when the PQ lost the 2014 elections. See Jonathan Montpetit, 'Quebec's charter of values, revisited', 5 September 2016, cbcnews.com.

14 See also the chapter by Maleiha Malik: 'It is also difficult to carry out policy about integration in legislative assemblies because of the electoral success of far right parties who may dominate and distort sensitive discussions about less powerful minorities. In this context, there is a potential role for expert independent commissions that have a mandate to gather facts, consult less powerful minorities who do not have a voice in legislative assemblies and make recommendations that may not be popular with voters'.

15 See footnote 7.

16 For more on this, see Lefebvre and Brodeur, *Public Commissions on Cultural and Religious Diversity* (n 6).

17 A brief reaction by Patrick Charlier, explaining the specifics of the follow-up of the Belgian Round Tables from the perspective of Unia, follows the chapter by Marie-Claire Foblets (Chapter 12).

18 *Kokkinakis v. Greece*, App No 14307/88 (ECtHR, 25 May 1993) para 31.

19 Such accommodations can, for example, benefit persons with physical limitations by making a building accessible or providing for an ergonomically appropriate work environment (see UN Convention on the Rights of Persons with Disabilities).

20 In order to allow future generations to access not only the reports, but also the public reactions and consultations, the PLURI database has collected thousands of documents related to various expert commissions across the globe (whose websites are routinely shut down once the work is wrapped up). The database includes briefs submitted, interviews, media reports and other related documents. www.pluri.gdcr.umontreal.ca.

21 Bouchard and Taylor, *Building the Future*, 151.

22 Chapter by Ralph Grillo in this volume.

23 Modood, in his chapter, writes about the immediate reception and the long-term effect of the Parekh Commission report: 'the reception of the CMEB report was a catastrophe for the multiculturalist cause which it espoused. However, what is interesting is that on this specific point of controversy, if we look only a few years forward we see that what was deemed by the press and politicians to be unacceptable has come to be thought of as necessary, even relatively uncontroversial, amongst senior politicians'. He notes, 'Actually most of the report's more than 140 recommendations were acted upon quite soon after its publication.' Runnymede Trust, *Realising the Vision: Progress and Further Challenges*. Runnymede Trust, 2004.

24 BBC News, 'Is refusing terrorists funerals a deterrent?' 8 June 2017.

25 See, e.g., Carola Burkert and Anette Haas, 'Investing in the future: Labor market integration policies for new immigrants in Germany', Migration Policy Institute, October 2014; OECD/European Union, *Indicators of Immigrant Integration 2015: Settling In* (OECD Publishing, 2015); Karin Borevi, Kristian Kriegbaum Jensen and Per Mouritsen, 'The civic turn of immigrant integration policies in the Scandinavian welfare states', *Comparative Migration Studies* (2017), 5:9 (special issue) (authors discussing the effects of 'civic integration policies').

26 Ralph Grillo, 'Some thoughts on the Halle Conference', 29 June 2015 (on file with authors). In R. Grillo, *Muslim Families, Politics and the Law: A Legal Industry in Multicultural Britain* (Ashgate, 2015), Grillo explored how Muslim families and gender relations have come under constant public scrutiny and political and social criticism.

Part I

The importance of Britishness

The Parekh Commission Report and beyond

2 Pointing to a multicultural future

Rethinking race, ethnicity, religion and Britishness

Tariq Modood[1]

Introduction

The purpose of this essay is to consider the place of religion in relation to post-immigration diversity in Britain and British national identity as reflected in three national commission reports stretching over thirty years. This means that the immigration I am concerned with is mainly the primary, non-white immigration from the Commonwealth of the 1950s–1970s, and the secondary or family reunification migration it led to in the 1970s–1980s, and of course the children and grandchildren of these settlements. The 1990s saw the beginning of a new wave of migration into the UK, mainly of refugees and asylum seekers from non-Commonwealth countries such as Bosnia, Somalia, Afghanistan, Kurds from Turkey and so on. From 2004 onwards large inflows of migration from the newly acceded EU states of Eastern Europe has also taken place. The latter two waves have certainly deepened the diversity, perhaps creating a 'superdiversity'[2] but the diversity I am interested in relates to not being white and/or not Christian and the core of that as a settled British population relates to the two earlier waves of migration indicated above.

As befitting an essay in a book on national expert commissions, I shall explore my theme in relation to such commissions in Britain. My original brief was to focus on the Commission on Multi-Ethnic Britain (CMEB), which published its report in 2000. I execute that brief by looking backwards and forwards from that point by discussing a commission which could be said to precede or point towards it, and one which picks up some of its work. I have identified the *Education for All: Report of the Committee of Inquiry into the Education of Children from Ethnic Minority Groups* (Swann Report, 1985) as its predecessor, and the Commission on Religion and Belief in British Public Life (CORAB, 2015) as one which picks up on an aspect of its work.[3] My interest is not on the specific policy recommendations of these commissions but on the general approach or understanding of the context exhibited in relation to the place of religion in a changing Britain.

While Swann was a government inquiry, CMEB and CORAB were run by an independent organisation (a campaigning organisation, The Runnymede Trust, and an educational institution, The Woolf Institute, Cambridge respectively). All three public inquiries were led by members of the House of Lords:[4] the scientist and former vice-chancellor, Michael Swann, the political philosopher and former vice-chancellor, Bhikhu Parekh and the former senior judge, Elisabeth Butler-Sloss.[5] Each commission focused on the changing nature of minorities and their place in British society and how they were changing British society and how British society should be further changed to accommodate the minorities. Having said this, it should, however, be noted that Swann was specifically confined to the disadvantages and the (lack of) progress of primarily black and Asian children in English schools,[6] and CORAB has no direct or exclusive minority-oriented focus.

Of course, the choice of the commissions prior to and subsequent to the CMEB involved some selection. As the predecessor, for example, I could have chosen the Scarman Report (1982),[7] the first major inquiry commissioned by the government. Yet, the above three commissions were of a reflective kind and not directly a response to an immediate crisis, while Lord Justice Scarman's brief was 'to inquire urgently into the serious disorder in Brixton on 10–12 April 1981'. The same point, namely an urgent focus on a specific set of events also applies to another judge-led inquiry commissioned by the British government, namely the Macpherson report,[8] which in 1999 reported on the murder of Stephen Lawrence and its investigation by the London Metropolitan Police. A candidate for a successor to the CMEB report could be *Our Shared Future*, the 2007 report of the government appointed Commission on Integration and Cohesion (COIE).[9] It is, however, not a promising candidate. It is a rare national report on race and ethnicity that is published without considerable debate and engagement, even controversy, but that was the case with this report, which was perceived as having been overly guided by the government minister which appointed it.[10] The Scarman Report was thought by many in the government, probably including the Prime Minister, Mrs Thatcher, and its right-wing supporters as excusing criminality and riotous behaviour by focusing on 'racial disadvantage' and blaming the riots on the police through its use of the concept of 'institutional discrimination'.[11] The Parekh/CMEB Report was vociferously attacked for being anti-British, anti-patriotic and for portraying Britain as inherently racist.[12] CORAB got attacked for suggesting that Christianity was no longer adhered to by the majority of Britons and so Britain was no longer 'a Christian country'.[13] Moreover, the COIE is much less referred to either generally or in terms of an intellectual impact than the reports mentioned above.[14]

In having chosen the commissions I have there is a further advantage of an even spread across time, as each is separated by the next nearest in time to it by exactly fifteen years, making a symmetrical triad.[15]

The Swann Report (1985)

Swann marks the early wave of British multiculturalism and includes some of the key ideas of racial disadvantage, discrimination and racism but explicitly tends towards a conceptualisation of the non-white minorities in terms of ethnic pluralism rather than racial dualism, despite the latter being the dominant perspective of the 1980s.[16] It reframes the concern with race into a conception of an ethnically plural Britain, which has been one of the dominant themes of British multiculturalism. Swann is, then, not just a pioneering expression of multiculturalism in general but in particular of this central feature of multiculturalism. Its deliberations on education were cast within a vision of accommodating cultural diversity within a framework of shared values and a pluralised and expanded sense of what it is to be British. While this vision was expressed in a brief introductory chapter, it has become the idea of multiculturalism in Britain or of multicultural Britishness. It argued:

> we are not looking for the assimilation of the minority communities within an unchanged dominant way of life, we are perhaps looking for the 'assimilation' of all groups within a redefined concept of what it means to live in British society today. We are not seeking to fit ethnic minorities into a mould which was originally cast for a society, relatively homogeneous in language, religion and culture, nor to break this mould completely and replace it with one which is in all senses 'foreign' to our established way of life. We are instead looking to recast the mould into a form which retains the fundamental principles of the original but within a broader pluralist conspectus – diversity within unity.[17]

It is a sentiment which perhaps harks back to Roy Jenkins's unelaborated but famous definition of integration in 1966 during his period as Home Secretary: '. . . not as a flattening process of assimilation but as equal opportunity, accompanied by cultural diversity, in an atmosphere of mutual tolerance'. It can be seen more clearly in an essay by Bhikhu Parekh, when in the early 1970s he had argued that 'pluralistic integration within the framework of a generally accepted conception of the good life should be the ideal governing Britain's relations with her immigrant population'.[18] While Parekh saw racism in its various manifestations as a significant obstacle to this integration, the aspect in which he was ahead of his time is his going on to say:

> In the ultimate analysis pluralistic integration entails that the Briton's perception of his identity should be revised . . . Only when it is acknowledged as a matter of course that a Briton is not by definition white but could be black, brown or yellow, that he might speak Swahili, Mandarin or Hindustani as his first and English as his second language, and that his 'kith and kin' might be found in Bombay, Barbados and Ibadan as well as in Salisbury and Wellington, can the non-white minority feel as authentically British as the native, and can be so accepted by the latter.[19]

Hence, it is perhaps not very surprising that a commission in which Parekh had been a member in its early period expressed a very similar idea. We shall see this theme was to become central to CMEB fifteen years later, under Parekh's chairmanship.

Swann included a discussion of religion (devoting to it one of its sixteen chapters[20]), but in a not altogether positive way. For example, despite considerable submissions to the contrary and open dissent from six ethnic minority members,[21] it explicitly argued against extending state funding from Christian and Jewish (voluntary-aided) state-funded faith schools to Islam, Hinduism and Sikhism.

The Commission on Multi-Ethnic Britain (2000)

According to its chair, Bhikhu Parekh, 'The Commission's remit was to analyse the current state of multi-ethnic Britain and to propose ways of countering racial discrimination and disadvantage and making Britain a confident and vibrant multicultural society at ease with its rich diversity' (p. iix). The report was summarised as: 'Building and sustaining a community of citizens and communities will involve:

- rethinking the national story and national identity;
- understanding that all identities are in a process of transition;
- developing a balance between cohesion, equality and difference;
- addressing and eliminating all forms of racism;
- reducing material inequalities;
- building a pluralistic human rights culture'.

(p. xii)

The Commission on Multi-Ethnic Britain had a number of unusual features for a national commission. It was created by an independent race relations think-tank, The Runnymede Trust, and while it was launched by the Home Secretary, Jack Straw, it was wholly independent of the government and included no members of the judiciary or representatives of the government. Of its twenty-five members (not all of whom served the full term) over a third were non-white and nearly a third were academics.[22] Besides its distinguished chair it included prominent public intellectuals and race equality professionals such as the late Professor Stuart Hall; the late Professor (later, Sir) Bob Hepple, who went on to become the Master of Clare College, Cambridge; the journalist and writer, Yasmin Alibhai-Brown; the Chair of the Greater London Assembly (and later of the Commission for Racial Equality, and after that, the Equality and Human Rights Commission), and broadcaster, Trevor Phillips; Sir (later Lord) Herman Ouseley, the Chair of the Commission for Racial Equality at the time; and Andrew Marr, Chief Political Editor at BBC Television at the time. As a result of this mix, some of the report had an academic character, such as Part I, 'A Vision for Britain', which began with 'Rethinking the National Story', included several sociological chapters and a political theory chapter entitled 'Cohesion, Equality and Difference'. These chapters attempted, as the report did as a whole, to be accessible to the general as well as the public policy reader, and wore their academic apparel lightly; nevertheless, they offered an intellectual framework for thinking about minority–majority relations in Britain and, so, for the more concrete analyses of the rest of the report. It may be the case, however, that the commission did not have the personnel composition and balance that people are accustomed to and that it had a more theoretical and academic orientation than journalists and politicians would expect. A consequence of this was the wilful misunderstanding of the report so that it could be accused by the right-wing press of being an overly minority perspective and unpatriotic and dismissed as an expression of an out of touch liberal intelligentsia.[23]

The CMEB broke new ground in a number of ways in relation to discourses and understanding of post-immigration minorities and how they should be accommodated in Britain. One of these was the inclusion of religious community identification and needs within the conceptualisation of the 'multi-ethnic'; although this marks a continuity with Swann, there is much development beyond it too. For example, in the 1990s it was clear that historic colour-racism was now also accompanied by new contemporary forms of 'cultural-racism'[24] and forms of anti-Muslim hostility or Islamophobia[25] and both of these were features of CMEB's conceptualisation of 'racism' (chapter 5). In this respect, the CMEB marks a transitional point, where, subsequent to the Rushdie Affair of 1988–1989, the effects of Muslim and other minority religious political assertiveness is seen to be changing the nature of multi-ethnicity, yet without becoming prominent in the report (the report being prior to the emergence of domestic terrorism in the name of Islam and the international 'war on terror'). 'The Future of Multi-Ethnic Britain' that the report desires consists of promotion of (in the key phrases of the report) 'race equality and cultural diversity' and 're-thinking the national story' in relation to 'black and Asian people'. The CMEB clearly – and up to that point unprecedentedly for a public report on race and ethnicity – acknowledges that religion has been neglected within the frame of racial equality. Its (self-)critical remarks include:

> Most race equality organisations are broadly secular, not religious. It is perhaps for this reason that they frequently appear insensitive to forms of racism that target aspects of religious identity. For example, they are widely perceived by British Muslims to be insensitive to distinctive Muslim concerns, by Jewish people to be uninterested in antisemitism, and by Irish people to be indifferent to sectarianism and anti-Catholicism. People with Indian cultural backgrounds and ties to Sikhism or Hinduism similarly feel that anti-racist organisations have little or nothing to say about their religious affiliations and identities.[26]

Yet the CMEB does not fully integrate the dimension of religion into its vision of equality and diversity. In the building of a 'community of communities', it is primarily ethno-racial and ethno-cultural individuals and communities that are in mind, rather than the ethno-religious. 'Religion and belief' is largely confined to one chapter and five recommendations in a report of twenty chapters and 135 recommendations. None are cross-referenced or mentioned in the recommendations under the headings of Education, Employment, Policing and so on. They are supplementary rather than integrated. The five specific recommendations are:

1 In all faith communities there should be closer connections between anti-racism and work to improve inter-faith relations.
2 Legislation should be introduced prohibiting direct and indirect discrimination on grounds of religion or belief.
3 A statement of general principles should be drawn up on reasonable accommodation in relation to religious and cultural diversity in the workplace and in schools, and case-study examples of good practice should be provided.
4 A study should be made of police responses to hate crimes containing a religious component.
5 A commission on the role of religion in the public life of a multi-faith society should be set up to make recommendations on legal and constitutional matters.

In relation to national identity the commission had an overarching message, which was then qualified in certain ways. The overarching message was that the rethinking and political action to make Britain more inclusive, which had begun, had to continue and had to focus on British identity itself if the country was to continue to progress towards an inclusive, non-racial, multicultural Britishness. The argument was that the inequalities and exclusions associated with racism, including material inequalities and disadvantages, could not be countered by merely materialist strategies but required 'rethinking the national story', our collective identity, in a plural way. A qualifying message was that there could be no complacency about the importance of anti-racism, which needed more political will, if Britishness was to be made inclusive in fact and not just rhetoric. Another qualifying message was that old-fashioned, monistic, assimilationist, majoritarian nationalism was past its usefulness and had to be replaced by a new, plural kind of national identity (sometimes referred to as 'post-national' in the report). It is argued that this was happening as ethnic minorities were assuming a British identity and qualifying it with a minority ethnicity, and thus not just passively accepting Britishness but making it their own; which, together with affirmation of their minority identity, is a mark of a new confidence (para 3.29). The report notes, however, that there is one major and so far insuperable barrier to this development. Britishness, as much as Englishness, has systematic, largely unspoken, racial connotations. Whiteness nowhere features as an explicit condition of being British, but it is widely understood that Englishness, and therefore by extension, Britishness is racially coded (para 3.30). Unfortunately, these few qualifying remarks about racism and how the perception that Britishness means whiteness unleashed a hysterical reaction in the right-wing press which influenced more generally the reception of the report. The overall message was ignored and the commission was condemned by many for being unpatriotic and even denying that Britain was or could be a unified country.[27]

The right-wing media was so successfully able to caricature the commission thus, that even the Home Secretary of the time, Jack Straw, who had given the commission its blessings at its launch, was now reported as saying, under the headline '"Proud to be British" Straw raps race report':

> he was appalled when he read part of the document suggesting that the term British had racial connotations and was no longer appropriate in a multicultural society. He ripped up a speech prepared for the launch of the document yesterday and instead delivered a strong attack on the part which he believed lacked intellectual rigour.
>
> 'Unlike the Runnymede Trust I firmly believe that there is a future for Britain and a future for Britishness,' Mr Straw declared. 'I am proud to be British and of what I believe to be the best of British values'.[28]

That these remarks were based on a false reading of the report is evidenced by an article that had been published in *The Guardian* on the previous day by Parekh as the chair of the CMEB, in which he argued:

> The report recognises that, while cherishing cultural diversity, Britain must remain a cohesive society with a shared national culture. That culture is based on shared values, including such procedural values as tolerance, mutual respect, dialogue and peaceful resolution of differences, as well as such basic ethical norms as respect for human dignity, equal worth of all and equal life chances.
>
> The common national culture includes shared symbols and a shared view of national identity, and these are best evolved through a democratic dialogue between our various communities. The report sees Britain both as a national community with a clear sense of collective purpose and direction and also made up of different communities interacting with each other within a shared moral framework.[29]

Despite such statements, the reception of the CMEB report was a catastrophe for the multiculturalist cause which it espoused. However, what is interesting is that on this specific point of controversy, if we look only a few years forward we see that what was deemed by the press and politicians to be unacceptable has come to be thought of as necessary, even relatively uncontroversial, among senior politicians.[30] Soon afterwards, cabinet members started expressing exactly the view that had been lambasted; or to put it more precisely, which was not sufficiently identified because of the lambasting. In 2001, John Denham argued that Britishness, as it existed, was insufficient, hence 'positive action must be taken to *build* a shared vision and identity'[31] and in 2007, Jack Straw himself said that it was necessary 'to *develop* an *inclusive* British story which reflects the past, takes a hard look at where we are now and *creates* a potent vision . . . to make sense of our shared future'.[32] Note the active verbs: 'build', 'develop', 'creates' – exactly the view that had motivated the CMEB. Nor were such views confined to New Labour. A project within a Leverhulme Trust Bristol-UCL Programme that took interviews from New Labour cabinet ministers and Conservative shadow cabinet ministers during 2007–2008 did not find a uniformity of views on this matter but found considerable cross-party agreement that British national identity had to be opened up to include minorities and that politicians and the state had a role to play in this process.[33] It is a point of view that is not always lived up to but the opening ceremony of the Olympic Games in London in July 2012 was an excellent expression of a multicultural Britishness that New Labour tried to articulate without ever quite succeeding and its positive reception in the British media – including the same papers that had lambasted the CMEB – shows what is possible.[34] An Australian political theorist opined that the Britain displayed at the Olympics meant that '[m]any countries are [now] looking to Britain as an example of a dynamic multicultural society united by a generous patriotism'.[35] Yasmin Alibhai-Brown, the left-wing journalist and a member of the CMEB, who returned her MBE as a protest against the Iraq War, wrote:

> these two weeks have been a watershed of true significance. There has been a visceral reaction among black and Asian Britons to what we have seen. For some, it has been perhaps the first time they have really felt a part of this country. For others, the promise of tolerance and integration has come true.[36]

So, while the initial media reception misunderstood the Parekh Report on British national identity – partly it should be added because of some ambivalent phrases in the report[37] – it underlined, even in the period before 9/11 and the 'multiculturalism has failed' discourse, how multiculturalists and race egalitarians approached the question of British national identity was one of the most charged topics that they could address. On the other hand, the report brings out how central the question of national identity, especially the idea of remaking or multiculturalising national identity, is to multiculturalism.

Commission on Religion and Belief in British Public Life (2015)[38]

In one way or another nearly all the CMEB recommendations in relation to religion (quoted in full above) have been pursued, with the last having now been realised in the existence of the Commission on Religion and Belief in British Public Life (CORAB).[39] Convened by the Woolf Institute, Cambridge and chaired by Baroness Butler-Sloss, the terms of reference of this commission which ran from 2013 to 2015 were to:

- consider the place and role of religion and belief in contemporary Britain, and the significance of emerging trends and identities;
- examine how ideas of Britishness and national identity may be inclusive of a range of religions and beliefs, and may in turn influence people's self-understanding;
- explore how shared understandings of the common good may contribute to greater levels of mutual trust and collective action, and to a more harmonious society;
- make recommendations for public life and policy.

CORAB is dual-aspected. On the one hand, it is not concerned directly with the position of minorities but the place of religion in public life generally. On the other hand, it has – I believe – come into being because of the fact of expanding and deepening religious diversity which is the principal cause of the new public salience of religion in Britain.[40] Specifically, while the terms of reference do not refer to race, ethnicity or even minorities, some of the mechanics of the origins of CORAB display a direct continuity with the Parekh Report. This was achieved through Mohammed A. Aziz, who had been a Commissioner in the Commission for Racial Equality and a consultant senior advisor on race and religion to ministers and cabinet secretaries. Subsequent to leaving government in 2011, he formulated the idea of the commission and proposed it to the Woolf Institute, Cambridge, whose director, and later also the Vice-Chair of CORAB, Dr Ed Kessler, warmly received the suggestion that the Woolf Institute host CORAB and Mohammed went on to raise the necessary funds, seek out suitable commissioners and head the secretariat. The link with the CMEB was directly present through the Chair of the CMEB, Lord Parekh becoming not just one of four patrons of the CORAB but the only one who participated actively in its deliberations throughout the life of the commission; Robin Richardson had drafted the CMEB report and was given the same responsibility in relation to the CORAB report; and I, having been an active part of the CMEB, was appointed a commissioner and member of the Steering Group of CORAB. More substantively, it should be clear that religion has primarily re-entered public life, political debates, mobilisation and policy making because of – in Britain as in the rest of Western Europe and beyond – the presence, assertiveness and terrorism of Muslims – even though the terrorism in question is supported, let alone engaged in, by a very small proportion of Muslims. While the reference '9/11' is the most relevant here in terms of public consciousness, its specificity in relation to terrorism means that it is less relevant in relation to the themes of this essay, namely those of equality, diversity and national identity. In relation to the latter, the pivotal events could be said to be those captured by the terms, 'the Rushdie Affair' and 'l'affaire du foulard', both occurring independently yet simultaneously in 1988–1989.[41]

A good way to approach the CORAB is through the argument that Britain consists of a Christian legacy, new faiths and the non-religious.[42] More specifically, that current debates about multiculturalism in relation to religion can I think be seen in terms of three contending forces,[43] which may be expressed in the contentions that Britain's public institutions and national identity should reflect that it is:

1 still predominantly a Christian country,
2 now a multi-faith country,
3 mainly a secular country.

In relation to the last we need to distinguish between several meanings of 'secular'. First, 'secular' may refer to philosophical and ethical perspectives; non-religion not as an absence but a presence; not as the opposite of religion but in terms of its status as a conversational partner of religion or even as one of many 'spiritual families' that constitute belief in modernity.[44] One of the most fundamental conceptual features of the CORAB report is the importance it gives to such secular worldviews, what it refers to, using the vocabulary of French legislation in relation to freedom of belief, as 'les convictions' (p. 14). While the CORAB report notes that such ideas developed during the Enlightenment and are at the core of ideas of equality and liberal democracy, and at times have been in an antagonistic relationship to religion, actually '[t]he two sets of values have also at times intermingled and converged, and have deeply influenced each other' (para 2.11). Yet the decision to place religion and secular convictions on the same plane was arrived at not by a philosophical or a theological discussion but by a resolution at the first meetings to fully embrace the legal non-discrimination concept 'religion or belief' and to express this in the title and terms of reference. The UK Equality Act, 2010, itself following the European Convention on Human Rights in relation to freedom of religion, expresses the legal offence of religious discrimination by reference to 'religion or belief', that is to say, religion and secular beliefs are treated on a par. CORAB took the legal category 'religion or belief', up to now only applied to in relation to non-discrimination and freedom of religion, and applied it to all aspects of the report, for example to education and the media and not just the law. Humanistic beliefs, principles and conscience are treated on a par with religion; one might say, though CORAB does not use this vocabulary, as an almost quasi-religion or alternative to religion.[45] It did not mean that everyone who in a Census or attitude survey said they had no religion had a humanist philosophy or participated in humanist meetings or rituals, but that was no disability as this simply paralleled the situation of those who identified as Christian or Muslim. It is possible, however, that CORAB did not fully appreciate the implications of generalising the equality between religious and non-religious beliefs. An example of this egalitarian inclusion, what I have elsewhere called 'equalising upwards',[46] is the recommendation that '[f]unding for chaplaincies in hospitals, prisons and higher education should be protected, but with equitable representation for those from non-Christian religious traditions and for those from humanist traditions' (para 3.27).

A second meaning of 'secular' can be understood in relation to political secularism. As a concept it might be said to have two alternative contemporary understandings of 'secular':

a) Religion is a private matter and the state may act to confine it thus.

Some British people often think that political secularism has the first of these meanings until they reflect on Anglican participation in the national rituals such as the remembrance of the dead in wars or a royal wedding or state-funded faith schools, which most British people have long supported and in the main continue to support. This then tends to lead British people to express the alternative view:

b) Religion has something to contribute to the public good but should do so within liberal democratic constitutionalist norms and framework.

What most British people approve of in political secularism is not an absolute separation between religion and the state but that religion as a feature of the public life of the country works within – indeed freely and actively supports – the liberal democratic constitutional character of the British state and politics.[47]

It is perhaps not surprising that the first of these is not represented in a commission investigating the place of religion in public life, but it should also be noted that judging by the composition of CORAB, secularist philosophies are not significantly represented.[48] Indeed, there is only one commissioner whose affiliation is indicated by reference to a secularist philosophy, namely, Andrew Copson, the chief executive of the Humanists UK. Given, though, the fact of multiple identities, it would be appropriate to recognise that at least some members with a religious community, if not a religious organisation identification, are likely to overlap with secularist as well as religious perspectives.[49] So even though the CORAB could be said to limit the scope of the secular *vis-à-vis* public religion relative to the concept of privatisation and radical secularism (as per (a) above), of more significance is how by embracing an extensive concept of 'religion and belief', the commission adjusted the balance within the triangular relationship in favour of secularism – not because of any animus to Christianity but in an inclusive spirit in the context of socio-demographic trends, to which I now turn.

According to the Census the number of people who identified themselves as Christian in England and Wales declined from 71 to 58 per cent between 2001 and 2011 (p. 86).[50] While, the Censuses are of the full population they are a relatively superficial measure of identification. In this regard it is best to supplement the Census with the British Social Attitudes (BSA) survey findings, which is what CORAB does. The BSA, which includes Scotland, only has a sample of 3,000 but it is a systematic purposive attitudes survey and has been measuring religious attitudes annually since 1984. It shows that Christian identification has dropped from 68 per cent in 1984 to 43 per cent in 2014 (p. 16). This shows a steeper drop than the Census and so we might suppose that Christian identification has different meanings in different contexts and/or the true picture is somewhere between the two figures. The BSA offers data by Christian denominations and shows that during 1984–2014 the numbers identifying as Anglicans declined from 40 to 17 per cent; and Catholics from 10 to 8 per cent. Other Christians remain steady at about 18 per cent of the British population. While there has been a decline in older Protestant churches such as Methodists and Baptists, there has been a growth in newer churches such as Pentecostals and Seventh Day Adventists, many of which are black majority-led. The findings are, then, that Christians are barely or not at all a majority in England and Wales, and Anglicans are no longer a majority among Christians; though of course Anglican is an elastic identity, as indeed is 'Christian' itself, as comparing the Census and the BSA shows. Yet Anglicanism does seem to have suffered a particularly steep decline.

On the other hand, those who say they have no religion have gone up from 15 to 25 per cent between the 2001 and 2011 Census (p. 86); and in the BSA from just under a third in 1984 to almost half in 2014 (p. 16) – the two datasets being more discrepant on this point than on Christians. The key point is that both datasets show the same trend and strongly suggest that it is not short-term.

As for religions other than Christianity, the BSA sample sizes are too small to be useful but the 2001 and 2011 Censuses show that a significant number of non-Christian religious minorities are to be found in all the major cities and towns and doubled in size, comprising nearly 10 per cent of England and Wales in 2011.[51] Being younger and having larger families, they will continue to grow. For example, while Muslims are about 5 per cent of the population, they are 8 per cent of those under 25. While Hindus, Sikhs, Jews and Muslims are about 12 per cent of 9–13 year olds (three-quarters being Muslims) they are double this proportion in London.

The new religious diversity and significance of religion

While for some right-wing newspaper commentators and editorials the big story of the commission allegedly was one of Christian marginalisation and secular majoritarianism, my interest here is in the third point of the triangular relationship, namely, the minority faiths.[52]

While each new generation across the last century seems to be less Christian than its predecessor and so Christianity is deemed to be important to their life by few young people today, this generational indifference is not found among post-immigrant groups. Indeed, among ethnic minorities expressions of commitment among the young can be exceptionally high: more than a third of Indians, and two-thirds of Pakistani and Bangladeshi 16–34-year-olds said in a national survey at the end of the twentieth century that religion was very important to how they led their lives compared to a fifth of Caribbeans and 5 per cent of whites.[53] In the case of young Muslims, the importance of religion has been rising and overtaking their elders.[54]

Beyond that, religion has a social importance for minorities. In South Asia, from where the majority of British Hindus, Sikhs and Muslims originate, religious identity has a salience much higher than in Britain, so it is not surprising that during the last few decades religion – rather than say colour or linguistic heritage or national origins – has risen in the individual and community self-identities of these minorities together with their sense of Britishness.[55] Of course groups are partly defined not by what they say they are but by what others – usually a dominant group – say they are. British society, as it happens, has been very receptive to ethnic minority self-definitions and re-definitions. For example, when West Indians said they were African-Caribbean; when non-whites said they were not 'coloured' but blacks; when Asians said they were not 'black'; ethnic minority collective self-projections quickly altered public discourses and prevailed over older nomenclature. So, similarly, when ethnic minority groups such as Pakistanis started dismissing 'Pakistani' and 'Asian' for themselves in preference for 'Muslim', the wider British public, especially the media and public organisations, fell in line. Since the 1980s–1990s onwards there seems to have been a constant rise in religious community identification among South Asians to supplement or demote race/ethnic identifications, and in a very few years they moved from 'black' to 'Asian' to Indian/Pakistani/Bangladeshi etc. to Sikh/Muslim/Hindu etc.[56]

This was then picked up and reinforced by the media, politicians and public discourse in relation to those groups or more widely. For example, the spring–summer 2001 riots in northern cities like Bradford and Oldham were originally reported by the national press as 'Asian' (though South Asians themselves were referring to the rioters as Muslim), but as the months passed that term shifted to 'Muslim', a transition completed by the events of '9/11' in September and discussions thereof.[57] This does not necessarily refer to religiosity but is a recent manifestation of the well-known phenomenon that Jews generally and Catholics in locations like Northern Ireland can call themselves and can be called by others as Jews and Catholics respectively even if they are not religious and may even be anti-religious. We are here clearly talking about group identity or ethno-religious community membership not belief.[58]

Of course, as indicated above, what minorities are usually unable to do is to control the meaning of terms. This again is most evident in the recent period, especially after 9/11 and in Britain especially after the July 2005 bombings in London ('7/7'), in relation to Muslims and Muslim identity or public discourses of Islam. Muslims may have demanded recognition *qua* Muslims and may have propelled that identity into public discourse and popular consciousness but very few Muslims have sought to have 'Muslim' mean fanatic, fundamentalist, misogynist, separatist or terrorist but anyone familiar with current affairs and how it is reported in the British media knows that this is what Muslim currently connotes in Britain.[59] This stereotyping of Muslims, part of the phenomenon generally called Islamophobia, can be understood as 'racialisation'. Not simply because that is what happens to groups designated as 'races', nor even because non-whiteness is closely associated with being a Muslim,[60] but it is to treat Muslims as if they were a single, racial or quasi-racial group. The dissonance that one might experience here in accepting the idea that a religious group is a 'race' can be eased by considering the general case of how the Jews have been racialised (indeed in continental Europe the Jews are the quintessential race), as well as the specific case of Catholics in Northern Ireland or Muslims in the 'ethnic cleansing' rampages in the former Yugoslavia.[61] In this regard it is worth noting that the CMEB recommendation that discrimination on grounds of religion or belief should be outlawed (quoted above) began to be fulfilled from 2003 onwards; initially only in relation to employment, following the EU directive 2000/78, later also covering services, and was fully implemented in the Equality Act, 2010. In that Act, religion or belief as a ground for discrimination was put on a par with all other grounds for discrimination in the strongest anti-discrimination legislation in Europe. The CORAB was very mindful of the phenomenon of religious discrimination and of how it was often based on a racialised view of religion and intertwined with racial discrimination, suggesting that the term 'ethno-religious' – a legal concept in Australia – captures an aspect of this (para 2.5). It did make a recommendation for the law to better understand this intertwining (paras 8.18–8.21) but in the light of the 2010 Act there was much less scope for it to make recommendations in relation to discrimination than might be the case in many other countries.

It should be clear from the above that the meaning of religion can vary between religions, in particular between Christians and non-Christians, or between being a member of a majority or a minority religion: for Jews, Muslims, Hindus and Sikhs religion is not just about belief but also sometimes primarily about community and cultural heritage or identity, including resisting stereotypes about one's own community or discriminatory treatment. Yet another way in which religion is not just about belief is that it often requires a public performance or a behaviour, e.g. in relation to codes of dress or food, and so is much more publicly visible and sometimes requires institutional adaptation in order to be accommodated.[62] While this is barely a feature of modern, especially Protestant, Christianity, where 'inner belief' can be considered sufficient and it is often deemed unnecessary, perhaps even inappropriate to display markers, even a cross, of one's faith, this is quite exceptional in global, and now British, terms. Most religions require the observance of rules of piety and Britain is experiencing such practice-based religions re-entering the public space – Muslim dress being the most visible and contentious – after quite a long period in which such religion has been eroded away or transformed into private belief. Institutions and areas of public life which have given up the need to accommodate Christians are now having to adjust to the needs of minority faiths, and sometimes stimulating Christian reappraisal of its retreat from public piety (e.g. the display of a wearer's cross, as in the *Eweida* case at ECHR). Dietary requirements, space for worship, and gender relations, besides dress, are also prominent as elements of religious praxis that institutions such as schools, hospitals and prisons, and even workplaces are being asked to adjust to. Adaptations of codes of dress or uniforms, or provision of vegan, vegetarian, kosher and halal meals, places for worship and time off to use them are the kinds of requests being made upon state institutions, universities, employers and so on even when no parallel provision exists for Christians and is not being requested for by Christians. This praxis-based accommodation is a significant multiculturalist challenge because it is not simply a matter of granting minorities provisions already enjoyed by the majority but a matter of respecting minority religions in ways that Christians may be indifferent too in relation to their own faith. And, of course, it is not just a symbolic recognition that is being requested as substantive provisions or institutional changes are sometimes necessary.

Partly because minority faiths are behaving in ways that most British people are not familiar with and, at least *prima facie*, sometimes may not approve of, and whose accommodation require some funding or others to change their behaviour, especially at the level of institutional rules and the shared public space, there is some majoritarian unease. Even when minority claims of public recognition and respect can be met with modest policy and resource commitments, some people, perhaps a growing number at the moment, are uncomfortable with and believe there is an over-religionising, specifically over-Islamising of the public sphere. Some of this cultural anxiety and antipathy to (too much public) religion merges with aspects of cultural hostility which may be forms of racism and Islamophobia. It may even be that in the standard case it overlaps with anti-minorities, especially anti-Muslim, sentiment. Nevertheless, the cultural anxiety I am referring to is not the same as cultural racism and needs to be identified and engaged with in its own right. CORAB partly acknowledged this in one of its grandest recommendations:

> A national conversation should be launched across the UK by leaders of faith communities and ethical traditions to create a shared understanding of the fundamental values underlying public life. It would take place at all levels and in all regions. The outcome might well be – in the tradition of Magna Carta and other such declarations of rights over the centuries – a statement of the principles and values which foster the common good, and should underpin and guide public life.
>
> (para 3.30)

Of course the anxieties – no less than the hostility and collective blaming – are considerably heightened when we are not simply talking about cultural difference, public religion or institutional accommodation but of terrorism and security. Some Muslims are involved in international terror campaigns and so, for this reason too, are a source of public anxiety and suspicion, and the object of state surveillance, which in turn leads Muslims to be alienated and distrustful. Here CORAB recommends a governmental approach that works with and not against the Muslim community and with academic research (paras 8.22–8.25).

The net result of what I have been describing in this section is that minority religions have come to have a significant – even if contested – public presence.[63] Public campaigns for inclusion and equality, conflicts over faith schools, women's dress and gender more generally, not to mention all the issues to do with the 'war on terror' and Islamist radicalism, has made religion much more politically prominent and public affairs generally. Public dialogue, representation and leadership is often sought and realised by those who define themselves in terms of religious community organisations.

While these are some of the reasons, religion has become more publicly salient and the meaning of religion has changed, so some appropriate public learning is required. It has also to be acknowledged that most people are losing touch with their own religious heritage, let alone understanding the new significance and variety of religion. At this juncture of simultaneous growth in ignorance about religions and their rising public, political, international and geo-strategic importance CORAB makes one of its central recommendations. Namely that there is a substantial need for raising the level of religion and belief literacy among the public, journalists, policymakers and various kinds of professionals (p. 9). The suggestion that the low level of public understanding of religion is a problem and needs to be remedied is not new to CORAB.[64] What I believe is distinctive about CORAB is the understanding of religion that it is said greater literacy of which is needed. Namely, religion which is extended to include secular belief systems; yet which is not reducible to belief for it includes pious practice, as for example, in relation to dress; which is a community or ethnic identity, especially for minorities; and which is intertwined with racism such as in the case of anti-Semitism or Islamophobic racialising of Muslims as an 'Other'. This complex, multi-dimensional understanding of religion, the lack of literacy in which is a serious problem of public life, is a thread across the report and is fundamental in the report's discussions of law, the media, education, social action and dialogue. Just as CORAB's generalisation of the legal concept of 'religion or belief' has extended religion in one direction, so the ideas of praxis, racialisation and identity in the concept of 'ethno-religious' has extended it in another. Moreover, religious literacy is required across society and especially by opinion-formers and policy-shapers and implementers because religion in this extended sense has something to contribute to the common good. But this is not a crudely pro-religion view. Nor just because secular beliefs and ethno-religious communities are folded into it. CORAB recognises that religion is 'a public good' but also 'a public bad' and so for both these reasons governments have a legitimate and indeed a necessary interest in it (para 2.6).[65]

If religion and belief literacy is primarily intended to remedy the educational deficits of adults, it is of course important to bring up the next generation with fewer of these deficits. In some of its educational recommendations, if taken together, the 'living with difference' (as the report was entitled) approach of CORAB comes out nicely. For example, it recommended the ending of the current statutory requirement of Christian worship in state schools – in reality more observed in the breach – and suggested the development of inclusive but non-compulsory school assemblies (para 4.28). This angered some Christian observers but it should not do so when it is taken together with the recommendation of 'publicly funded schools to be open for the provision of religion- or belief-specific teaching and worship on the school premises outside of the timetable for those who request it and wish to participate' (para 4.28). Moreover, it pressed for all schools to teach a subject dealing with religious and non-religious worldviews, which should have the same status as other humanities subjects (p. 83). It also affirmed state-funded faith schools and while recognising that they can increase segregation and have a divisive effect that needs to be actively addressed by seeking a mix of pupils, it resisted the imposition of any form of quota of non-faith pupils upon some schools, as it was urged to do. In this way I think it avoided being either pro or anti any one part of the Christian-secular-diversity triangle that I said characterises contemporary Britain, but sought to achieve a reasonable balance and interdependence between the points of the triangle. With a judicious mix of the compulsory and the encouraging it avoided a one-size-fits-all solution and ensured space for difference while making sure that there be a commonality, a learning together of all religions and belief systems in mixed class settings.[66] This set of recommendations is clearly trying to strike a balance between inclusivity and allowing some space for difference; recognising that the latter can be divisive, at least in state schools it would be in a context of everyone taught together to acknowledge and respect difference as well as commonality. Beyond that, all faith traditions may continue to run independent schools and state-supported schools but encouraged to include diversity too. I therefore do not accept, as one-sided commentators have claimed, that in its educational recommendations CORAB erased the idea of Britain as a Christian nation. It was not a turning back on the meaning that Christianity has in the formation of an evolving national identity but a weaving together of all three points of the national triangle in a way that is inclusive of each and balances the cultivation of a common life with the nourishing of difference for whom that is important.

Conclusion

It has been remarked that the CMEB had some ambivalence about Britishness. While it was wrongly perceived as suggesting that to be British or wanting to be British was racist, and it firmly put 'rethinking the national story' on the race equality agenda, it was ambivalent about whether Britain was a nation and could achieve the goal of an anti-racist, multicultural belonging, or whether it was a 'post-nation'.[67] CORAB's terms of reference, cited above, indicate a much more affirmative view of Britain by making 'ideas of Britishness and national identity' central to its purpose. A dialogical or multilogical Britishness seems to suggest itself as a goal, the solution to the problems discussed by CORAB. It is interesting that at a time when academic attention has been on globalisation, the post-national and transnationalism, British diversity has been centred on the national. In many ways, this is a more general European trend in which when politicians talk of 'integration' they are not usually talking about European integration but about how to make the new minorities truly French, more German etc. In so far as Britain differs it is that the national identity is seen as something which itself should be understood in terms of diversity. This is present in the central message of the two latter commissions discussed, though perhaps it is carried further in CORAB. While the CMEB top bullet-pointed summary item above emphasised 'rethinking the national story and the national identity', CORAB explicitly foregrounded in its terms of reference the dialogical relationship between commonality and difference: 'examine how ideas of Britishness and national identity may be inclusive of a range of religions and beliefs, and may in turn influence people's self-understanding'. The difference is perhaps slight but while CMEB emphasised – too one-sidedly in the view of its critics – the interrogation of dominant ideas of British national identity, CORAB emphasised the dialogical remaking of inherited national identities which are both productive of minority inclusion and reflective of difference. Perhaps CORAB was merely trying to state the message of the CMEB more clearly and avoid the brickbats received by the latter. If CMEB was interpreted as over-emphasising the deconstructive attitude towards received notions of national identity, CORAB was careful to project a more constructive tone.

The thirty-years trajectory from Swann via CMEB to CORAB shows not just effort to think through equality and diversity in a period in which the minorities were growing in size and becoming more significant in relation to British national identity. It is also a thinking about the nature of multiculturalism in a context which changes in ways that few if any would have predicted. This is certainly to do with how religion and not just colour becomes the marker of minority status and of otherness. Most importantly it is about the interaction of race/ethnicity and religion in the 'ethno-religious', which when combined with an extensive concept of 'religion and belief' displays in the British context a reworked concept of religion. At the same time there is – despite some ambivalence in the CMEB – a line of affirmation in an evolving, plural, British national identity – one which cannot be taken for granted but must be revisited as a work in progress. This happens at a time, 1985–2015, when many have been anticipating a fading away of the national in favour of the global and the local.[68]

Notes

1 I would like to thank Geoff Levey, Bhikhu Parekh, Robin Richardson and Varun Uberoi for their comments on this essay and for numerous conversations over the years on some of the topics of the essay. They of course are not responsible for the views expressed here.

2 Vertovec, Steven. "Super-diversity and its implications." *Ethnic and Racial Studies* 30.6 (2007): 1024–1054.

3 Declaration of interest: I was the academic adviser and contributor to CMEB and an active part of the collective authorship of its report (also known as 'The Parekh Report'), and a commissioner and steering group member of CORAB. Of course, all views expressed in this essay are my own and no doubt others from CMEB and CORAB might express some points differently or perhaps disagree in some instances.

4 Though Parekh became a peer only towards the end of the life of his commission.

5 Actually Bhikhu Parekh served on all three commissions – as a member (resigned November 1981 to become Vice Chancellor, University of Baroda), as a chair, and as an active patron respectively.

6 The final report also contained chapters on the educational needs of Italian, Cypriot, Traveller and Ukrainian children as well.

7 Scarman, Leslie. *The Scarman Report: The Brixton Disorders 10–12 April 1981: Report of an Inquiry*. Penguin, 1981.

8 Macpherson, Sir William. *The Stephen Lawrence Inquiry Report*, Presented to Parliament by the Secretary of State for the Home Department by Command of Her Majesty. February 1999 Cm 4262-I.

9 http://resources.cohesioninstitute.org.uk/Publications/Documents/Document/Default.aspx?recordId=18.

10 Hunter, Alistair, and Christina Boswell. "Research-policy dialogues in the United Kingdom." In Peter Scholten et al. (eds), *Integrating Immigrants in Europe*. Springer, 2015: 233–251.

11 https://en.wikipedia.org/wiki/1981_Brixton_riot.

12 Richardson, Robin. "Children will be told lies: Distortions, untruths and abuse in the media coverage." *Runnymede Trust Bulletin* 324 (2000): 12–13; McLaughlin, Eugene, and Sarah Neal. "Who can speak to race and nation? Intellectuals, public policy formation and the Future of Multi-ethnic Britain Commission." *Cultural Studies* 21.6 (2007): 910–930; Modood, Tariq. "Multiculturalism and Britishness: Provocations, hostilities and advances." In Romain Garbaye and Pauline Schnapper (eds), *The Politics of Ethnic Diversity in the British Isles*. Palgrave Macmillan, 2014: 21–37.

13 Bingham, John and Steven Swinford. "Britain is no longer a Christian country and should stop acting as if it is, says judge." *Daily Telegraph*, 7 December 2015. www.telegraph.co.uk/education/12036287/Britain-is-no-longer-a-Christian-country-and-should-stop-acting-as-if-it-is-says-judge.html; Gledhill, Ruth. "New Butler-Sloss religion report destroys our nation's defence against evil." *Christian Today*, 7 December 2015.

14 Ralph Grillo in his essay in this volume also identifies these three reports as having a special significance and shares the point just made about COIE. In relation to the latter, see Hunter and Boswell. "Research-policy dialogues". For more general discussions about how research on immigrant integration is used in public or policy discourses across Europe, see Scholten, Peter, et al. *Integrating Immigrants in Europe: Research-Policy Dialogues*. Springer, 2015; and Husband, Charles (ed.). *Research and Policy in Ethnic Relations: Compromised Dynamics in a Neoliberal Era*. Policy Press, 2015.

15 It is worth mentioning a certain continuity of personnel: Bhikhu Parekh was present in all three (see note 5), I was present in the two most recent (as per note 3) and Robin Richardson was the editor of those two commission reports.

16 Modood, Tariq. "'Black', racial equality and Asian identity." *Journal of Ethnic and Migration Studies* 14.3 (1988): 397–404; Modood, Tariq. "Political blackness and British Asians." *Sociology* 28.4 (1994): 859–876.

17 Swann, Lord. *Education for All: Report of the Committee of Inquiry into the Education of Children from Ethnic Minority Groups*, 1985: 8.

18 Parekh, Bhikhu C. *Colour, Culture and Consciousness: Immigrant Intellectuals in Britain*. No. 3. G. Allen & Unwin, 1974: 230.

19 Parekh, *Colour, Culture and Consciousness*: 230–231.

20 Chapter 8: "Religion and the role of the school: Religious education and the 'separate' schools debate."

21 Their dissenting view is on p. 515 of the report.

22 CMEB Report: 366–371. I include the Report's editor, Robin Richardson, who is not listed on pp. 366–371 but who played a key role in the shaping and drafting of the report.

23 Parekh, Bhikhu. "Reporting on a Report." *The Runnymede Bulletin*, June 2000: 1–8; Richardson, "Children will be told lies"; Modood, "Multiculturalism and Britishness."

24 Modood, Tariq. "Difference, cultural-racism and anti-racism." In Pnina Werbner and Tariq Modood (eds), *Debating Cultural Hybridities: Identities and the Politics of Anti-Racism*. Zed Books, 1997: 154–172; Modood, Tariq. *Multicultural Politics: Racism, Ethnicity and Muslims in Britain*. University of Minnesota Press and University of Edinburgh Press, 2005: 11–18, 27–45.

25 Runnymede Trust, *Islamophobia: A Challenge for Us All*. London: Runnymede Trust, 1998.

26 CMEB Report: 237.

27 Richardson, "Children will be told lies."

28 Ford, Richard. *The Times*, 12 October 2000.

29 Parekh, Bhikhu. "A Britain we can all belong to." *The Guardian*, 11 October 2000. For a fully philosophically elaborated statement, see Parekh, B. *Rethinking Multiculturalism: Cultural Diversity and Political Theory*. Harvard University Press, 2000. For a discussion of Parekh's understanding of multiculturalism and national identity and how it is present in the CMEB report, see Uberoi, Varun. "National identities and moving beyond conservative and liberal nationalism." In Varun Uberoi and Tariq Modood, (eds), *Multiculturalism Rethought*. Edinburgh University Press, 2015: 75–94, and Uberoi, Varun. "The 'Parekh Report': National identities without nations and nationalism." *Ethnicities* 15.4 (2015): 509–526.

30 Actually most of the report's more than 140 recommendations were acted upon quite soon after its publication. Runnymede Trust. *Realising the Vision: Progress and Further Challenges*. Runnymede Trust, 2004.

31 Denham, John. *Building Cohesive Communities: A Report of the Ministerial Group on Public Order and Community Cohesion*. Home Office, 2001 (italics added).

32 Straw, Jack. Cyril Foster Lecture, 25 January 2007 (italics added).

33 Uberoi, Varun, and Tariq Modood. "Inclusive Britishness: A multiculturalist advance." *Political Studies* 61.1 (2013): 23–41. For more on the Leverhulme programme see

www.bristol.ac.uk/ethnicity/projects/leverhulme/ and Modood, Tariq, and John Salt. "Global migration, ethnicity and Britishness." In Tariq Modood and John Salt (eds), *Global Migration, Ethnicity and Britishness*. Palgrave Macmillan, 2011: 248–268.

34 Katwala, S. "An island story: Boyle's Olympic opening was irresistibly British." *Open Democracy*, 31 July 2012. www.opendemocracy.net/ourkingdom/sunder-katwala/island-story-boyles-olympic-opening-was-irresistibly-british.

35 Soutphommasane, Tim. *The Virtuous Citizen: Patriotism in a Multicultural Society*. Cambridge University Press, 2012.

36 Alibhai-Brown, Y. "Mo's joyful embrace of Britishness and why these Games mark a truly historic watershed." *Mail Online*, 12/13 August 2012. www.dailymail.co.uk/debate/article-2187469/Mo-Farahs-joyful-embrace-Britishness-Games-mark-truly-historic-watershed.html. As CORAB later commented: 'What the ceremony portrayed, essentially, was a multinational nation that was rooted in but not stuck in the past, proud yet also self-deprecating, open to the future, open to the wider world, and concerned with the common good' (para 3.8).

37 Modood, "Multiculturalism and Britishness."

38 CORAB is also discussed by Maleiha Malik in her essay in this volume and set in the context of a useful discussion of the role of commissions in Britain and of non-legal regulation.

39 'Reasonable accommodation' in the workplace has not been made into law as in the US and Canada. CORAB argued that where a legal offence of indirect discrimination is in place, the additional right to reasonable accommodation achieves no material benefit (para 8.10).

40 Modood, Tariq. "2011 Paul Hanly Furfey Lecture: Is there a crisis of secularism in Western Europe?" *Sociology of Religion* 73.2 (2012): 130–149.

41 Modood, "Crisis of secularism."

42 Weller, Paul. *Time for a Change: Reconfiguring Religion, State and Society*. Bloomsbury, 2005: 117.

43 Modood, "Crisis of secularism."

44 Taylor, Charles. "Foreword: What is secularism?" In Geoffrey Brahm Levey and Tariq Modood (eds), *Secularism, Religion and Multicultural Citizenship*. Cambridge University Press, 2009: xii.

45 Interestingly, in 2013 two Britons, Sanderson Jones and Pippa Evans, started The Sunday Assembly because they 'wanted to do something that was like church but totally secular and inclusive of all – no matter what they believed'. There are now over seventy Sunday Assembly chapters in eight different countries where people sing songs, hear inspiring talks and create community together: www.sundayassembly.com/story. Meanwhile, there is a 'growing network of over 300 celebrants trained and accredited by Humanists UK to conduct non-religious weddings, funerals and other ceremonies': https://humanism.org.uk/ceremonies/. In July 2016, following the likes of Harvard, Westminster University created the first British paid post of secular advisor to join its multi-faith chaplaincy: www.independent.co.uk/student/news/westminster-university-secular-humanist-adviser-isabel-millar-jane-flint-a7137196.html.

46 Modood, Tariq. *Multiculturalism, Muslims and the British State*. British Association for the Study of Religions, 2003: 164.

47 Modood, Tariq. "Moderate secularism, religion as identity and respect for religion." *The Political Quarterly* 81.1 (2010): 4–14.

48 Members of the commission can be seen at www.corab.org.uk/background-information.

49 A point not taken into account by some who commented on the membership of the commission (e.g. Voas, David. "Can the commission justify religious privilege?" *Public Spirit*, 14 October 2014: www.publicspirit.org.uk/can-the-commission-justify-religious-privilege/; Morris, Bob. "'Living with difference': The Butler-Sloss Commission's report reflects the interests of its members rather than the public interest." UCL Constitution

Unit. https://constitution-unit.com/2016/01/05/living-with-difference-the-butler-sloss-commissions-report-reflects-the-interests-of-its-members-rather-than-the-public-interest/). The steering group Commissioner, Lord Harries, the former Bishop of Oxford and the most senior Christian on the commission was happy to describe himself as a Christian humanist. The final report was clear that while religious and humanistic values may sometimes conflict, and certainly have done so historically, they may also overlap and indeed be intertwined and influence each other (paras 2.11–2.12).

50 A religion question was only introduced in 2001 and so there is no earlier Census data on this. There is less diversity in Scotland, although a higher proportion are non-religious. In Northern Ireland, Christian affiliation is much higher (with Protestants and Catholics about 40 per cent each).

51 Or 7.4 per cent of the UK population, consisting of Muslims (4.8%), Hindus (1.5%), Sikhs (0.8%), Jews (0.5%) and others (0.8%), and double these proportions in many urban areas.

52 The reception of the report included some anger in certain Christian quarters, who interpreted it as anti-Christian and a secularist hollowing out of a Christian legacy and especially objected to the proposal to end compulsory Christian worship in all schools. This point of view was reflected in the *Daily Telegraph, Daily Express, Daily Mail* and *Spectator*. More favourable were the centrist/mainstream responses, as in *The Times, Independent* and *Guardian* and the BBC, who saw it as appropriately recognising the growth of non-religion as well as diversity.

53 Modood, Tariq, Richard Berthoud, Jane Lakey, James Nazroo, Patten Smith, Satnam Virdee and Sharon Beishon. *Ethnic Minorities in Britain: Diversity and Disadvantage*. No. 843. Policy Studies Institute, 1997.

54 GfK, N.O.P. "Social Research, (2006) Attitudes to Living in Britain: A Survey of Muslim Opinion"; see also Mirza, Munira, Abi Senthilkumaran and Zein Ja'far. *Living Apart Together: British Muslims and the Paradox of Multiculturalism*. Policy Exchange, 2007.

55 It is doubtful for example that most South Asians in Britain ever thought of themselves in terms of colour identities such as black or brown as much as some observers thought to be the case (Modood, "Political blackness and British Asians"; Modood et al. *Ethnic Minorities in Britain*: 291–297). In relation to Britishness see Modood et al. *Ethnic Minorities in Britain*: 338–331, and Heath, Anthony, and Neli Demireva. "Has multiculturalism failed in Britain?" *Ethnic and Racial Studies* 37.1 (2014): 161–180.

56 The first South Asian group to assert a religious rather than an ethnic (Indian) identity were the Sikhs in relation to workplace bans on the wearing of turbans in the 1960s–1980s and in favour of a Sikh separatist movement in the Punjab or how it was combatted by the Indian state in the 1980s.

57 I rely here on memory as interestingly this does not seem to have been documented in detail, though see Lewis, Phillip. *Islamic Britain, Religion, Politics and Identity among British Muslims*. 2nd edn. I.B. Tauris, 2002: 210–211.

58 Modood, Tariq. "Anti-essentialism, multiculturalism and the recognition of religious groups." *Journal of Political Philosophy* 6 (1998): 378–399. Of course historically speaking it could be said that the Jews were a people who had a religion (which came to be called Judaism) rather than a religious group; the same could perhaps be said of Hindus and Hinduism. The term 'ethnoreligious' here is therefore most apt.

59 Morey, Peter, and Amina Yaqin. *Framing Muslims*. Harvard University Press, 2011.

60 That 'Muslim' is racially coded (as colour, cultural alienness and not being European) in the way the CMEB argued Britishness is racially coded as whiteness.

61 Modood, *Multicultural Politics*; Meer, Nasar, and Tariq Modood. "Refutations of racism in the 'Muslim question'." *Patterns of Prejudice* 43.3–4 (2009): 335–354. Note, however, the point made in footnote 56. Jews may be considered as a racialised religious group or as a religionised ethnic group or 'nation'.

62 Modood, Tariq. "Religion in Britain today and tomorrow." *Public Spirit*, 26 January 2015. www.theosthinktank.co.uk/comment/2015/01/29/religion-in-britain-today-and-tomorrow; cf. Nilsson DeHanas, Daniel. *London Youth, Religion, and Politics: Engagement and Activism from Brixton to Brick Lane*. Oxford University Press, 2016. It is often said of some religions, for example Judaism and Hinduism, that they do not really have (many) core beliefs but are lived out and transmitted through core practices. This relates to the limitations of the word 'religion' as alluded to in footnote 56.

63 Modood, *Multicultural Politics*; Dinham, Adam, and Vivien Lowndes. "Faith in the public realm." In Adam Dinham, Robert Furbey and Vivien Lowndes (eds), *Faith in the Public Realm: Controversies, Policies and Practices*. Policy Press, 2009: 1–20.

64 Dinham, Adam, and Matthew Francis (eds), *Religious Literacy in Policy and Practice*. Policy Press, 2015.

65 Modood, "Moderate secularism"; as applied to the role of universities, see: Modood, Tariq and Craig Calhoun. *Religion in Britain: Challenges for Higher Education*, Stimulus Paper. Leadership Foundation in Higher Education, 2015.

66 Modood, Tariq. "Schools need to do more to improve children's religious literacy." *The Conversation*, 7 December 2015. https://theconversation.com/schools-need-to-do-more-to-improve-childrens-religious-literacy-51926.

67 Modood, "Multiculturalism and Britishness"; Uberoi, "The 'Parekh Report'."

68 Antonsich, Marco, and Tatiana Matejskova. "Immigration societies and the question of 'the national'." *Ethnicities* 15.4 (2015): 495–508.

Bibliography

Alibhai-Brown, Y. "Mo's joyful embrace of Britishness and why these Games mark a truly historic watershed." *Mail Online*, 12/13 August 2012. www.dailymail.co.uk/debate/article-2187469/Mo-Farahs-joyful-embrace-Britishness-Games-mark-truly-historic-watershed.html.

Antonsich, Marco, and Tatiana Matejskova. "Immigration societies and the question of 'the national'." *Ethnicities* 15.4 (2015): 495–508.

Commission on Multi-Ethnic Britain (CMEB/Parekh Report). *The Future of Multi-Ethnic Britain*. London: Profile Books, 2000.

Commission on Religion and Belief in British Public Life (CORAB). *Living with Difference: Community, Diversity and the Common Good*. Cambridge: The Woolf Institute, 2015.

DeHanas, Daniel Nilsson. *London Youth, Religion, and Politics: Engagement and Activism from Brixton to Brick Lane*. Oxford: Oxford University Press, 2016.

Denham, John. *Building Cohesive Communities: A Report of the Ministerial Group on Public Order and Community Cohesion*. London: Home Office, 2001.

Dinham, Adam, and Matthew Francis (eds), *Religious Literacy in Policy and Practice*. Bristol: Policy Press, 2015.

Dinham, Adam, and Vivien Lowndes. "Faith in the public realm." In Adam Dinham, Robert Furbey and Vivien Lowndes (eds), *Faith in the Public Realm: Controversies, Policies and Practices*. Bristol: Policy Press, 2009: 1–20.

Gledhill, Ruth. "New Butler-Sloss religion report destroys our nation's defence against evil." *Christian Today*, 7 December 2015.

Heath, Anthony, and Neli Demireva. "Has multiculturalism failed in Britain?" *Ethnic and Racial Studies* 37.1 (2014): 161–180.

Hunter, Alistair, and Christine Boswell. "Research-policy dialogues in the United Kingdom." In P. Scholten, H. Entzinger, R. Penninx and S. Verbeek (eds), *Integrating Immigrants in Europe: Research-Policy Dialogues*. Cham: Springer, 2015: 233–251.

Husband, Charles (ed.). *Research and Policy in Ethnic Relations: Compromised Dynamics in a Neoliberal Era*. Bristol: Policy Press, 2015.

Katwala, S. "An island story: Boyle's Olympic opening was irresistibly British." *Open Democracy*, 31 July 2012. www.opendemocracy.net/ourkingdom/sunder-katwala/island-story-boyles-olympic-opening-was-irresistibly-british.

Lewis, Phillip. *Islamic Britain: Religion, Politics and Identity among British Muslims*. Second edition. London: I.B. Tauris, 2002.

McLaughlin, Eugene, and Sarah Neal. "Who can speak to race and nation? Intellectuals, public policy formation and the Future of Multi-ethnic Britain Commission." *Cultural Studies* 21.6 (2007): 910–930.

Macpherson, Sir William. *The Stephen Lawrence Inquiry Report*, Presented to Parliament by the Secretary of State for the Home Department by Command of Her Majesty. February 1999 Cm 4262-I.

Meer, Nasar, and Tariq Modood. "Refutations of racism in the 'Muslim question'." *Patterns of Prejudice* 43.3–4 (2009): 335–354.

Modood, Tariq. "'Black', racial equality and Asian identity." *Journal of Ethnic and Migration Studies* 14.3 (1988): 397–404.

Modood, Tariq. "Political blackness and British Asians." *Sociology* 28.4 (1994): 859–876.

Modood, Tariq, "Difference, cultural-racism and anti-racism." In Pnina Werbner and Tariq Modood (eds), *Debating Cultural Hybridities: Identities and the Politics of Anti-Racism*. London: Zed Books, 1997: 154–172.

Modood, Tariq. "Anti-essentialism, multiculturalism and the recognition of religious groups." *Journal of Political Philosophy* 6 (1998): 378–399.

Modood, Tariq. *Multiculturalism, Muslims and the British State*. British Association for the Study of Religions, 2003.

Modood, Tariq. *Multicultural Politics: Racism, Ethnicity and Muslims in Britain*. Minneapolis, MN and Edinburgh: University of Minnesota Press and University of Edinburgh Press, 2005.

Modood, Tariq. "Moderate secularism, religion as identity and respect for religion." *The Political Quarterly* 81.1 (2010): 4–14.

Modood, Tariq. "2011 Paul Hanly Furfey Lecture: Is there a crisis of secularism in Western Europe?" *Sociology of Religion* 73.2 (2012): 130–149.

Modood, Tariq. *Multiculturalism: A Civic Idea*. Second edition. Cambridge: Polity Press, 2013.

Modood, Tariq. "Multiculturalism and Britishness: Provocations, hostilities and advances." In Romain Garbaye and Pauline Schnapper (eds), *The Politics of Ethnic Diversity in the British Isles*. Houndmills: Palgrave Macmillan, 2014: 21–37.

Modood, Tariq. "Religion in Britain today and tomorrow." *Public Spirit*, 26 January 2015. www.theosthinktank.co.uk/comment/2015/01/29/religion-in-britain-today-and-tomorrow.

Modood, Tariq. "Schools need to do more to improve children's religious literacy." *The Conversation*, 7 December 2015. https://theconversation.com/schools-need-to-do-more-to-improve-childrens-religious-literacy-51926.

Modood, Tariq and Craig Calhoun, *Religion in Britain: Challenges for Higher Education*, Stimulus Paper. London: Leadership Foundation in Higher Education, 2015.

Modood, Tariq, and John Salt. "Global migration, ethnicity and Britishness." In Tariq Modood and John Salt (eds), *Global Migration, Ethnicity and Britishness*. Houndmills: Palgrave Macmillan, 2011: 248–268.

Morey, Peter, and Amina Yaqin. *Framing Muslims*. Cambridge, MA: Harvard University Press, 2011.

Morris, Bob. "'Living with difference': The Butler-Sloss Commission's report reflects the interests of its members rather than the public interest." UCL Constitution Unit. https://constitution-unit.com/2016/01/05/living-with-difference-the-butler-sloss-commissions-report-reflects-the-interests-of-its-members-rather-than-the-public-interest/.

Parekh, B. *Colour, Culture and Consciousness: Immigrant Intellectuals in Britain*. London: Allen & Unwin, 1974.

Parekh, B. *Rethinking Multiculturalism: Cultural Diversity and Political Theory*. Cambridge, MA: Harvard University Press, 2000.

Parekh, B. "Reporting on a Report." *The Runnymede Bulletin*, June 2000: 1–8.

Parekh, B. "A Britain we can all belong to." *The Guardian*, 11 October 2000.

Richardson, Robin. "Children will be told lies: Distortions, untruths and abuse in the media coverage." *Runnymede Trust Bulletin* 324 (2000): 12–13.

Runnymede Trust, *Islamophobia: A Challenge for Us All*. London: Runnymede Trust, 1998.

Runnymede Trust, *Realising the Vision: Progress and Further Challenges*. London: Runnymede Trust, 2004.

Scarman, Leslie. *The Scarman Report: The Brixton Disorders 10–12 April 1981: Report of an Inquiry*. London: Penguin, 1981.

Scholten, P., H. Entzinger, R. Penninx and S. Verbeek (eds). *Integrating Immigrants in Europe: Research-Policy Dialogues*. Cham: Springer, 2015.

Soutphommasane, Tim. *The Virtuous Citizen: Patriotism in a Multicultural Society*. Cambridge: Cambridge University Press, 2012.

Straw, Jack. Cyril Foster Lecture, 25 January 2007.

Swann, Lord. *Education for All: Report of the Committee of Inquiry into the Education of Children from Ethnic Minority Groups*, 1985.

Uberoi, Varun, "National identities and moving beyond conservative and liberal nationalism." In Varun Uberoi and Tariq Modood (eds), *Multiculturalism Rethought*. Edinburgh: Edinburgh University Press, 2015: 75–94.

Uberoi, Varun. "The 'Parekh Report': National identities without nations and nationalism." *Ethnicities* 15.4 (2015): 509–526.

Uberoi, Varun, and Tariq Modood. "Inclusive Britishness: A multiculturalist advance." *Political Studies* 61.1 (2013): 23–41.

Vertovec, Steven. "Super-diversity and its implications." *Ethnic and Racial Studies* 30.6 (2007): 1024–1054.

3 National identities and the Parekh Report[1]

*Varun Uberoi**

'Multiculturalists' often advocate the value of cultural diversity, intercultural dialogue, intercultural learning and policies of multiculturalism like teaching children about a polity's multicultural nature.[2] But such 'multiculturalists' also advocate the importance of national identities,[3] yet many claim that 'multiculturalists' do not do so or that they have an 'anti-national bias'.[4] Likewise, scholars frequently examine why and how 'multiculturalists' advocate policies of multiculturalism.[5] But scholars seldom examine why and how 'multiculturalists' advocate national identities. This essay helps to fill this gap by examining how the Commission for Multi-Ethnic Britain (CMEB) advocated national identities. I will show that the CMEB's report reflects a previously unnoticed way of thinking about the nature and worth of national identities that its chairman, and prominent political theorist, Bhikhu Parekh, had been devising since the 1970s. This way of thinking about national identities is shown to be distinct as it differs from how other prominent thinkers discuss national identities, and valuable as it avoids the questionable ways in which these thinkers discuss the nature and worth of national identities.

Those familiar with the CMEB's report may think it strange to claim that this report reflects a valuable way of thinking about national identities as they recall how the media depicted the report. For 'several days' after the report was published, its claims about why cultural minorities should also be seen as 'British' were distorted by journalists who claimed, for example, that the report said 'being British' is racist.[6] Such 'damning headlines' did not prevent some of the report's ideas being used by various politicians.[7] But scholars who remember these headlines may think that the report is unlikely to reflect a valuable way of thinking about national identities. Yet I will show that the report does so, as it reflects Parekh's way of thinking about such identities.

Parekh's texts[8] relating to national identities began to appear in the 1970s when membership of the European Economic Community, mass immigration and Scottish and Welsh nationalism led him to ask who 'the British' were any longer. Parekh's texts relating to national identities continued.[9] But his work is not discussed by prominent scholars who debate the value of national identities.[10] These scholars may assume that as Parekh devotes limited attention to national identities in well-known books like *Rethinking Multiculturalism* and *A New Politics of Identity*, he also devotes limited attention to national identities in general. This assumption is understandable, but it is also mistaken.

This is because Parekh's various works relating to the nature and worth of national identities from the 1970s onwards can be found among his less well-known book chapters, review articles and published lectures. These pieces are not like peer reviewed journal articles as they contain ideas and arguments that Parekh was 'trying out' and that provoke thought but are, at times, unclear. The relevant texts by Parekh thus require some interpretation to clarify the nature and significance of the ideas and arguments in them or there is a danger that these ideas and arguments will just be ignored or deemed implausible. I will clarify some of Parekh's ideas and arguments using my own terms, examples and ideas just as Parekh explicitly does with other thinkers[11] and others scholars do too.[12] I will focus primarily on texts that Parekh wrote prior to the CMEB's report to show that the report reflects how Parekh had long discussed the nature and worth of national identities, while avoiding the difficulties that other prominent thinkers encounter when they discuss the nature and worth of national identities.

These other prominent thinkers are not Isaiah Berlin,[13] Elie Kedourie[14] or Ernest Gellner,[15] who all focus on the intellectual, political and functional sources of *nationalism* and barely mention *national identities*. Nor are the thinkers I am referring to ones who want to substitute ideas of nations and nationalism with ideas of the 'post-national' or thinkers who want to explain when, why and how nations and nationalism emerged as such thinkers do not argue specifically for the worth of national identities.[16] Instead, the prominent thinkers I am referring to are similar to Parekh as they too discuss at length both what national identities are and why we should value them. These prominent thinkers are those political theorists such as David Miller and Yael Tamir,[17] who are often known as 'liberal nationalists', and those such as Roger Scruton[18] who are often known as 'conservative nationalists'.[19]

Each of these nationalist thinkers has different presuppositions, philosophical approaches, arguments and, thus, political positions. But each of them refers to themselves as a 'conservative' or a 'liberal' and these traditions of thought shape their descriptions of ideas of nationhood like nations and nationalism. For example, the Burkean idea of polities being a 'partnership'[20] between the living, the dead and the unborn shapes how a conservative such as Scruton[21] understands what nations are as he depicts them as those sharing not only descent but a culture bequeathed from one generation to the next. Likewise, the liberal ideal that comes from Immanuel Kant,[22] J.S. Mill[23] and others of an individual choosing his life 'plan for himself' makes Ernest Renan's[24] idea of nations constituted by '*consent*' and a '*desire* to live a common life' attractive to liberals such as Miller and Tamir[25] who use Renan's ideas. After describing conservative and liberal ideas of nationhood, these thinkers endorse such ideas of nationhood so much so that Miller, Tamir and Scruton[26] argue that nations should be preserved through forms of 'self-rule' or by constituting states.[27] In doing so, these conservative and liberal thinkers advocate different forms of national*ism*[28] and thus unsurprisingly, they are often classified as conservative or liberal national*ists*.

As conservative nationalists think nations share descent and a single common culture they unsurprisingly do not claim that cultural diversity, intercultural dialogue or policies of multiculturalism are valuable.[29] But some well-known liberal nationalists elegantly advocate policies of multiculturalism and are seen as 'multiculturalists' too,[30] while other liberal nationalists are not.[31] I will show that the CMEB's report avoided the questionable ways in which all such conservative and liberal nationalists discuss national identities yet it reflected Parekh's distinct and valuable way of doing so. And I take two steps to achieve this.

First, I will show that the way in which the CMEB's report discusses the nature and worth of national identities differs from how conservative and liberal nationalists do. Second, I show how the report's discussion of national identities reflects Parekh's work, which also helps to illuminate why these nationalist thinkers discuss the nature and worth of national identities in questionable ways. I conclude by discussing why scholars and British politicians can use the ways in which Parekh and the CMEB advocate national identities.

The CMEB's report

A 'sense of national identity' was said in the Report to entail mental images or understandings of what Britain is.[32] Such understandings of Britain were disturbed after the Second World War as Britain's imperial, military and economic might disappeared and membership of the European Economic Community, mass immigration and devolution appeared.[33] The CMEB said that Britain can respond by trying 'to turn the clock back . . . digging in, defending . . . old values' or it can try to 'create more . . . inclusive' and 'cosmopolitan' images of itself. The report endorses creating such inclusive and cosmopolitan images of Britain as 'one nation'. But this 'nation' is not described as conservative and liberal nationalists might describe a 'nation'.[34]

This more inclusive and cosmopolitan nation is thus not, as conservative nationalists claim, a 'natural' 'organism' whose members share descent and fate and are of the same 'kind'.[35] This is neither inclusive nor cosmopolitan and the CMEB also did not describe a nation as a liberal nationalist might. Liberal nationalists, we saw, follow Renan[36] and claim members of a nation share a belief that they want to live a 'collective life' and liberal nationalists also suggest that 'forgetting' divisive historical episodes can aid this.[37] Yet for the CMEB, divisive historical episodes about, for example, discrimination or Empire often cannot just be forgotten by minorities and British people must tell 'truths' about these episodes.[38] Likewise, unlike conservative and liberal nationalists, the report, at times, also referred to Britain as a 'state' and a 'post nation' and thus the CMEB seemed, at times, ambiguous and sceptical about Britain being a nation.[39] Indeed, when 'one nation' was defined, the report said it must be 'understood as a community of communities and . . . of citizens'.[40] This would ensure British people included in their understandings of Britain not just 'individual' citizens but those who see themselves as part of racial, religious, cultural, regional and other groups.[41] It is difficult to find conservative and liberal nationalists who make similar claims about what nations are.

The CMEB also claimed that people's national identities had value for two reasons. First, when British people have such understandings of Britain, they have understandings of 'a social unity' or a collective of which they are part. This then helps British people to conceive collective challenges, goals and interests with other members of this collective.[42] Such understandings of Britain often, of course, exclude racial, cultural and religious minorities and can help to cultivate fear and discrimination against them too.[43] Yet if British people had 'mental images' of Britain that include minorities, this would help to normalise minorities in Britain and reduce fear of them and discrimination against them, which is then a second reason to value national identities, at least when they include minorities.

Such mental images of Britain are thus not said to be valuable, as conservative nationalists claim, because they are an 'eternal' understanding of the nation – one that is the same for current members as it was for their ancestors and will be for their children.[44] The report suggests, as we saw, how such understandings should change. Likewise, the CMEB does not suggest that national identities are valuable because they convey 'life's meaning'.[45] National identities are not said to make sense of life, or reconcile a person to the world in Hegel's famous sense. Likewise, the report does not suggest, as liberal nationalists can, that our national identities aid our ability to make choices.[46] But national identities are said in the report to help people to see themselves as a 'community', and liberal nationalists claim this too. Yet liberal nationalists claim such a community is one of 'solidarity', 'obligation' and 'trust' in which people willingly redistribute their income and the CMEB focus on other features of a community.[47] The CMEB thus note how such communities contain individuals and groups who do not exhibit 'blind loyalty' to the community. Instead, members of the community are so attached to it that they criticise it so as to improve it. Members of the community also disagree about many things but just as disagreements and disputes among family members do not stop them being a family, the same is said to be true in such a community whose internal disputes are thus called 'family quarrels'.[48]

Not all understandings of Britain necessarily include minorities and the CMEB wanted this to occur, thus the CMEB also discussed 're-imagining' understandings of Britain so that these understandings come, over time, to include minorities. This might occur through at least four means. First, public disputes and controversies such as the Rushdie Affair force people to think about Britain's multicultural nature, which is a necessary step if their views of it are to include different minorities.[49] Second, policies of multiculturalism can be introduced. Such policies can declare the polity to be multicultural as Canada did, or can mandate teaching children about the polity's multicultural nature. Such policies legitimise and promote understandings of Britain that include different minorities.[50] Third, perhaps through publicly funded education and the arts, British people must tell 'the truth' about episodes of discrimination and Empire as doing so reveals the implications of seeing British people as solely white or superior to others. Fourth, again using perhaps publicly funded education and the arts, British people must portray how different regional, religious, cultural and other groups exist and often understand Britain differently. Thus, we might say there are 'several' understandings of Britain, all of which must come to include minorities.[51]

Where conservative nationalists think national identities cannot be remade as they are natural and can be conserved or can die,[52] clearly the CMEB disagreed as it discussed national identities becoming more inclusive. The latter is something that liberal nationalists discuss too yet only some of them might agree with the CMEB that policies of multiculturalism can aid this process[53] while other liberal nationalists who are more sceptical about these policies might disagree.[54] Indeed, where both types of liberal nationalists distinguish between policies of multiculturalism and 'nation-building' policies that promote understandings of a polity through, for example, history classes in schools, the CMEB did not make this distinction.[55] Declaring a polity to be multicultural and teaching children about its multicultural nature will promote understandings of a polity too and thus these sorts of policies of multiculturalism may also be seen as nation-building policies.[56] Likewise, where liberal nationalists follow Renan and advocate 'forgetting' divisive episodes the CMEB advocated 'truth telling'. The report thus exhibits a way of thinking about national identities that differs from how conservative and liberal nationalists think about such identities. I will now show that the report reflects Parekh's distinct and valuable way of thinking about national identities

The report's Parekhian approach to national identities

Parekh often discusses national identities in two ways.[57] First, he discusses them as the identity of a polity and thus for example, Britain, France, Canada or America's identity and second as an identity that a person might exhibit when they say they feel 'British', 'French', 'Canadian' or 'American'. We often think and talk about national identities in both ways yet conservative and liberal nationalists do not distinguish between them. Nor did the report explicitly, but it discussed national identities in both ways.

The report thus discussed understandings of what Britain is and in doing so it was referring to a polity's identity. A polity is seen by Parekh as 'a territorially concentrated group of people' who accept 'a common mode of conducting their collective affairs' which might include common institutions, values, norms, procedures and so on.[58] Understandings of what a polity is often focus on a territorially concentrated group of people and can refer to their history, race, language, their political ideals, values, norms, institutions, the traditions of thought and behaviour that they use to conduct their collective affairs, and so on. But understandings of what a polity is can be held by those who are not its members, thus de Tocqueville offered an understanding of America, and Rousseau had an understanding of Poland when writing about it.

Yet the report said members of a polity have such understandings themselves and publicly funded education can help to provide these understandings. Thus children can be taught in schools about the language, norms, values, ideals, traditions of behaviour, history and so on of their polity. But such understandings of a polity often remain vague and thus politicians, historians, philosophers and other intellectuals often try to clarify them. Such people interpret and relate their polity's ideals, procedures, history and so on in different ways according to their own experiences, religious, political and other views. Other members of the polity might repeat the different understandings of their polity that are offered in schools or by politicians or by intellectuals while interpreting and relating their polity's features in their own ways too. Thus a polity comes to be understood by its members in a range of different ways hence some in Britain understand it as England, some in Canada think it is multicultural and some in America think it has a 'civil religion'.

However, the report, we saw, also referred to people's 'sense of national identity' and thus their feeling, for example, British, French, American or Canadian and so on. When a person feels this way, they feel their polity shapes what they are and thus they may also say that they are British, French, American or Canadian and so on. And they might feel shaped by their polity because its legal, political and educational institutions regulate their behaviour and condition their notions of what is acceptable and normal. They are partly, yet unavoidably, shaped by their polity. Thus we might say that while a polity is shaped by its members, its members are also shaped by it. Hence, when someone says they feel British they are not just saying they feel part of Britain, as they are also saying that they sense how it shapes them, thus they may discuss their 'Britishness' too. Yet such a person is now exhibiting the other way of discussing a national identity that I referred to above which is not the identity of a polity, but an identity that a person has.

Such an identity does not depict what a person is as a whole, but part of what they are, just as their sexual or religious identities might. Such identities usually only refer to parts of us that help to make us the person we are, like being 'a man', 'a Muslim' or 'an artist'. Thus, it is strange for a person to say they have an identity as a 'reading group member' or a 'commuter', as it is unlikely that these aspects of them make them who they are. But it is not strange for an immigrant to say they have become more British, French, Canadian and American and so on when they sense how their new polity has come to shape them. And this immigrant must have some, albeit usually vague, understanding of their polity, or they cannot identify it as being something that shapes them. Indeed, the same is true for all those who say that they feel British, French, Canadian, American and so on. They must also have some understanding of Britain, France, Canada or America or they cannot identify it as being something that shapes them. Thus the two ways in which we usually think and talk about national identities are not unrelated. Hence, the CMEB's report said, as we saw, that a person's sense of national identity entails an understanding of what their polity is.

The report also reflected Parekh's scepticism about whether polities need to be understood as being relatively culturally uniform and thus being what we often call 'nations'.[59] Few polities are so culturally uniform and those that were have altered and become what Habermas calls 'post-national'.[60] We saw how the report described Britain as a 'post-nation', but so did Parekh,[61] who refers to Habermas when saying this. Indeed, the report's hesitancy to describe Britain as a nation reflects Parekh's.[62] He suggested that Britain is a polity that is too culturally diverse to be a nation. But as many in Britain have long been shaped by its legal and political institutions, it is unsurprising that so many also see themselves, in part, as British. This suggests how people in a polity like Britain can have national identities and see themselves as 'British', but without being part of a nation called Britain.[63]

England, Scotland and Wales approximate much more closely than Britain to how Parekh conceives of nations. A history of separate educational institutions in Scotland, England and Wales, separate legal institutions in Scotland and now devolution means that many also see themselves as English, Scottish and Welsh. These are national identities too, thus a person can have more than one national identity and be, for example, British and English. However, those who see themselves as Scottish or Welsh *but not* British increased just after devolution and since,[64] and this is unsurprising. As legal and political institutions across Britain reduce, so does the extent to which people feel shaped by Britain and so does the extent to which they see themselves as British.

Conservative nationalists are unlikely to challenge Parekh's claim that nations are relatively culturally uniform as they say the same.[65] But liberal nationalists think about nations differently to Parekh. Hence liberal nationalists might question the idea that what we often call nations are relatively culturally uniform by saying that nations need only be culturally uniform at the 'public' level in the sense of a language, norms, values, beliefs and institutions regulating the collective affairs of a citizenry.[66] Such a 'public culture' originates from a cultural group but leaves space for norms, beliefs, traditions of other cultural groups including minority ones who over time can also shape a 'public culture'.[67] And describing a nation as relatively culturally uniform hides how 'thin' and thus accommodating a nation's culture can and should be. But while it is difficult to deny that some polities, like America and Canada, have 'thin' and accommodating public cultures, it remains unclear whether we do, and should, think of these polities as 'nations', and there is also another problem.

It is unclear how the cultures that liberal nationalists discuss can be both 'thin' and offer what they call 'meaningful ways of life across the full range of human activities . . . encompassing both public and private spheres'.[68] For some liberal nationalists, such cultures are 'the defining feature of nations' and as liberal nationalists refer to these cultures as 'pervasive cultures' and 'encompassing cultures' it is unclear why these cultures are also 'thin'.[69] Likewise, if the cultures that nations do or should have are so accommodating, it is unclear why all prominent liberal nationalists claim that immigration, in large numbers, threatens such cultures.[70] Such cultures change anyway, but liberal nationalists claim a nation should be able to 'control' how its culture changes and preference should be given to potential immigrants 'whose cultural values are closer to those of the existing population'.[71] These claims suggest that liberal nationalists are not always thinking about 'thin' and accommodating cultures and often have 'thick' and somewhat unaccommodating cultures in mind. Despite their claims then, liberal nationalists do not always themselves think that a nation's culture can and should be 'thin' and accommodating, thus it is unclear why we should think this either.

But liberal nationalists might instead query how Parekh can conceive of national identities without nations.[72] Liberal nationalists may note that something must make an identity 'national' and this is surely a nation, which is seemingly why liberal nationalists also often clarify what national identities are by first defining and defending what nations are.[73] But 'national' need not imply a nation when we refer to 'national negotiations' over a trade treaty or a 'national representative' like a prime minister. In such examples, 'national' can merely imply a polity and the same is true with what we usually call 'national identities'. This is why we can disagree about whether America, Canada and Britain are nations,[74] but this does not deter discussions about America, Canada and Britain's identity or people's American, Canadian and British identities. When referring to national identities, we do not necessarily have nations in mind, but we are often thinking of either the identity of a polity and, for example, America's identity, or the identity of a person when they claim, for example, they are American. Parekh's work illuminates both ways of discussing national identities and shows why specifying the natures of both need not require us to define and defend what nations are.

Yet Parekh also saw how understandings of a polity often focus on the ethnic and cultural characteristics of an ethnic and cultural majority.[75] Ethnic or cultural minorities are thus seen as outsiders even though they are often citizens and this exacerbates their exclusion and discrimination. Thus, at the time of the report's publication, Parekh argued that understandings of a polity should not be 'ethno-cultural'. This is because the polity too easily becomes understood as ethno-culturally uniform, and ethno-cultural minorities are excluded from such understandings.[76] But part of opposing an ethno-culturally uniform understanding of a polity is opposing a culturally uniform understanding of it. And a culturally uniform understanding of a polity is what Parekh calls a nation. This means that Parekh thought not only that understandings of a polity *need not* depict a nation, but also that they *should not* depict a nation.[77]

Hence Parekh claimed, at the time of the report, that people's understandings of their polity must be solely 'politico-institutional'.[78] Such understandings could, for example, focus on a constitution and a bill of rights as this occurs in America and does not exclude ethno-cultural minorities. But Tariq Modood noted that understandings of a polity that are solely 'politico-institutional' cannot logically include 'ethno-cultural' minorities *in* them precisely because these understandings are solely politico-institutional in nature.[79] If the aim is to include ethno-cultural minorities in popular understandings of the polity it is unclear why 'politico-institutional' understandings achieve this aim. Understandably, Parekh stopped claiming that understandings of a polity should be only 'politico-institutional' and discussed how members of a polity are also understood using the other features I referred to earlier.[80] But note that the report did not claim that Britain should be understood in solely 'politico-institutional' terms, perhaps because Modood was the CMEB's academic advisor.[81]

Parekh also argued that national identities were valuable for people individually and for their polities too.[82] Individually, these identities depict, as we saw earlier, part of what makes a person who they are. And a person acknowledging they are, for example, 'American' aids their self-understanding. Indeed, even people who see themselves as citizens or workers of the world will over time usually be shaped by their polity's legal, political and educational institutions and they will thus often have something, for example, 'British', 'French', 'American', 'Canadian' and so on about them.

The value of national identities for a polity stems from how its members must, from time to time, conceive collective goals and challenges that they will meet together.[83] To conduct such activity, it helps if members of a polity think of themselves, *inter alia*, as a group, but why would they? Members of a polity often differ by race, religion, class, education, occupation and so on, and are mostly unknown to one another, but their national identities can help them to think of themselves as a group. This is because those who, for example, think they are 'American', cannot do so without understandings of America that may be vague. But if these understandings of America are clear they often help Americans to visualise themselves as a group. Likewise, those who think of themselves as, for example, 'American' often think of themselves as a group just as those sharing a religious identity and are 'Muslim', or those sharing a sexual identity and are 'gay', might. People's national identities are thus valuable as they *help* members of a polity to think of themselves as a group; recall that the CMEB described national identities as being important in a remarkably similar way.

Likewise, those who refer, for example, to the 'truly British' or 'naturally British' have understandings of their polity in which they think it to be solely 'white'. This helps to generate fear of racial, religious and other minorities and the exclusion and discrimination of minorities too. Thus, from 1974 onwards, Parekh argued as the CMEB's report later would, that people must alter their understandings of their polity to include minorities, as this can help to remove a source of discrimination and exclusion against minorities.[84]

Parekh had also long promoted the measures that the report discussed to encourage people's understandings of their polity to include minorities. He thus noted how public controversies like the Rushdie Affair alert people to how their understanding of their polity excludes minorities who are citizens and, as citizens, they are entitled to request changes just as other citizens are.[85] Equally, Parekh noted that polities should declare themselves multicultural, as Canada did.[86] Schools should teach children about the polity's multicultural nature as this will help children to develop understandings of it that include minorities. Parekh also rejects Renan's advice to forget divisive episodes in the polity's past, as these episodes can teach us much about the discrimination and exclusion that follows when popular understandings of a polity suggest it is solely 'white', Christian and so on.[87]

The above reasons for national identities being valuable differ to claiming that people's understandings of their nation are valuable for being an 'eternal' view of it as, we saw, conservative nationalists suggest. Indeed, it is unclear why something that is eternal is also valuable and Parekh notes how understandings of a polity change and thus understandings of Britain are not today what they were even fifty years ago.[88] Parekh also rejects the conservative nationalist view that national identities have value by conveying 'life's meaning', as this is asserted not argued. And it is unclear why, for example, Canada, Britain, France or America's identity or being Canadian, British, French or American would convey life's meaning.[89]

But liberal nationalists are right to suggest that choice requires something to choose with.[90] Yet, at times, liberal nationalists suggest that we choose using our national *identities* even though they refer to the norms, values, traditions of thought and so on of a national *culture*.[91] Equally, it is unclear how a person's national identity could be used to make choices and the following example will illustrate the point: a person might say that as an Englishman he chooses to fight for England or to study English history. But it is unclear whether this person is using his English culture, English identity, English loyalty or an English education to make such choices as all of them may be being used simultaneously. Likewise, none of them may be being used as this person may fight for England to protect his own life and those of his family and friends. And a person may choose to study English history simply because he has an interest in it as so many academics around the world do. We might go further and note that a person might choose not to attend an elite university. This is because he is 'working class' and fears that he will not 'fit in', or he might choose not to drink alcohol as he is a Muslim. Such choices do not relate to national identities but rather to other identities, and perhaps with such points in mind, Parekh does not connect national identities and choice.[92]

Further, for Parekh[93] national identities need not cultivate the 'solidarity', 'obligation' and 'trust' that people require to redistribute their income as liberal nationalists hope.[94] This is because, whether national identities aid such redistribution is an empirical issue and liberal nationalists accept this, thus they subtly show how national identities must be conceived to examine this empirical issue.[95] But liberal nationalists have said little about what 'solidarity', 'trust' and 'obligation' are, what forms of them are needed, what our indicators for them are or how we can disaggregate national identities as stimulating them from other stimuli.[96] Liberal nationalists themselves concede that what they can say about the 'sources' of 'solidarity' is 'meagre and speculative'.[97] Likewise, liberal nationalists themselves note that while empirical evidence does not disprove what they claim it does not support it either.[98] Unsurprisingly, Parekh has long criticised this way of arguing that national identities are valuable and describes the value of such identities differently.[99]

The above arguments from conservative and liberal nationalists are questionable and Parekh argues for the value of national identities in ways that seem less so. But Parekh is not a nationalist in the sense that I referred to earlier. Parekh does not discuss the importance of nations or the importance of nations having 'self-rule' or constituting states as, for example, Scottish and Welsh nationalists might. Parekh thus does not advocate the nationalism that we often think is necessary to call someone a nationalist.[100] This does not mean that Parekh is unconcerned with the goals of those seeking secession, independence or greater 'autonomy' in a more federal political structure. Parekh simply thinks that such groups or communities need not always be nations, that in some cases arguments can be offered not to secede and that nationalism is not the only way to think about their goals. Indeed, Parekh showed why rectifying previous injustice, preventing further abuse, the need for 'recognition' and better representation could all be used in different ways to plausibly justify the goals of such groups. He also showed why counter arguments were often implausible.[101] In this way, Parekh and the CMEB's report *outline* or *sketch* for us how we can begin to think about the importance of national identities without nations, nationalism and without being nationalists.

Conclusion

This essay has shown that the CMEB's report reflected Parekh's distinct and valuable way of thinking about national identities which has previously gone unnoticed. Parekh thus does not assume that explaining what national identities are requires us to explain what nations are, but it involves discussing two ways in which we often think about national identities, as for example, America's identity, or being American. Parekh also shows why national identities are valuable but not, as we saw above, by claiming that they convey an 'eternal' understanding of the polity, life's meaning, or that they can aid choice or the redistribution of income. Such arguments are questionable and thus Parekh avoids them. The report, we have seen, avoids these arguments too. The report was thus repeatedly criticised in the media for its discussion of national identities, even though it reflects a way to think about national identities that avoids the questionable arguments of rival ways to think about such identities.

Indeed, within and beyond the academy, Parekh, as a prominent 'multiculturalist', has used his academic and non-academic writing to articulate a distinct and valuable way to discuss the nature and worth of national identities for forty years. Yet political theorists examine only the parts of his work that do *not* relate to national identities.[102] When examining discussions about national identity some assume that liberal nationalism is either the only way of advocating national identities or the most plausible way.[103] Others assume that conservative nationalism is the only alternative to liberal nationalism when advocating national identities.[104] Yet we have seen there is another way of thinking about national identities that has been present in scholarly and public discussions for forty years that comes from Parekh and much can be gained from studying it. For example, liberal nationalists depict what national identities are and why they have value, as does Parekh, but he avoids some of their more questionable arguments. Liberal nationalists may thus benefit from expanding upon Parekh's work to address the relationships between national identities, citizenship, what rights immigrants who are not yet citizens are entitled to and so on.[105] Parekh, and other 'multiculturalists', say surprisingly little about these issues. But as national identities are central to what liberal nationalists say about them, clearer notions of what national identities are and why they are important are needed, and Parekh may help to provide them.

Parekh's work may also be important for those who seek to escape the need for national identities by arguing that citizens only need to 'possess a sense belonging to a polity' in the sense that they 'identify with it' and see it as theirs.[106] This is because a person cannot sense that he belongs, for example, to America without at least some understanding of America, and such an understanding is one way in which Parekh describes national identities. Equally, if citizens should 'identify' with their polity and see it as theirs, this might occur if they think their polity helps to shape what they are, and makes them, for example, British, French, Canadian, American and so on, which is the second way in which Parekh describes national identities. Similarly, those who criticise liberal nationalists and who endorse 'post-national' or 'post-nationalist' polities may benefit from Parekh's work.[107] This is because, like them, Parekh accepts that polities are often no longer nations and he does not advocate nationalism or being a nationalist. But Parekh does not take the step of rejecting national identities, as he recognises why what we usually call national identities need not be as connected to nations and nationalism as some think.

Parekh's work and the report may also be of use to British politicians. For sure, Parekh himself claimed 'the report was born an orphan as no one wanted it'.[108] Yet, as the Runnymede Trust argues, two-thirds of the report's 120 recommendations were implemented within three years of its publication.[109] The recommendations that were implemented did not relate to the subject of national identities thus British politicians can still use the report and Parekh's way of thinking about such identities in at least the two following ways.

First, many senior British politicians already discuss promoting more inclusive understandings of Britain, as Tariq Modood and I showed elsewhere.[110] Indeed, Modood and I showed that these politicians advocate promoting more inclusive understandings of Britain in a way that is remarkably similar to the way that the report did. Liberal nationalist reasons can be used to justify this practice as some public intellectuals suggest, and this may accord with the 'muscular forms of liberalism' that some British politicians discuss too.[111] But if British politicians use liberal nationalist reasons to justify promoting more inclusive understandings of Britain, it is unclear how they avoid the questionable arguments that we saw liberal nationalists can make. The CMEB report and 'multiculturalists' like Parekh thus offer British politicians another way to justify the importance of understandings of Britain being more inclusive.

Second, the CMEB and Parekh show which policies are likely to help British people to have more inclusive understandings of Britain. Hence, they discuss how anti-discrimination measures, over time, alter what many think is permissible in a polity, thus suggesting that such measures, over time, alter people's understandings of their polity. Likewise, a government can promote more inclusive understandings of a polity by declaring the polity to be, *inter alia*, multicultural, as Canada did, and authorise school curricula to teach children about the different cultures in the polity. Equally, a government need not hide from episodes in the polity's past in which many were discriminated against and excluded. This is because a government can instead use such episodes to promote the implications of not having inclusive understandings of a polity. In these ways the CMEB report and Parekh show what politicians can actually do to try to achieve more inclusive understandings of Britain. There is then some scholarly and non-scholarly value in the way that 'multiculturalists' advocate national identities and more research into the many different ways in which different 'multiculturalists' do so[112] is needed.

Notes

* Senior Lecturer in Political Theory and Public Policy, Brunel University London, UK.

1 This essay draws heavily on V. Uberoi, 'Parekh Report: National identities without nations and nationalism', *Ethnicities* (2015) 15(4), 509–526.

2 Tariq Modood, 'Multiculturalism and Britishness: Provocations, hostilities and advances', in R Garbaye and P Schnapper (eds) *The Politics of Ethnic Diversity in the British Isles* (Palgrave Macmillan 2014) 21, 31; Bhikhu Parekh, *Rethinking Multiculturalism* (Macmillan 2000) 6.

3 Tariq Modood, *Multiculturalism* (Polity Press 2013) 135; Bhikhu Parekh, *Rethinking Multiculturalism* 230.

4 Brian Barry, *Culture and Equality* (Polity 2001) 77; David Goodhart, *The British Dream: Successes and Failures of Post-war Immigration* (Atlantic Books 2013) 201.

5 Brian Barry, *Culture and Equality* 315; George Crowder, *Theories of Multiculturalism* (Polity 2013) 50; P Jones, 'Liberal equality and accommodation', in Varun Uberoi and Tariq Modood (eds) *Multiculturalism Rethought* (Edinburgh University Press 2015) 128.

6 Eugene Mclaughlin and Sarah Neal, 'Misrepresenting the multicultural nation', *Policy Studies* (2004) 25(3); Modood, 'Multiculturalism and Britishness' 24.

7 Modood, 'Multiculturalism and Britishness' 25, 32; Varun Uberoi and Tariq Modood, 'Has multiculturalism in Britain retreated?', *Soundings* (2013) 53, 135–136.

8 Bhikhu Parekh, 'The spectre of self-consciousness', in B Parekh (ed) *Colour, Culture and Consciousness: Immigrant Intellectuals in Britain* (George Allen & Unwin 1974); Bhikhu Parekh, 'Asians in Britain: Problem of opportunity?', in Commission for Racial Equality (ed) *Five Views of Multi-racial Britain* (Commission for Racial Equality 1978).

9 Bhikhu Parekh, 'The New Right and the politics of nationhood', in *The New Right: Image and Reality* (Runnymede Trust 1986); Bhikhu Parekh, 'Discourses on national identity', *Political Studies* (1994) 42; Bhikhu Parekh, 'Ethnocentricity of the nationalist discourse', *Nations and Nationalism* (1995) 1(1); Bhikhu Parekh, *Rethinking Multiculturalism*; Bhikhu Parekh, 'Being British', in A Gamble and T Wright (eds) *Britishness Perspectives on the British Question* (Wiley 2009).

10 Arash Abizadeh, 'Does liberal democracy presuppose a cultural nation? Four arguments', *American Political Science Review* (2002) 96(3); Andrew Mason, *Community Solidarity Belonging* (Cambridge University Press 2000); Andrew Mason, *Living Together as Equals: The Demands of Citizenship* (Oxford University Press 2012); David Miller, *On Nationality* (Clarendon Press 1995); David Miller, *Citizenship and National Identity* (Polity Press 2000).

11 Bhikhu Parekh, *Marx's Theory of Ideology* (Johns Hopkins University Press 1982) ii.

12 For example, Gerald A Cohen, *Karl Marx's Theory of History: A Defence* (Oxford University Press 1978).

13 Isaiah Berlin, 'The bent twig: On the rise of nationalism', in I Berlin (ed) *The Crooked Timber of Humanity* (Pimlico 2003).

14 Elie Kedourie *Nationalism* (Oxford 1998).

15 Ernest Gellner, *Nations and Nationalism* (Blackwell 1983).

16 Jürgen Habermas, *The Inclusion of the Other* (MIT Press 1998) 119; Anthony Smith, *Nationalism* (Polity Press 2001) 43–61. Likewise, Benhabib (2008), Nussbaum (1996)

and Viroli (1995) only briefly mention national identities and discuss instead replacing ideas of nations and nationalism with ideas of 'republican patriotism' or 'cosmopolitanism'. See Seyla Benhabib, *Another Cosmopolitanism* (Oxford University Press 2008); Martha Nussbaum, 'Patriotism and cosmopolitanism', in M Nussbaum (ed) *For Love of Country* (Beacon Press 1996); Maurizio Viroli, *For Love of Country* (Oxford University Press 1995).

17 Miller, *On Nationality*; Miller, *Citizenship and National Identity*; Yael Tamir, *Liberal Nationalism* (Princeton University Press 1993).

18 Roger Scruton, 'In defence of the nation', in J Clark (ed) *Ideas and Politics in Modern Britain* (Macmillan 1990); Roger Scruton, *Arguments for Conservatism: A Political Philosophy* (Continuum 2006).

19 Crowder, *Theories of Multiculturalism* 83–91.

20 Edmund Burke, 'Reflection on the revolutions in France', in *Modern Political Thought: Readings from Machiavelli to Nietzsche* (Hackett 1996) 572.

21 Scruton, 'In defence of the nation' 305; Roger Scruton, *How to Be a Conservative* (Bloomsbury 2014) 82.

22 Immanuel Kant, *Selections* (Charles Schreiber 1929) 335.

23 John Stuart Mill, *On Liberty and Other Essays* (Oxford University Press 1991) 65.

24 Earnest Renan, 'What is a nation?', in V Pecora (ed) *Nations and Identities: Classic Readings* (Blackwell 2001) 175 (emphasis added).

25 Miller, *On Nationality* 22–5; Tamir, *Liberal Nationalism* 33.

26 Miller, *On Nationality* 81; Tamir, *Liberal Nationalism* 75; Scruton, *Arguments for Conservatism* 25.

27 Scruton's 'In defence of the nation' rejects nationalism but later Scruton (*Arguments for Conservatism* 16, 21, 25) accepts the importance of nation-states and endorses the importance of nations having states and is thus a nationalist in the sense described.

28 Gellner, *Nations and Nationalism* 1; Kedourie, *Nationalism* 67; Smith, *Nationalism* 22.

29 Scruton, 'In defence of the nation' 305; Scruton, *How to Be a Conservative* 83.

30 Will Kymlicka, *Multicultural Citizenship* (Clarendon Press 1995) 113; Will Kymlicka, *Can Liberal Pluralism Be Exported?* (Oxford University Press 2001) 54–58.

31 Miller, *On Nationality* 149, 154.

32 CMEB, *The Parekh Report* (Profile Books 2000) 15–16.

33 CMEB, *The Parekh Report* 22–26.

34 CMEB, *The Parekh Report* 15, 22, 103.

35 Scruton, 'In defence of the nation' 314, 317; Roger Scruton, *England: An Elegy* (Continuum 2001) 11; Scruton, *Arguments for Conservatism* 27.

36 Renan, 'What is a nation?' 166.

37 Kymlicka, *Multicultural Citizenship* 189; Miller, *On Nationality* 22–25; Tamir, *Liberal Nationalism* 67.

38 CMEB, *The Parekh Report* 103.

39 CMEB, *The Parekh Report* 15, 36–39; Miller, *On Nationality* 173–174; Scruton, *Arguments for Conservatism,* 16. By saying the report seemed at times sceptical about Britain being a nation I am, I hope, echoing Modood's claim ('Multiculturalism and Britishness' 28) that the report was unclear in places about Britain being a nation.

40 CMEB, *The Parekh Report* 56.

41 CMEB, *The Parekh Report* ix.

42 CMEB, *The Parekh Report* ix, 16.

43 CMEB, *The Parekh Report* 16.

44 Scruton, 'In defence of the nation' 306.

45 Scruton, 'In defence of the nation' 310.

46 David Miller, 'Nationalism', in J Drysek, B Honig and A Phillips (eds) *The Oxford Handbook of Political Theory* (Oxford University Press 2009) 535.

47 Miller, *On Nationality* 83, 91; David Miller *Justice for Earthlings* (Cambridge University Press 2013) 158; Tamir, *Liberal Nationalism* 110.

48 CMEB, *The Parekh Report* 50.
49 CMEB, *The Parekh Report* 52.
50 CMEB, *The Parekh Report* xviii, 276.
51 CMEB, *The Parekh Report* 103.
52 Scruton, 'In defence of the nation' 305–306; *Nationalism* 244.
53 Will Kymlicka, 'The rise and fall of multiculturalism? New debates on inclusion and accommodation in diverse societies', in S Vertovec and S Wessendorf (eds) *Multiculturalism Backlash* (Routledge 2010) 39.
54 Miller, *On Nationality* 149, 154; David Miller, 'Immigrants, nations and citizenship', *Journal of Political Philosophy* (2008) 16(4), 380.
55 Kymlicka, *Can Liberal Pluralism Be Exported?* 19, 49; Will Kymlicka and Keith Banting, 'Immigration, multiculturalism and the welfare state', *Ethics and International Affairs* (2006) 20(3), 300; Miller, 'Immigrants, nations and citizenship' 380. Kymlicka and Banting ('Immigration, multiculturalism and the welfare state' 301) note how policies of multiculturalism can help to foster national solidarity yet they retain the distinction between policies of multiculturalism and nation-building policies. See Varun Uberoi, 'Social unity in Britain', *Journal of Ethnic and Migration Studies* (2007) 33(1); also Varun Uberoi, 'Do policies of multiculturalism undermine national identities?', *Political Quarterly* (2008) 79(3); Varun Uberoi, 'Multiculturalism and the Canadian charter of rights and freedoms', *Political Studies* (2009) 57(4).
56 The fact that liberal nationalists conceive of nation-building policies as promoting understandings of a polity can be seen by considering how, for example, celebrating symbols, heroes and so on are listed as potential nation-building policies. See Will Kymlicka, *Politics in the Vernacular* (Oxford University Press 2001) 34; Kymlicka and Banting, 'Immigration, multiculturalism and the welfare state' 300.
57 Bhikhu Parekh, 'Postscript', in B Parekh (ed) *Colour, Culture and Consciousness: Immigrant Intellectuals in Britain* (George Allen & Unwin 1974) 230; Bhikhu Parekh, *Ghandi's Political Philosophy: A Critical Examination* (Macmillan 1989) 74–75; Bhikhu Parekh, *A New Politics of Identity: Political Principles for an Interdependent World* (Palgrave Macmillan 2008) 56.
58 Parekh, 'Discourses on national identity' 501; Bhikhu Parekh, 'Citizenship and political obligation', in P King (ed) *Socialism and the Common Good* (Frank Cass 1996) 263; Parekh, *Rethinking Multiculturalism* 181–84.
59 Elsewhere, I suggest that Parekh thinks nations are morally, ethnically and culturally uniform but this presupposes that ethnicity and morality are unrelated to culture, and as Parekh does not take this position I describe his understanding of a nation differently here.
60 Habermas, *The Inclusion of the Other* 119.
61 Bhikhu Parekh, 'The future of multi-ethnic Britain: Reporting on a report. The Round Table: The Commonwealth', *Journal of International Affairs* (2001) 90(362).
62 Bhikhu Parekh, 'The concept of national identity', *New Community* (1995) 21(2), 38; Parekh, 'The future of multi-ethnic Britain' 695; Parekh, 'Response to David Goodhart', in D Goodhart, *Progressive Nationalism* (Demos 2005) 76.
63 Parekh, 'The concept of national identity' 255–256.
64 Sundas Ali and Anthony Heath, *Future Identities: Changing Identities in the UK: The Next Years* (Government Office for Science 2013) 6–8; Iain McLean and Alistair McMillan, *State of the Union* (Oxford University Press 2005) 4.
65 Scruton, 'In defence of the nation' 315; Scruton, *How to Be a Conservative* 82.
66 Kymlicka, *Can Liberal Pluralism Be Exported?* 55–56; Miller, *On Nationality* 26.
67 Kymlicka, *Multicultural Citizenship* 88–89; Kymlicka, *Politics in the Vernacular* 56–57; Miller, *On Nationality* 25–26; Miller, 'Immigrants, nations and citizenship' 388.
68 Kymlicka, *Multicultural Citizenship* 76.
69 Kymlicka, *Multicultural Citizenship* 80.

70 Kymlicka, *Politics in the Vernacular* 219; David Miller, 'Immigration: The case for limits', in A Cohen and H Wellman (eds) *Contemporary Debates in Applied Ethics* (Blackwell 2005) 199; Tamir, *Liberal Nationalism* 127.
71 Miller, 'Immigration: The case for limits' 200, 204; David Miller, 'Justice and immigration', *European Journal of Political Theory* (2015) 10; Tamir, *Liberal Nationalism* 127.
72 Parekh, 'The concept of national identity' 255–256.
73 Will Kymlicka, *Contemporary Political Philosophy* (Oxford University Press 2002) 265; Miller, *On Nationality* 18; Tim Soutphommasane, *The Virtuous Citizens* (Cambridge University Press 2012) 71.
74 See Frank Bechhoffer and David McCrone, 'Being British: A crisis over identity', *Political Quarterly* (2007) 78(2), 253; Varun Uberoi and Iain McLean 'Britishness: A role for the state', *Political Quarterly* (2009) 79(s1), 41; and Miller, *On Nationality*, 141.
75 Parekh, 'Postscript' 230.
76 Parekh, *Rethinking Multiculturalism* 231.
77 Indeed, if a nation is defined as a homogenous ethno-cultural group, Parekh's entire philosophical system is about rejecting this. See Varun Uberoi, 'Introduction', in V Uberoi, and T Modood (eds) *Multiculturalism Rethought* (Edinburgh University Press 2015).
78 Parekh, *Rethinking Multiculturalism* 231.
79 Tariq Modood, 'Their liberalism our multiculturalism', *British Journal of Politics and International Relations* (2001) 3(2), 249.
80 Parekh, *A New Politics of Identity* 59; Parekh, 'Being British' 36.
81 CMEB, *The Parekh Report* 369. Modood ('Multiculturalism and Britishness' 24) 'took an active part in the collective authorship' of the CMEB report.
82 Parekh, *A New Politics of Identity* 57.
83 Parekh, *Rethinking Multiculturalism* 196.
84 Parekh, 'Postscript' 230; Parekh, 'Discourses on national identity' 502.
85 Bhikhu Parekh, 'Britain and the social logic of pluralism', in Commission for Racial Equality (ed) *Britain: A Plural Society. Report of a Seminar* (Commission for Racial Equality 1990) 1, 70.
86 Parekh, 'Britain and the social logic of pluralism' 70; Parekh, *Rethinking Multiculturalism* 229, 235.
87 Parekh, 'The concept of national identity' 264.
88 Parekh, 'Ethnocentricity of the nationalist discourse' 504.
89 Bhikhu Parekh, 'The incoherence of nationalism', in R Beiner (ed) *Theorizing Nationalism* (State University of New York Press 1999) 308.
90 Will Kymlicka, *Liberalism, Community and Culture* (Clarendon Press 1989) 166; Miller, *On Nationality* 44.
91 Miller, 'Nationalism' 535. Kymlicka, *Multicultural Citizenship* 77.
92 Parekh, *Rethinking Multiculturalism* 103.
93 Parekh 'The incoherence of nationalism' 313–314.
94 Kymlicka, *Multicultural Citizenship* 77; Miller, *On Nationality* 83, 91, 92; *Justice for Earthlings* 90; Tamir, *Liberal Nationalism* 110.
95 David Miller and Sundas Ali, 'Testing the national identity argument', *European Political Science Review* (2014) 6(2).
96 Only now has David Miller begun to clarify the concept of solidarity. See. David Miller, 'Solidarity and its sources', in K Banting and W Kymlicka, *The Stains of Commitment* (Oxford University Press 2017).
97 Will Kymlicka, 'Solidarity in diverse societies: Beyond neoliberal multiculturalism and welfare chauvinism', *Comparative Migration Studies* (2015) 3(17), 8, 9.
98 Miller and Ali, 'Testing the national identity argument' 240, 254, 257.

99 Parekh, 'The incoherence of nationalism'; Bhikhu Parekh, 'Multicultural society and the welfare state', in J Connelly and J Hayward (eds) *The Withering of the Welfare State* (Palgrave Macmillan (2012).
100 Kedourie, *Nationalism* 67; Smith, *Nationalism* 9; Gellner, *Nations and Nationalism* 1.
101 Parekh, 'Britain and the social logic of pluralism' 191; Bhikhu Parekh, 'Liberal democracy and national minorities', in F Requejo and M Caminal (eds) *Political Liberalism and Plurinational Democracies* (Routledge 2011) 33–35.
102 Brian Barry, 'The muddles of multiculturalism', *New Left Review* (2001) (8) (March–April); Paul Kelly, 'Situating Parekh's multiculturalism: Bhikhu Parekh and twentieth-century British political theory', in V Uberoi and T Modood (eds) *Multiculturalism Rethought* (Edinburgh University Press 2015) 29–55.
103 Mason, *Living Together as Equals* 181.
104 Crowder, *Multiculturalism* 83–91.
105 Miller, 'Immigration: The case for limits'; David Miller, *National Responsibility* (Oxford University Press 2007).
106 Mason, *Community Solidarity Belonging* 127; Mason, *Living Together as Equals* 181.
107 Abizadeh, 'Does liberal democracy presuppose a cultural nation?' 495; Jan-Werner Müller, *Constitutional Patriotism* (Princeton University Press 2007) 63.
108 I heard Parekh say this to an audience at the Political Studies Association conference 2010 in Edinburgh.
109 L. Holloway, 'The Future of Multi-Ethnic Britain at 15 years', http://www.runnymedetrust.org/blog/the-future-of-multi-ethnic-britain-15-years-on.
110 Modood, 'Multiculturalism and Britishness' 31; Uberoi and Modood, 'Has multiculturalism in Britain retreated?'; Varun Uberoi and Tariq Modood, 'Inclusive Britishness: A multiculturalist advance', *Political Studies* (2013) 61(1).
111 Goodhart, *The British Dream* 285.
112 Modood, *Multiculturalism*; Uberoi, 'Social unity in Britain'.

Bibliography

Abizadeh A, 'Does liberal democracy presuppose a cultural nation? Four arguments', *American Political Science Review* (2002) 96(3).

Ali S and Heath A, *Future Identities: Changing Identities in the UK: The Next Years* (Government Office for Science 2013).

Barry B, *Culture and Equality* (Polity Press 2001).

Barry B, 'The muddles of multiculturalism', *New Left Review* (2001) 8 (March–April).

Bechhoffer F and McCrone D, 'Being British: A crisis over identity', *Political Quarterly* (2007) 78(2).

Benhabib S, *Another Cosmopolitanism* (Oxford University Press 2008).

Berlin I, 'The bent twig: On the rise of nationalism', in I Berlin (ed) *The Crooked Timber of Humanity* (Pimlico 2003).

Burke E, 'Reflection on the revolutions in France', in *Modern Political Thought: Readings from Machiavelli to Nietzsche* (Hackett 1996).

Cohen GA, *Karl Marx's Theory of History: A Defence* (Oxford University Press 1978).

Commission on the Future of Multi-Ethnic Britain (CMEB), *The Parekh Report* (Profile Books 2000).

Crowder G, *Theories of Multiculturalism* (Polity Press 2013).

Gellner E, *Nations and Nationalism* (Blackwell 1983).

Goodhart D, *The British Dream: Successes and Failures of Post-war Immigration* (Atlantic Books 2013).

Habermas J, *The Inclusion of the Other* (MIT Press 1998).
Jones P, 'Liberal equality and accommodation', in V Uberoi and T Modood (eds) *Multiculturalism Rethought* (Edinburgh University Press 2015).
Kant I, *Selections* (Charles Schreiber 1929).
Kedourie E, *Nationalism* (Oxford 1998).
Kelly P, 'Situating Parekh's multiculturalism: Bhikhu Parekh and twentieth-century British political theory' in V Uberoi and T Modood (eds) *Multiculturalism Rethought* (Edinburgh University Press 2015).
Kymlicka W, *Liberalism, Community and Culture* (Clarendon Press 1989).
Kymlicka W, *Multicultural Citizenship* (Clarendon Press 1995).
Kymlicka W, *Can Liberal Pluralism Be Exported?* (Oxford University Press 2001).
Kymlicka W, *Politics in the Vernacular* (Oxford University Press 2001).
Kymlicka W, *Contemporary Political Philosophy* (Oxford University Press 2002).
Kymlicka W, 'The rise and fall of multiculturalism?: New debates on inclusion and accommodation in diverse societies', in S Vertovec and S Wessendorf (eds) *Multiculturalism Backlash* (Routledge 2010).
Kymlicka W and Banting B, 'Immigration, multiculturalism and the welfare state', *Ethics and International Affairs* (2006) 20(3).
McLaughlin E and Neal S, 'Misrepresenting the multicultural nation', *Policy Studies* (2004) 25(3).
McLean I and McMillan A, *State of the Union* (Oxford University Press 2005).
Major J, Speech to the Conservative Group for Europe (1993), available at http://www.johnmajor.co.uk/page1086.html, last accessed 10 May 2014.
Mason A, *Community Solidarity Belonging* (Cambridge University Press 2000).
Mason A, *Living Together as Equals: The Demands of Citizenship* (Oxford University Press 2012).
Mill JS, *On Liberty and Other Essays* (Oxford University Press 1991).
Miller D, *On Nationality* (Clarendon Press 1995).
Miller D, *Citizenship and National Identity* (Polity Press 2000).
Miller D, 'Immigration: The case for limits', in A Cohen and H Wellman (eds) *Contemporary Debates in Applied Ethics* (Blackwell 2005).
Miller D, *National Responsibility* (Oxford University Press 2007).
Miller D, 'Immigrants, nations and citizenship', *Journal of Political Philosophy* (2008) 16(4).
Miller D, 'Nationalism', in J Drysek, B Honig and A Phillips (eds) *The Oxford Handbook of Political Theory* (Oxford University Press 2009).
Miller D, *Justice for Earthlings* (Cambridge University Press 2013).
Miller D, 'Justice and immigration', *European Journal of Political Theory* (2015), published online first, 8 May. DOI: 10.1177/1474885115584833.
Miller D and Ali S, 'Testing the national identity argument', *European Political Science Review* (2014) 6(2).
Modood T, 'Their liberalism our multiculturalism', *British Journal of Politics and International Relations* (2001) 3(2).
Modood T, *Multiculturalism* (Polity Press 2013).
Modood T, 'Multiculturalism and Britishness: Provocations, hostilities and advances', in R Garbaye and P Schnapper (eds) *The Politics of Ethnic Diversity in the British Isles* (Palgrave Macmillan 2014).
Müller JW, *Constitutional Patriotism* (Princeton University Press 2007).

Nussbaum M, 'Patriotism and cosmopolitanism', in M Nussbaum (ed) *For Love of Country* (Beacon Press 1996).

Parekh B, 'Postscript', in B Parekh (ed) *Colour, Culture and Consciousness: Immigrant Intellectuals in Britain* (George Allen & Unwin 1974).

Parekh B, 'The spectre of self-consciousness', in B Parekh (ed) *Colour, Culture and Consciousness: Immigrant Intellectuals in Britain* (George Allen & Unwin 1974).

Parekh B, 'Asians in Britain: Problem of opportunity?', in Commission for Racial Equality (ed) *Five Views of Multi-racial Britain* (Commission for Racial Equality 1978).

Parekh B, *Marx's Theory of Ideology* (Johns Hopkins University Press 1982).

Parekh B, 'The New Right and the politics of nationhood', in *The New Right: Image and Reality* (Runnymede Trust 1986).

Parekh B, *Ghandi's Political Philosophy: A Critical Examination* (Macmillan 1989).

Parekh B, 'Britain and the social logic of pluralism', in Commission for Racial Equality (ed) *Britain: A Plural Society. Report of a Seminar* (Commission for Racial Equality 1990).

Parekh B, 'Discourses on national identity', *Political Studies* (1994) 42.

Parekh B, 'Ethnocentricity of the nationalist discourse', *Nations and Nationalism* (1995) 1(1).

Parekh B, 'The concept of national identity', *New Community* (1995) 21(2).

Parekh B, 'Citizenship and political obligation', in P King (ed) *Socialism and the Common Good* (Frank Cass 1996).

Parekh B, 'The incoherence of nationalism', in R Beiner (ed) *Theorizing Nationalism* (State University of New York Press 1999).

Parekh B, 'Defining British identity', *Political Quarterly* (2000) 71(1).

Parekh B, *Rethinking Multiculturalism* (Macmillan 2000).

Parekh B, *Integrating Minorities* (Institute of Contemporary Arts 2001).

Parekh B, 'The future of multi-ethnic Britain: Reporting on a report. The Round Table: The Commonwealth', *Journal of International Affairs* (2001) 90(362).

Parekh B, 'Response to David Goodhart', in D Goodhart, *Progressive Nationalism* (Demos 2005).

Parekh B, *A New Politics of Identity: Political Principles for an Interdependent World* (Palgrave Macmillan 2008).

Parekh B, 'Being British', in A Gamble and T Wright (eds) *Britishness Perspectives on the British Question* (Wiley 2009).

Parekh B, 'Liberal democracy and national minorities', in F Requejo and M Caminal (eds) *Political Liberalism and Plurinational Democracies* (Routledge 2011).

Parekh B, 'Multicultural society and the welfare state', in J Connelly and J Hayward (eds) *The Withering of the Welfare State* (Palgrave Macmillan 2012).

Renan E, 'What is a nation?', in V Pecora (ed) *Nations and Identities: Classic Readings* (Blackwell 2001).

Scruton R, 'In defence of the nation', in J Clark (ed) *Ideas and Politics in Modern Britain* (Macmillan, 1990).

Scruton R, *England: An Elegy* (Continuum 2001).

Scruton R, *Arguments for Conservatism: A Political Philosophy* (Continuum 2006).

Scruton R, *How to Be a Conservative* (Bloomsbury 2014).

Smith A, *Nationalism* (Polity Press 2001).

Soutphommasane T, *The Virtuous Citizens* (Cambridge University Press 2012).

Tamir Y, *Liberal Nationalism* (Princeton University Press 1993).

Uberoi V, 'Social unity in Britain', *Journal of Ethnic and Migration Studies* (2007) 33(1).

Uberoi V, 'Do policies of multiculturalism undermine national identities?', *Political Quarterly* (2008) 79(3).

Uberoi V, 'Multiculturalism and the Canadian charter of rights and freedoms', *Political Studies* (2009) 57(4).

Uberoi V, 'Introduction', in V Uberoi and T Modood (eds) *Multiculturalism Rethought* (Edinburgh University Press 2015).

Uberoi V and McLean I, 'Britishness: A role for the state', *Political Quarterly* (2009) 79(s1).

Uberoi V and Modood T, 'Has multiculturalism in Britain retreated?', *Soundings* (2013) 53.

Uberoi V and Modood T, 'Inclusive Britishness: A multiculturalist advance', *Political Studies* (2013) 61(1).

Viroli M, *For Love of Country* (Oxford University Press 1995).

Yack B, *Nationalism and the Moral Psychology of Community* (Chicago University Press 2012).

4 Religion, belief and diversity in transition

The Commission on Religion and Belief in British Public Life

Maleiha Malik

Introduction

The Commission on Religion and Belief in British Public Life (CORB) was a non-statutory commission that undertook its work between November 2013 and June 2015.[1] CORB is a civil society initiative that was funded by the Woolf Institute, University of Cambridge, the Open Society Institute, the Joseph Rowntree Charitable Trust, the David Foundation and the Dominion Trust. Its main aim was to examine, analyse, consult and hear public evidence on the 'place and role of religion and belief in British public life'. I was a CORB Commissioner, and in this essay I start by setting the context within which there is increasing use of commissions to manage racial, religious and cultural diversity. I then move on to consider the specific case of the UK by developing a descriptive overview of CORB that draws out wider issues of process and substance that are relevant to debates about the use of statutory and non-statutory commissions to regulate increasing diversity in liberal democracies. I then set out the aims, design and work of CORB. I identify some processes and design techniques for commissions that can be used to analyse the work of CORB. I then discuss the implications of using commissions in relation to a number of key issues with a special focus on concerns about minorities and gender. Finally, I develop a comparison between CORB and other commissions, such as those in France. The contrast between the UK and France is significant because it points to the different motivations and functions of commissions, as well as differences in their impact and success as a form of regulating religious diversity. I also explore the wider lessons that CORB can offer about how commissions can more successfully respond to increasing diversity in liberal democracies.

The context for CORB: diversity in liberal democracies

The Commission on Religion and Belief in Public Life in the UK (CORB) was established to address how to re-imagine religion and belief in public life rather than issues of migration or minorities. Nevertheless, increasing diversity that is the outcome of migration was significant to the context for discussions.

The CORB's official terms of reference included: to consider the place of religion and belief in contemporary Britain, and the significance of emerging trends and identities; to examine how ideas of Britishness and national identity may be inclusive of a range of religions and beliefs, and may in turn influence people and their self-understanding; to explore how shared understandings of the common good may contribute to greater levels of mutual trust and collective action, and a more harmonious society. CORB's remit also included making recommendations for public life and policy.

In the UK, there had been a number of commissions similar to CORB. The Swann Commission, which issued the Swann Report *Education for All* (1985) was a statutory 'Committee of Enquiry into the Education of Children from Ethnic Minority Groups'.[2] The Parekh Commission (2000) was the Commission on the Future of Multi-Ethnic Britain, a non-statutory commission set up by the Runnymede Trust.[3] The Parekh Commission considered the present state of multi-ethnic Britain and made suggestions for countering racial discrimination and disadvantage and for making Britain a 'multicultural society' at ease with 'diversity'.

The Swann and Parekh Commissions had also focused on the topics of integration of minorities, but CORB's terms of reference were different. CORB moved away from a predominant focus on race, ethnicity and socio-economic issues towards addressing the increasing significance of national identity that involved majorities as well as minority communities. This shift from Swann and Parekh to CORB reflected the social changes that had made a British national identity becoming more important as an issue for majorities as well as ethnic minorities.

Moreover, the focus on religion and belief rather than ethnicity reflected the increasing relevance of religion and belief in debates about diversity. The post-9/11 shift from a sole focus on colour racism and ethnicity to cultural racism and religion was reflected in CORB.[4] The reasons for this shift are partly a result in demographic changes but they are also related to the rise of the religious identity of groups such as British Muslims with the resulting majoritarian response of Christians and secularists that is often expressed as a cultural anxiety about the public presence of Islam. The increasing relevance of religious diversity to debates about diversity was reflected in the composition of commissioners and the process for CORB that included Christian, other religious and non-religious viewpoints.

In the summer of 2015, the Syrian war and the subsequent migration of large numbers of refugees seeking asylum in the European Union dominated public debate in most EU states. Although there has always been migration into and across the EU, the post-Second World War movement of populations from non-Western into Western nations means that it has become now undeniably important to also tackle the diversity that emerges because of the presence of non-Western religious and cultural traditions. Developing a sharp binary between Western and non-Western contexts can be problematic, but the Western–non-Western contrast is relevant because it increases the complexity of understanding and accommodating cultural and religious practices. The increasing presence of non-Western communities expands the 'geographical reach' of religious and cultural practices that originate in the Middle East, Africa or South Asia.[5]

Unlike previous patterns of migration into the UK in the nineteenth century that were from Ireland or Eastern Europe, this recent pattern of migration is significant because it has changed the nature of the challenge of diversity: not only does the UK now have to address issues of diversity within the Western religious, ethnic and cultural tradition, it also has to confront the challenge of non-Western traditions. Moreover, the increasing presence of non-Western communities exaggerates the 'geographical reach' of diversity. For instance, UK nationals may refer to non-UK law in the Middle East or Pakistan to resolve their disputes, thereby increasing the likelihood of the 'import' of foreign legal concepts as well as the co-existence of state and non-state legal concepts, norms and institutions.[6] The import of norms from other countries and legal systems, which may not give the same definition or weight to values such as freedom or equality, creates the possibility of conflict or tension between the minority legal order and liberal constitutional norms. Accommodating non-Western cultural practices, either through a minority legal order or through policies of multiculturalism raises the possibility of a conflict between practices and state law, especially in areas such as equality and human rights norms.

As the introduction to this volume makes clear, this new pattern of migration from third non-Western countries gives new urgency to the existing challenges of the integration of new migrants as well as the management of increasing diversity. Understanding the nature of diversity will be important especially if there are very few members of minority communities involved in political and policy-making forums. There is an understandable focus on established minority communities identified by language, ethnicity, culture or religion. There is fluidity and hybridity of cultural exchange with constant movement of individuals between different spheres. Individuals and groups are constantly introducing new elements into their identity, while at the same time maintaining older recognisable traditions. However, this assumption of stability needs to be balanced by greater attention to individuals who often move in and out of more than one group. It is, therefore, important to ask questions about the impact of legal and social policies on the individuals who choose to remain members of that community. How is culture or religion being defined? Who has the power to decide? Who is being excluded from this process? Will state intervention promote values such as individual autonomy or democratic participation? Will individuals have the power to choose the leaders who represent them or the norms that govern them? Will decisions be made impartially through processes of accountability, deliberation and transparency that involve all the participants rather than through the imposition of the will and authority of a few individuals?[7]

Often, it will be difficult for 'outsiders' to have the knowledge to evaluate the impact of a particular state intervention or policy. In these situations, neglect or a badly designed intervention may cause a religious practice to become ossified in ways that continue to harm their individual members, especially 'minorities within minorities'. Cultural change can take different directions that need to be tracked at both the individual and the group level. On the one hand, group representatives may refuse to adapt their norms to reflect social change or to share power. For instance, religious practice may function as a static non-negotiable marker for group 'identity' and may cease to be a dynamic system of normative social regulation that responds to social change. On the other hand, group representatives may be willing to take an imaginative approach towards developing their norms in order to meet the needs of their individual users, including 'minorities within minorities' as well as adjust to the limits on being religious in liberal societies.

The legal regulation of migration and diversity is now a complex field and it takes the form of security law, immigration law as well as human rights and discrimination law. As security concerns become more important throughout the EU, non-legal forms of regulation are also increasingly relevant to manage violent and non-violent extremism associated with Muslim minorities such as the UK Prevent strategy.[8]

The changing political context within which public debates about cultural diversity are carried out is also relevant. During earlier historical periods, policies of assimilation of minorities were not considered to be problematic. For example, there was an assumption that majoritarian values were the basis for organising legal and political institutions, as well as access to public life. Now, however, the increasing legitimacy of policies of multiculturalism or the accommodation of cultural diversity means that minority groups are willing to claim that their legal norms or legal system should be 'accommodated' by the state system. One consequence is that the power of state law to declare a matter 'settled' may be more readily resisted by those whose identities and values are different, and those who view the law as a sign of cultural identity rather than command. In nationalist narratives it is assumed that there is a clear delineation between territorial jurisdiction, which is in the control of the state, and traditions and cultures. In reality, however, these boundaries are becoming more fluid.

Another political development is the shift in the countries of origin of migrants. In the UK, the public hysteria that follows high-profile incidents of political violence associated with Muslims, such as the London 7/7 bombings in 2005, makes it more difficult to have a reasonable public discussion about accommodating Muslim religious norms in an atmosphere in which Islam and Muslims are routinely singled out for ridicule in public debates.[9] The over-reaction to proposals to accommodate Muslim legal norms in Britain was also echoed in the Canadian experience and is also now influencing the public debates about the use of foreign laws in US courts. Public debates about how to accommodate Muslim minorities, such as their legitimate family law practices of marriage or divorce, may be hijacked by those who are using legitimate concerns about the impact of religious practices on women as a means of persecuting social groups such as Jews and Muslims. For example, some version of anti-Islam ideology has been adopted by all the far right parties in Europe, including Pegida in Germany, Geert Wilders' Party for Freedom in the Netherlands, as well as the English Defence League and the British National Party in the UK. Far right groups in the United States, such as

Stop the Islamisation of America, have also now adopted Anti-Islam ideology. All these parties now include fighting against the introduction of Islamic law (sharia) as a central political goal, and anti-sharia politics has become a defining feature of far right anti-Islam ideology. This adoption of anti-Islam ideology by far right parties who are posing a challenge to mainstream democratic parties makes it especially difficult to have a public debate about the integration of European Muslims through normal democratic processes.[10] Within this complex framework it is increasingly difficult to design forms of legal and non-legal regulation that are objective, transparent and independent. It is also difficult to carry out policy about integration in legislative assemblies because of the electoral success of far right parties who may dominate and distort sensitive discussions about less powerful minorities. In this context, there is a potential role for expert independent commissions that have a mandate to gather facts, consult less powerful minorities who do not have a voice in legislative assemblies and make recommendations that may not be popular with voters. This raises a core issue about democratic governance and minorities. How can minorities secure their best interests if they are a permanent minority especially in the context of present pressures on not only legislatures and electoral politics but also on consistency within European constitutional and human rights law?

Royal Commissions and the Commission on Religion and Belief in British Public Life

While CORB was a non-statutory commission, it included many of the features that are typical of the UK tradition of Royal Commissions.

In the UK, there is a historical tradition of using Royal Commissions to address issues of public significance. The definition of a 'Royal Commission' is not clear but it was described by Sir Edward Coke – an authority on English legal history – as "a delegation by warrant of an Act of Parliament or of the common law whereby jurisdiction, power and authority is conferred on others".[11] The UK tradition of Royal Commissions dates back to the later middle ages around the eleventh century, but they became especially important in the Victorian late nineteenth century period as part of an increasing interest by Parliament and the executive to know more about British social life from a non-partisan expert enquiry rather than a politically motivated select committee.[12]

The use of the term 'Royal' denoted that the commission was created by a delegated royal prerogative and that it was a body of the executive although it was also partly maintained at a distance from the executive branch of government. Clokie noted that Royal Commissions could include a number of functions. They could prepare the ground for a predetermined government policy. Their ad hoc nature meant that they could focus more precisely on a particular issue and appoint experts who were independent from the political process to verify accurate facts and to provide an analysis to offer a solution for a policy problem that the government wanted to address or act as a delegated representative body to address a major social, economic or political controversy that the executive (in the UK the Cabinet) did not feel able to settle.[13] More cynically, Clokie notes that Royal Commissions could also perform the function of deflecting from criticism or political pressure or postponing a question that the executive did not want to address in order to pacify a section of the public.[14]

By way of contrast to Royal Commissions, non-statutory commissions such as CORB are initiated and funded through private or civil society stakeholders. In the case of CORB, for example, the initiative came from the Woolf Institute, which is affiliated to the University of Cambridge. The Woolf Commission has as its main aim the study of relations between Christians, Jews and Muslims. The general goal of the Woolf Institute is to use research and education to explore the relationship between religion and society and to reduce intolerance. The Woolf Commission built on its idea for a public commission by acting as a partial funder and steer for the initiative, while at the same time selecting a wide range of key commissioners who were independent and from different faiths and backgrounds. These independent commissioners acted as the main analysts and decision-makers who met frequently to consider public evidence and expect advice before reaching their conclusions.

CORB replicated the Royal Commission practice of leadership by prominent judges. It was chaired by the Rt Hon Baroness Butler-Sloss who was formerly a Lord Justice of Appeal of the UK Court of Appeal, and the President of the Family Division of the High Court, with Dr Edward Kessler MBE, founder and executive director of the Woolf Institute acting as vice-chair.

Regulating religion: commissions as 'soft regulation'

In this essay I describe and evaluate the work of CORB within the paradigm of 'regulating religious diversity' as a point of entry to discuss not only constitutional, human rights and discrimination law but also soft techniques for the control and management of religious diversity. Although a non-statutory commission cannot directly legislate, it acts as a form of 'soft regulation' that influences the development of law and policy. I use the idea of 'regulation' rather than just focusing on human rights treaties, legislation or case law. This allows a discussion of how religion is a subject of control in a liberal democracy by going beyond conventional legal sources such as legislation and case law.

Definitions of regulation vary, but at their core they include the idea that the state can use its power to influence socially valuable behaviour that has adverse side effects through three clusters of techniques: first, it can set standards to distinguish between preferable and less preferable behaviour; second, it can introduce a system for information or monitoring that produces accurate knowledge about the current or future state of the behaviour; and third, it can use tools for behaviour modification to change the nature of the behaviour towards a more preferable outcome.[15] The core idea is that the state can consciously use its power to influence socially valuable behaviour but this may have adverse side effects by introducing rules and establishing systems of control such as information gathering, monitoring or standard setting. This way of approaching the relationship between the liberal state and religious diversity has a number of advantages. It challenges the pre-occupation with human rights law and the assumption that state law is the only source for articulating the collective goals of a political community. This, in turn, allows a wider analysis that takes into account the ways in which non-state institutions such as CORB may be a source of social control and influence as well as being a valuable forum in which public deliberations about religion take place. Shifting attention away from an understanding of regulation as solely focused on legislation, positive law and state power also allows an analysis of the extent to which the state has the power to control religious diversity where it is emerging in sites in the private, public or civil society sector in overlapping and complex ways. Finally, once the focus shifts away from state law and power, it is possible to include a more detailed analysis of the precise non-legal techniques for managing or influencing religious diversity such as alternative dispute resolution, mediation, bargaining and negotiation.

CORB, therefore, has potential regulatory effects by creating information about religious communities, as well as through its recommendations. It is an interesting example of regulating religion because it is a non-statutory commission that, nevertheless, has the potential to have significant effects. CORB also needs to be understood within the context of a UK approach to using commissions and public enquiries to supplement law-making processes. This tradition has tended to supplement the legislative process by appointing prominent public figures to consider an issue of controversy or public interest, often chaired or steered by a leading judge. In the context of diversity and minorities there are a number of advantages with commissions as the preferred form of regulation. The challenge in these circumstances is that majoritarian populist opinions will be tainted by prejudice against minorities. Yet, the paradox is that these viewpoints will often have power within democratic processes that favour majorities in order to secure electoral success. A commission can take 'one step back' from stereotypes that can distort the production of knowledge and policies in majoritarian democratic processes that are vulnerable to populist pressures from far right politics. A commission can create high standards of accuracy that apply to the production of knowledge about minorities that are based on reliable facts and analysis. A commission can also ensure a wider range of consultation and participation by minorities that can be more inclusive than political processes that are often dominated by majorities. In the case of CORB, for example, majoritarian viewpoints were included, but these were also balanced by ensuring there was representation by not only Jews and Muslims, but also Sikhs, Hindus and Humanists. In this way, commissions can perform a valuable function in creating a framework and public process for fact-based and inclusive analysis of how to address the challenges of democracy.

In the context of diversity the UK tradition of commissions have a number of further advantages. They provide a more diffuse form of factual enquiry and governance that can address the issues in a space that is slightly removed from the centralised nationalised form of governance by the executive. This is important because it may provide a more sensitive answer to populist pressures of voters who are hostile to migrants and minorities. It is also important because the more diffuse flexible structure of commissions can allow for the terms of reference and enquiry to be based on a wider dialogue with stakeholders who may lack power to access mainstream democratic processes.

Commissions can play a crucial role in establishing accurate facts and giving policy guidance about minorities because they can address the distortions of the democratic process and the public sphere. Commissions can be designed to be objective, independent and transparent and also involve or hear evidence from minority communities as important stakeholders in the decision-making process. A commission can step back and create a more neutral space when public and political debates about migrants become acrimonious. At one level, commissions can be an important source of gathering accurate facts about diversity and minority against a background of moral panics and distortions about migration and minority communities. They can also be designed to give voice to less powerful minorities who are marginalised in public debates and in democratic processes. For example, they can provide an alternative to a short-term focus on community cohesion as a proxy for anti-terrorism policy by focusing on the integration of minorities as an important democratic rather than purely security aim in liberal societies. One example of this at CORB was the consultation with Muslim communities in Leicester and Leeds that gave voice to the concerns of these communities outside the paradigm of concern with national security. The 'neutral space' within which a commission operates can correct the powerlessness of minorities in mainstream political processes, correct factual distortions that may distort public debates and provide a decision-making process that is based on accurate facts, public reason and that takes a longer view.

CORB identified three core issues that were of public significance. First, it considered "the place and role of religion in contemporary Britain, and the significance of emerging trends and identities". Second, it examined how "ideas of Britishness and national identity may be inclusive of a range of religions and beliefs, and may in turn influence people's self-understanding". CORB's third role was more diffuse than most Royal Commissions because it encouraged an exploration of how "shared understandings of the common good may contribute to greater levels of mutual trust and collective action, and to a more harmonious society".

CORB had a complex structure that ensured that there was a mix of technical expertise of academics and professionals as well as a diversity of religious and gender perspectives. There were four patrons: Professor Lord Parekh of Kingston upon Hull, emeritus professor of political philosophy at the universities of Hull and Westminster; Sir Iqbal Sacranie OBE, formerly secretary-general of the Muslim Council of Britain; the Rt Revd and Rt Hon Lord Williams of Oystermouth, master of Magdalene College, Cambridge, and formerly Archbishop of Canterbury; and the Rt Hon Lord Woolf, CH, formerly Lord Chief Justice of England and Wales.

The Chair, Baroness Butler-Sloss, was assisted by a steering group whose members were Dr Shana Cohen, deputy director of the Woolf Institute; the Rt Revd Professor Lord Harries of Pentregarth, emeritus Gresham professor of divinity, honorary professor of theology at King's College, London, and formerly Bishop of Oxford; Dr Edward Kessler MBE, founder and executive director of the Woolf Institute and fellow of St Edmund's College, University of Cambridge; and Professor Tariq Modood MBE, professor of sociology, politics and public policy at the University of Bristol, and founding director of the Centre for the Study of Ethnicity and Citizenship, University of Bristol. There were also commissioners who reflected a range of professional skills, religion and belief perspectives (including humanism) and different UK regions such as Scotland and Northern Ireland. Crucially each member serves in an individual capacity and not as an official representative or delegate of an institution. In this way, the commission's membership reflected a range of backgrounds in terms of religion, conviction, occupation, age, gender and geographical location.

The process through which substantive issues were debated and agreed was designed to ensure that decision-making was based on accurate facts but that, at the same time, there was an opportunity for stakeholder engagement. The commission met for six weekends over a period of two years, November 2013–June 2015. A pre-commission day event was held on 30 September 2013 to finalise terms of reference and modes of operation, and agree a preliminary schedule of themes and issues to be discussed at the weekend meetings, as well as agree on an outline programme of work.

As with Royal Commissions, the terms of reference for CORB were drawn to focus on a specific issue of religion and belief in British public life, and the way in which the issue was investigated included academic research, consultation with internal experts (who were appointed to the steering group as commissioners) and outside (government, NGO and academic) experts. Each weekend meeting was centred on specially commissioned papers that ensured that the deliberations of the commissioners were based on accurate facts. This technical, factual enquiry was supplemented by consultation with key stakeholders including minority groups.

CORB also made use of the process of holding public consultations with the general public in a number of locations, although unlike a Royal Commission it lacked the formal investigatory powers to summon witnesses under oath, in camera hearings, offers of indemnity, or seizure of evidence or classified information. This wider consultation took place through five local public hearings arranged in Belfast, Cardiff, Glasgow, Leeds and Leicester. In addition, there were national public hearings – one in London and the other, focusing on young people, in Birmingham. At each public meeting the chair was accompanied by 3–4 commissioners many of whom had knowledge of the local communities in the hearing locations. On each occasion, there were also private meetings at which individuals or organisations would be invited to give evidence in confidence including raising issues on sensitive matters, followed by a larger meeting with people involved in an open discussion and airing of views. Both kinds of meeting were by private invitation. A broad range of views was sought, including those of – for example – editors of local papers and borough or city councillors with lead responsibility for community cohesion. The commission also issued a general call for evidence from individuals and organisations.

In a marked contrast to French commissions, whose establishment was triggered by a very specific problem with one religion (e.g. veils, women and Islam) the CORB approach was to remain open-minded about the issues that would arise but to steer the work by identifying some key areas for analysis. To that end, the discussions and the final report focused on key substantive areas: vision, education, media, dialogue, social action and law. Moreover, the analysis of each of these themes was organised around issues of context and background, challenges and concerns and ways forward.

All evidence taken from the hearings and the general call fed into the final weekend meeting of the commission before the drafting of the final report that was an outcome of on-going dialogue within CORB. Each section was drafted by a small team. The final draft was an outcome of debate and dialogue with the whole CORB team who met to finalise the final version of the draft. There are always negotiations and compromise in all collective drafts that are an outcome of this type of process. In the case of CORB, the process of regular meetings and on-going dialogue between the team ensured that differences were resolved in an on-going iterative process rather than being stored up until the end. The process for producing the final CORB report, therefore, did not lead to substantial or acrimonious disagreement. The process sought to reach a balance between a bottom-up approach (through listening to evidence from public hearings and written consultations) and a top-down approach (drawing on the expertise and experience of the patrons, steering group and commissioners).

Public reception

The CORB report was launched in December 2015 to a mixed media reception. There was accurate reporting of the CORB recommendation that mandatory religious worship in public schools should be reconsidered in the light of social changes in the nature of religion and that positive steps should be taken to ensure that selection by faith did not lead to segregation.[16] There was also reporting of the wider issues raised in the report such as the need to open up the UK House of Lords to a wider range of religious minority groups and to focus religious literacy especially for public officials.[17]

Some commentators questioned the objectivity of the CORB report. They offered ad hominem critiques that linked the CORB recommendations to the personal religious and personal viewpoints of the commissioners.[18] Stakeholder groups such as Christian Concern rejected the central premise of CORB that Britain was now a pluralist country that required a shift in public institutions and culture. Andrea Williams, chief executive of Christian Concern, said that there was a climate of "vandalism" against Christianity and stated:

> All around us we are seeing the outworking of the cultural vandalism of our nation's Christian foundation. It's time to reverse that destruction. Who is it that really wants to diminish Christianity's place in our national life? Those who are most vocal in calling for further pursuit of the failed experiment of pluralism, are usually those who really see it as a vehicle for robbing us of the benefit of our Christian heritage.[19]

The Church of England also criticised the CORB report, especially in relation to its recommendation that faith schools should take measures to reduce selection of pupils and staff on grounds of religion.

Nigel Genders, Church of England chief education officer, stated:

> No school is neutral and in our schools we want heads who can develop and promote the Christian character of the school, which is hugely popular with parents. But we don't recognise the implication in this report that schools are involved in mass discrimination.[20]

Malcolm Brown, the Church of England's head of mission, expressed even stronger criticism by suggesting that the commission had been "hijacked". He said: "Religion shows no sign of going away or allowing itself to be relegated to the private sphere." He added: "The common assumption that religion is in decline and can safely be relegated to the margins of our cultural life is simply wrong."

A different set of criticisms by the UK National Secular Society focused on the fact that the report gave too much rather than too little weight to faith in British public life in the light of evidence that religion is less important to British citizens. Both religious and non-religious groups criticised the CORB recommendations for very different reasons.

The negative responses indicated a deep resistance to the factual findings and normative approach of the CORB report. These critiques were ad hominem, in that they focused on the personal characteristics of the commissioners rather than the facts or arguments that were developed in the final report. Moreover, the critics did not provide any balancing arguments that address the social facts and the changing demographics in the UK that CORB sought to address. The critiques provided no alternative responses to the social changes in the UK that were a main impetus for CORB that, the commissioners and experts argued, required a transformation of British public institutions and public life.

CORB, diversity and gender

The CORB commissioners included women from a number of backgrounds. The Chair, Baroness Butler-Sloss, was a prominent woman judge who is an Anglican. The other participants included Professor Gwen Griffith-Dickson, Dr Jagbir Jhutti-Johal, Professor Francesca Klug and myself. The CORB commissioners were not proxies or representatives or delegates for any one religious group. Nevertheless, this selection did ensure both specialist knowledge and personal experience of women from a number of religious perspectives including that of Christians, Jews, Muslims and Sikhs. In addition, women from the Charedi Jewish community and other religious communities were invited to give evidence at the CORB weekend seminars. A special effort was made to send out the public consultation documents to individual women and women's groups as well as ensuring that women were invited to give evidence to the public enquiry hearings. In this way, the design of the CORB process ensured that women were involved at each stage and also had the opportunity to raise their concerns, without specifically identifying gender as a problem for religion or for any particular religious community.

In terms of the substantive issues, CORB included gender as one of a number of questions that the commissioners, those presenting at the weekend seminars and the general public could raise. The questions for consultation did not raise a specific question about one religion (Islam) or a religious symbol (the headscarf or face veil). Instead, CORB approached its terms of reference in a more general way without replicating stereotypes about any one particular religion or religious symbol. Islamic veils are given significant popular media coverage. The public feedback to the CORB consultation, including feedback from Muslim women's groups, did not raise the Islamic veil as a significant topic. The feedback to public consultations did make clear that many stakeholders were concerned about the problems of gender segregation in public spaces such as UK universities. The issue of gender segregation emerged organically through a process of public consultation and reflection. It was discussed at length by the members of the commission, who sought expert opinion to inform their deliberations.

Two strategies were deployed to minimise these risks and steer a balanced approached between these two aspects. First, CORB encouraged women to raise issues but it did not raise problems on their behalf. That is, CORB sought to give voice to women when they raised concern rather than acting as a proxy seeking to speak on their behalf. One example of this is that CORB did not ask any questions about gender segregation at Islamic Society meetings at UK universities but, nevertheless, this was an issue that was raised by a number of stakeholders who submitted evidence and it was, therefore, considered, discussed and analysed. Another example is that CORB did not start with a view as to whether or not religious tribunals were good or bad for religious women, but specific site visits to religious tribunals and conversations with the general public did raise questions and comments about gender and women.

This cautious/explorative approach to issues of religion and belief and gender is significant because, as the regulatory paradigm confirms, CORB has – despite its non-statutory status – the power to create knowledge about religious individuals and communities. It is of crucial importance that this knowledge is based on the voices of actual religious individuals rather than replicating stereotypes about religion. As I argue in the next section, one distinct feature of CORB was precisely that it sought to be pro-active rather than reactive. CORB sought to avoid the approaches of other commissions, such as those in France, that were triggered by a specific panic about the religious practices of one particular religious community. In the next section, a contrast between the CORB and French commissions illustrates this difference.

CORB: comparing the UK and French approach

The CORB approach is in marked contrast to the French commissions that have been convened to address issues of diversity, gender and religious culture. The Stasi and the Gerin commissions were convened with a narrow remit to consider the problems of Islamic headscarves and face veils.

One crucial difference between CORB and the French Stasi and Gerin commissions was the political context within which the commissions were created. Both the Stasi and Gerin commissions were convened following pre-existing concerns and anxiety about diversity in a context of majoritarian resistance to the accommodation of Islamic veils in the public sphere.

The UK also has a tradition of setting up commissions and enquiries that react to conflicts that arise from increasing diversity, for example the Scarman Report (commissioned by the UK government after the 1981 Brixton riots) and the Macpherson Enquiry (commissioned by the UK government after the murder of Stephen Lawrence in 1993). In contrast, CORB was formed in response to a widely held perception in the academic and policy community that social changes in the nature of religion and belief required a deeper, more considered public response. Yet unlike these UK commissions, and the Stasi and Gerin commissions, there was no urgent 'trigger event' that preceded the formation of CORB.

A brief overview of the political context of the Stasi and Gerin commissions illustrates that they were 'reactive' rather than 'proactive' responses to the challenges of diversity in liberal societies. On 15 March 2004, the law banning the Islamic headscarf in public schools was passed through an ostensibly neutral legal provision prohibiting the wearing of conspicuous religious symbols in public primary and secondary schools.[21] Despite its seemingly neutral application, however, the main impetus for the ban was an anxiety about Islam and Muslim difference. The controversy over the Islamic headscarf (*hijab* in Arabic; *foulard* in French) was triggered in September 1989 when three female students were suspended for refusing to remove their headscarves in class at a secondary school in Creil, a town in France. Later that year, in November 1989, the Conseil d'Etat held that it was possible to reconcile the Islamic headscarf with the French principle of secularism (*laïcité*). Lionel Jospin, then Minister of Education, decided that schools had the responsibility for decisions about the permissibility of headscarves in public schools. The French Conseil d'Etat took a nuanced, contextualised approach and confirmed that the decision about the Islamic headscarf should be left to individual schools.[22]

L'affaire de foulard remained a point of friction between French Muslim migrants who were becoming increasingly visible in public and the French state authorities. French feminists, as well as the public more generally, were divided about the Islamic headscarf. On the one hand, prominent feminist intellectuals such as Elisabeth Badinter stated that the veil was a symbol of the oppression of women.[23] Sylviane Agacinski, who led calls for *parité*, also opposed the veil and supported its legal prohibition.[24] Minority women's groups such as Ni Putes, Ni Soumises (Neither Whores nor Downtrodden) were prominent in arguing in favour of a ban on the Islamic headscarf as an aspect of protecting young Muslim girls. Other French feminists disagreed. Francoise Gaspard opposed essentialist definitions of gender that informed the *parité* campaign and opposed the veil bans. Christine Delphy, another leading French feminist, signed a petition to protest the law to ban conspicuous religious symbols and the Islamic headscarf.[25]

By June 2003, in the face of continuing public criticism of the veil, prominent members of the Socialist Party and conservative nationalists of the Union for a Popular Movement (UMP) had proposed a ban in public schools. President Jacques Chirac established a commission chaired by Bernard Stasi, comprising twenty-eight members representing intellectuals, politicians and community representatives, to investigate the principle of secularism and make suitable proposals. The Stasi Commission recommended a ban on conspicuous religious symbols in public schools that was subsequently enacted as *Loi (law) 2004-228 of 15 March 2004.*[26] Their recommendations were not limited to the ban on the veil and they included a wide range of integration policies including progressive policies on the recognition of the religious holidays of minorities. Nevertheless, the politicised context in which the Stasi Commission was established and in which it reported ensured that the Islamic veil ban was the predominant focus of the reception and implementation of their report.

Turning to the Gerin Commission, on 11 April 2011, less than a decade after legal regulation of the Islamic headscarf in public schools, the French law criminalising the face veil came into force. It had its immediate origin in the political intervention of André Gerin, a Communist Party deputy from Vénissieux, who voiced anxieties about face veils after a high-profile legal decision that denied French citizenship to a woman (Faiza M) in 2008 because her choice to wear a face veil was interpreted as indicative of her failure to adopt the Republican value of gender equality.[27] Following recommendations of the Gerin Commission, a French criminal law was enacted that prohibited the wearing of all face-covering gear in public including masks, niqabs and burqas. The face-covering ban imposed a fine starting from Euros 150.00 for each violation or, in the alternative, citizenship education. The relevant provisions also included fines and up to one year's imprisonment for anyone who forced a woman to wear a face veil.[28]

The reactive nature of the French commissions had an unfortunate consequence on their ability to reach out to a wider range of actors who had a stake in *L'affaire de foulard* as well as issues of diversity. The Stasi Commission, for example, did not give substantive weight to the voices of the young school girls who had been excluded from public schools because they wore an Islamic headscarf. Instead, they accepted that the public disturbances that were felt by public officials at the sight of young women wearing Islamic headscarves were de-stabilising to the school environment. Moreover, the Stasi Commission referred to young girls being 'pressured' by North African Muslim migrant men into wearing the headscarf and that permitting Islamic headscarves in public schools encouraged or enabled men, families and communities to coerce the young women into wearing the Islamic headscarf, Yet, no evidence was put forward to justify these claims that replicated and reinforced stereotypes about migrant men at the same time as undermining the actual choices of the young girls who had not significant opportunity to confirm their own choices and desires.[29] Coercing women to wear a headscarf was already an offence that attracted a penalty, yet this stereotype that the young schoolgirls were coerced into wearing the Islamic headscarf was also used to justify a blanket ban even in those situations where there was no coercion by men and a free choice by women.

The Gerin Commission heard from a range of representatives (including liberals and feminists) but there was almost no consultation with the women who actually wear the face veil. One Muslim woman (Kenza Drider) wrote to the commission to say she wanted to give evidence. She was given permission to give evidence but she was also told to remove her face veil while giving evidence before the Gerin Commission.[30]

The failure to consult widely with Muslim women during the Stasi and Gerin commissions, as well as the use of the criminal law as the preferred form of legally regulating the face veil, has obscured the need to initiate wider dialogue about the Islamic headscarf and the face veil. This dilemma is accurately summarised by Ralph Grillo and Prakash Shah who conclude that the current approaches impose a single normative meaning and solution to the issue of the face veil, thereby "stifling or impeding what might otherwise be a 'natural' conversational and dialogical development among Muslims, and with non-Muslims, about the significance of the face veil".[31]

CORB started with the viewpoints of the subjects themselves rather than assuming that a particular religious practice, such as the veil, is problematic. Moreover, in relation to religious practices, CORB used the technique of dialogue with the majority and the minority to identify the viewpoints and solutions to diverse religious practices. This technique is crucial for democratic inclusion. In the French and Belgian context, for example, the failure to address religious diversity through dialogue before moving on to the use of the criminal law has been problematic from the point of view of democratic inclusion. In Belgium for example, Muslim women objected to the fact that they had not been invited to participate in consultations and debates. One Belgium woman objected: "They have not asked for our advice, . . . they have not asked our opinion, they have not asked the proposals we could make on this, they have forbidden us."[32] Case studies from Belgium and France confirm that these Muslim women are not only fully capable, but they are in fact eager to participate in such a dialogue, as well as fully participate in democratic processes. It is not insignificant that the current criminal bans on face veils in public did not ask Muslim women to give their viewpoint or participate in the creation of the laws that would have such a dramatic impact on their personal freedom. As the Belgian case study concludes, the hasty adoption of the ban, without any effort to enter into contact with the women concerned is experienced as a breach of democratic principles.[33]

The key problem with the French commissions – Stasi and Gerin – was that they were established in response to a political controversy that was itself an outcome of the increasing power of anti-Islam politics and the electoral success of the French far right. In this context, it is not surprising that the focus of the Stasi and Gerin commissions was on top-down regulation that sought to re-establish republican values and public order rather than a wider dialogue that produced long-term solutions to the challenges of diversity. The Stasi and Gerin commissions deployed systems of regulation and a process of speaking 'on behalf of' women rather than a dialogue that allowed an exchange of information and viewpoint. By way of contrast, CORB was established in response to academic and policy-makers identifying a need for more sustained analysis about religion and belief in British public life rather than as a reaction to a political 'trigger event'. This meant that CORB was able to steer a balance between two types of risks. An advantage of this approach was that the public hearings and public evidence was received and interpreted without preconception in support of one rather than another political position. CORB was able to hear the voices of women from religious communities without preconceptions about what they considered to be crucial issues or social problems. One example of this was a direct hearing from women representatives of the Charedi Jewish community in north London who gave evidence that their main concern was access to public services on equal terms rather than veiling. On the one hand, there are clearly core issues about how religion and belief impacts on gender equality (women, LGBT groups). On the other hand, CORB needed to avoid approaches where a commission targets a particular religion or religious community/culture thereby reproducing false knowledge and stigmatising a particular religion and community. CORB addressed this risk by allowing women themselves to raise the issues that were of concern to them rather than framing the 'problems' that they faced on their behalf.

Conclusion

CORB reported in December 2015 so it is too early to evaluate the wider reception and policy impact of its recommendations. Its central recommendations on the need for a review and renewal of British public institutions in response to social changes in the nature of religion and belief are not controversial, although there has been resistance to the need for change among some religious stakeholders. CORB was a significant commission in the context of discussion of diversity in the UK because it reflected a shift in the academic, political and policy discourse from a predominant focus on colour racism and ethnicity towards a greater focus on cultural racism, religion and belief. Moreover, CORB incorporated an analysis of minorities into a wider discussion of national identity that was also relevant to minorities. Although it is too early to evaluate the full impact of CORB, it is possible to make some concluding observations. CORB's process of incorporating evidence from experts, some of whom had alternative viewpoints, ensured that there was factual evidence to inform the deliberation of commissioners and the public consultations. This process ensured that CORB's deliberations and recommendations were rooted in the present reality about social life and diversity. CORB's public hearings were crucial because they were a dialogue about British identity rather than a top-down imposition of a static definition of identity on a minority. CORB's approach to the intersection of race, religion and culture with gender was to treat any potential tension as a problem for all religions rather than focusing on one particular religious minority or symbol. In this way, CORB's recommendations about religion and belief in British public life were developed dialectally in response to an understanding of social change and through hearing public evidence. CORB's design and processes, then, were promising ways of addressing the challenges of diversity during a period when the UK is going through a process of considerable transformation in the nature of race, religion and belief.

Notes

1 The Commission on Religion and Belief in British Public Life, *Living with Difference: Community, Diversity and the Common Good.* Report of The Commission on Religion and Belief in British Public Life. The Woolf Institute, 7 December 2015.
2 Anthony H. Richmond, "The Swann Report: Before and after the Swarm Report," *Journal of Ethnic and Migration Studies* 12, no. 2.
3 The Runnymede Trust, *The Future of Multi-Ethnic Britain. The Parekh Report* (The Runnymede Trust, 2000).
4 Tariq Modood, *Multiculturalism: A Civic Idea* (Polity Press, 2007). Maleiha Malik, "Introduction," in *Anti-Muslim Prejudice in the West: Past and Present*, ed. Maleiha Malik (Routledge, 2008).
5 See page 19 of Maleiha Malik, "Minority Legal Orders in the UK." The British Academy, 2012. See also Maleiha Malik, "Minorities and Law: Past and Present," *Current Legal Problems* 67, no. 1 (2014). Maleiha Malik, "Family Law in Diverse Societies," in *Routledge Handbook of Family Law and Policy*, ed. John Eekelaar and Robert George (Routledge, 2014).
6 Malik, "Minority Legal Orders in the UK."
7 Avigail Eisenberg and Jeff Spinner-Halev, eds. *Minorities within Minorities: Equality, Rights and Diversity* (Cambridge University Press, 2005).
8 Charlotte Heath-Kelly, "Counter-Terrorism and the Counterfactual: Producing the 'Radicalisation' Discourse and the UK Prevent Strategy," *The British Journal of Politics and International Relations* 15, no. 3 (2013).
9 Hans-Georg Betz and Susi Meret, "Revisiting Lepanto: The Political Mobilisation against Islam in Contemporary Western Europe," in *Anti-Muslim Prejudice in the West: Past and Present*, ed. Maleiha Malik (Routledge, 2008).
10 Ralph Grillo, *Muslim Families, Politics and Law: A Legal Industry in Multicultural Britain* (Routledge, 2015).
11 Thomas J. Lockwood, "A History of Royal Commissions," *Osgoode Hall Law Journal* 5, no. 2 (1967), 174.
12 Ibid., 181–182.
13 Hugh McDowall Clokie and J. William Robinson, *Royal Commissions of Inquiry* (Stanford University Press, 1937).
14 Lockwood, "A History of Royal Commissions," 172; Clokie and Robinson, *Royal Commissions of Inquiry.*
15 Robert Baldwin, Martin Cave and Martin Lodge, eds. *The Oxford Handbook of Regulation* (Oxford University Press, 2010).
16 Harriet Sherwood, "Top Judge Leads Calls to Scrap Mandatory Daily Christian Worship in UK Schools," *The Observer*, 6 December 2015.
17 Caroline Wyatt, "We Should Do God, Says Report into Religion in Public Life," BBC News, www.bbc.com/news/the-reporters-35005026.
18 Bob Morris, "'Living with Difference': The Butler-Sloss Commission's Report Reflects the Interests of Its Members Rather Than the Public Interest," UCL Constitution Unit, https://constitution-unit.com/2016/01/05/living-with-difference-the-butler-sloss-commissions-report-reflects-the-interests-of-its-members-rather-than-the-public-interest/.
19 Ruth Gledhill, "New Butler-Sloss Religion Report Destroys Our Nation's Defence against Evil," *Christian Today*, 7 December 2015.

20 Ibid.
21 Joan Wallach Scott, *The Politics of the Veil* (Princeton University Press, 2007).
22 Malcolm Cook and Grace Davie, eds. *Modern France: Society in Transition* (Routledge, 1999).
23 Nusrat Choudhry, "From the Stasi Commission to the European Court of Human Rights: L'affaire du foulard and the Challenge of Protecting the Rights of Muslim Girls," *Columbia Journal of Gender and Law* 16, no. 1 (2007).
24 Sylviane Agacinski, *Politique des sexes: précédé de mise au point sur la mixité* (Points, 2009).
25 Scott, *The Politics of the Veil*; Eleonore Lepinard, "Parity Laws and Their Implementation in French Local Politics," *Social Politics: International Studies in Gender, State and Society* 13, no. 1 (2006); Mayanthi L. Fernando, *The Republic Unsettled: Muslim French and the Contradictions of Secularism* (Duke University Press, 2014).
26 Choudhry, "From the Stasi Commission to the European Court of Human Rights."
27 John R. Bowen, "How the French State Justifies Controlling Muslim Bodies: From Harm-Based to Values-Based Reasoning," *Social Research* 2 (2011).
28 Maleiha Malik, "The Return of a Persecuting Society? Criminalizing Facial Veils in Europe," in *The Experiences of Face Veil Wearers in Europe and the Law*, ed. Eva Brems (Cambridge University Press, 2014).
29 Maleiha Malik, "Complex Equality: Muslim Women and the Headscarf," *Droit et société, revue internationale de théorie du droit et de sociologie juridique* 68, no. 1 (2008).
30 For more on the Gerin Commission in France, see Jennifer Selby, "Islam in France Reconfigured: Republican Islam in the 2010 Gerin Report," *Journal of Muslim Minority Affairs* 31 (2011), 387. See also the essay by Nadia Fadil in this volume.
31 See R. Grillo and P. Shah, "Reasons to Ban? The Anti-Burqa Movement in Western Europe," MMG Working Paper 12-05 (Germany: Max Planck Institute for the Study of Religious and Ethnic Diversity, 2012).
32 "Wearing the Face Veil in Belgium: Views and Experiences of 27 Women Living in Belgium concerning the Islamic Full Face Veil and the Belgian Ban on Face Covering," Report of the Human Rights Centre, Ghent University, 2012, at page 35. Eva Brems, "Introduction to the Volume," in *The Experience of Face Veil Wearers in Europe and the Law*, ed. Brems, 1–16.
33 Brems, ed. *The Experiences of Face Veil Wearers in Europe and the Law*.

References

Agacinski, Sylviane. *Politique des sexes: précédé de mise au point sur la mixité*. Points, 2009.

Baldwin, Robert, Martin Cave and Martin Lodge, eds. *The Oxford Handbook of Regulation*. Oxford University Press, 2010.

Betz, Hans-Georg and Susi Meret. "Revisiting Lepanto: The Political Mobilisation against Islam in Contemporary Western Europe." In *Anti-Muslim Prejudice in the West: Past and Present*, edited by Maleiha Malik. Routledge, 2008.

Bowen, John R. "How the French State Justifies Controlling Muslim Bodies: From Harm-Based to Values-Based Reasoning." *Social Research* 78(2) (2011): 325–348.

Brems, Eva, ed. *The Experiences of Face Veil Wearers in Europe and the Law*. Cambridge University Press, 2014.

Choudhry, Nusrat. "From the Stasi Commission to the European Court of Human Rights: L'affaire du foulard and the Challenge of Protecting the Rights of Muslim Girls." *Columbia Journal of Gender and Law* 16(1) (2007): 199–206.

Clokie, Hugh McDowall, and J. William Robinson. *Royal Commissions of Inquiry*. Stanford University Press, 1937.

Commission on Religion and Belief in British Public Life. *Living with Difference: Community, Diversity and the Common Good*. Report of the Commission on Religion and Belief in British Public Life. The Woolf Institute, 7 December 2015.

Cook, Malcolm and Grace Davie, eds. *Modern France: Society in Transition*. Routledge, 1999.

Eisenberg, Avigail and Jeff Spinner-Halev, eds. *Minorities within Minorities: Equality, Rights and Diversity*. Cambridge University Press, 2005.

Fernando, Mayanthi L. *The Republic Unsettled: Muslim French and the Contradictions of Secularism*. Duke University Press, 2014.

Gledhill, Ruth. "New Butler-Sloss Religion Report Destroys Our Nation's Defence against Evil." *Christian Today*, 7 December 2015.

Grillo, Ralph. *Muslim Families, Politics and Law: A Legal Industry in Multicultural Britain*. Routledge, 2015.

Heath-Kelly, Charlotte. "Counter-Terrorism and the Counterfactual: Producing the 'Radicalisation' Discourse and the UK Prevent Strategy." *The British Journal of Politics and International Relations* 15(3) (2013): 394–415.

Lepinard, Eleonore. "Parity Laws and Their Implementation in French Local Politics." *Social Politics: International Studies in Gender, State and Society* 13(1) (2006): 30–58.

Lockwood, Thomas J. "A History of Royal Commissions." *Osgoode Hall Law Journal* 5(2) (1967): 172–209.

Malik, Maleiha. "Complex Equality: Muslim Women and the Headscarf." *Droit et société, Revue internationale de théorie du droit et de sociologie juridique* 68(1) (2008): 127–152.

Malik, Maleiha. "Introduction." In *Anti-Muslim Prejudice in the West: Past and Present*, edited by Maleiha Malik. Routledge, 2008.

Malik, Maleiha. "Minority Legal Orders in the UK: Minorities Pluralism and the Law." The British Academy, 2012.

Malik, Maleiha. "Family Law in Diverse Societies." In *Routledge Handbook of Family Law and Policy*, edited by John Eekelaar and Rob George. Routledge, 2014.

Malik, Maleiha. "Minorities and Law: Past and Present." *Current Legal Problems* 67(1) (2014): 67–98.

Malik, Maleiha. "The Return of a Persecuting Society? Criminalizing Facial Veils in Europe." In *The Experiences of Face Veil Wearers in Europe and the Law*, edited by Eva Brems. Cambridge University Press, 2014.

Modood, Tariq. *Multiculturalism: A Civic Idea*. Polity Press, 2007.

Morris, Bob. "'Living with Difference': The Butler-Sloss Commission's Report Reflects the Interests of Its Members Rather Than the Public Interest." UCL Constitution Unit, https://constitution-unit.com/2016/01/05/living-with-difference-the-butler-sloss-commissions-report-reflects-the-interests-of-its-members-rather-than-the-public-interest/.

Richmond, Anthony H. "The Swann Report: Before and after the Swarm Report." *Journal of Ethnic and Migration Studies* 12(2) (1985): 223–227.

The Runnymede Trust. *The Future of Multi-Ethnic Britain. The Parekh Report*. The Runnymede Trust, 2000.

Scott, Joan Wallach. *The Politics of the Veil*. Princeton University Press, 2007.

Sherwood, Harriet. "Top Judge Leads Calls to Scrap Mandatory Daily Christian Worship in UK Schools." *The Observer*, 6 December 2015.

Wyatt, Caroline. "We Should Do God, Says Report into Religion in Public Life." BBC News, www.bbc.com/news/the-reporters-35005026.

5 British multiculturalism

From 'Parekh' to 'PREVENT', and beyond

Ralph Grillo

Introduction

Initially I was somewhat sceptical of the value of comparing the four commissions investigated in the *Religious and Cultural Diversity in Four National Contexts* project. After all, they were each very much of their own time and place; the Parekh Report,[1] for one, seemed fifteen years after its publication in 2000 to come from a far-away country where (or when) they did things differently. Nonetheless, such commissions (whether statutory or based in civil society), though having a long history, are a distinctive phenomenon of our time, and their institutional nature deserves investigation. They are also relevant to, indeed an integral component in, another phenomenon of our time, the continuous detailed scrutiny (in the public sphere and in the academy) of diversity and diverse populations, and what to do about them. This in itself deserves thorough, comparative, socio-scientific investigation: why, where, when, how, who, and with what effect? My book *Muslim Families, Politics and the Law*, subtitled *A Legal Industry in Multicultural Britain*,[2] explored aspects of this 'commission phenomenon' so far as the UK is concerned and the present essay hopefully takes the investigation a stage further.

The essay begins with an overview of what may be called the 'crisis of difference', globally and specifically within Europe and the UK. Although not everyone agrees, the idea that religious and cultural diversity constitutes a problem which must be resolved (perhaps by suppressing it) now figures prominently in the politics of most Western countries. The second section then focuses on the British experience, rehearsing historic developments in the multiculturalist agenda (associated initially with the so-called 'Jenkins formula' of 1967), which established a relatively benign framework for the accommodation of cultural and religious minorities – a common public sphere with distinctive beliefs and practices tolerated in the private sphere and to an extent the public sphere, too: minorities would be 'Here', that is part of British society, but 'Different', with their own distinctive cultures. Nonetheless, although there was some (considerable?) improvement, by the turn of the millennium members of minority communities still faced many economic and other disadvantages and discrimination, as the Parekh Report (hereafter *Parekh*) demonstrated and the reaction to what the report's assertions about 'Britishness' made abundantly clear.

The third section of the essay continues the discussion of *Parekh* by situating it among the numerous policy interventions by government and NGOs concerning immigrant and ethnic minorities of migrant background. Their sheer volume is striking as is their change in tone and orientation, especially through the 2000s. *Parekh*, a report by the long-standing Runnymede Trust, was very much of its epoch, but the benign tolerance it represented was overtaken, notably after 2005, by 'muscular liberalism', and an increasing emphasis on 'security', on combatting 'radicalisation', on 'British values', and on measures to monitor and 'reform' Islamic practices.

If the third section ends on a pessimistic note, the fourth offers some glimmers of hope by focusing on how in 'super-diverse' cities such as London quotidian accommodation is still fostered on the ground, and, sometimes surprisingly, in judicial responses to cultural diversity. This may prove naively optimistic, but I believe that the tradition of British multiculturalism as a 'negotiated order', which *Parekh* represented and which reappeared in the CORAB Report in 2015,[3] still has a role to play.

A crisis of difference?

Although this essay is principally about Britain's historic experience of diversity, I begin by reflecting on the current global situation, which is one of multiple crises: economic, political, social, cultural, religious, environmental, medical. Among these, and to an extent emanating from them, is a 'crisis of difference', with religious and cultural diversity major sites of contestation. How does this manifest itself?

From around the 1960s Western nation-states seemed more accommodating of ethnic, religious, and linguistic diversity. For example, in Britain there was growing recognition of the right to difference, and legislation enhancing minorities' freedom from discrimination. After 2000, however, there was a widespread 'backlash' against such 'multiculturalism', which was said to privilege ethno-linguistic cultural blocs, encourage 'parallel lives', and promote cultural and religious principles and practices unacceptable in societies espousing liberal, democratic, individualistic, and secular values. It is now widely believed that contemporary societies are characterised by what Giovanni Sartori called an 'excess of alterity',[4] and more often than not, the overriding (hegemonic) narrative is that diversity, especially that associated with immigrant minorities, is bad, or at least potentially damaging to existing social relations, and must be restrained.[5]

Why have diversity and difference become so problematic? Why is anxiety about 'our' culture and what is happening to it[6] seemingly ubiquitous? There are several interconnected reasons.

First, inequality between rich and poor countries, coupled with political instability and civil conflict (often associated with ethnic and religious difference), foreign armed interventions, and long-term environmental degradation, have driven millions of migrants and refugees to seek work and asylum across borders, within the global South, and between South and North. This has been of growing importance since the early 1990s, notably so in 2015–2016 with, in Europe and North America, a veritable moral panic over what has become the largest flow of 'displaced persons' since the Second World War. This reaction may have xenophobic, indeed racist undertones, which anti-immigrant parties and politicians articulate, but there are also genuine (if often misplaced) worries about the impact of immigration (and the settlement of families of immigrant background) on access to welfare, housing, education, health care, jobs ('Ils mangent notre pain', as the saying goes[7]), and on ways of life. There is also great anxiety about what is happening to nation-states in a period of 'disembedding',[8] when societies are undergoing drastic economic and political change within a neo-liberal world.

Second, there have been major changes in the demographic composition of the populations of migrant origin, with minority families increasingly significant. Hence, matters routinely affecting family life and relations of gender and generation have grown in importance, with immigration and settlement catalysts for changing perceptions of self and others, forcing both incomers and members of receiving societies, to re-assess and perhaps re-assert cherished values.

Third, although many are now long-term migrants, or born and brought up in the countries where they now reside, relationships with societies of origin have not diminished. The transnationalisation of relationships is a major factor in the contemporary scene.

Fourth, *some* people from such backgrounds seek to maintain *some* practices seemingly at odds with those of the societies in which they have settled, and therefore 'problematic' so far as public policy is concerned. I emphasise 'some', and add that issues may arise from what is occurring *within* minority families, in the changing dynamics of gender and generation, as much as from any disjunction between minority and majority practices.

Fifth, we are in what Habermas[9] has called a post-secular world, with people globally turning to religion to guide their conduct and seek advice on how to comport themselves in societies often seen as secular, individualistic, and immoral. In consequence, for many people (outsiders, insiders, believers, non-believers) a person's essence is seemingly encapsulated in their religion.

The combination of such factors (along with the globalisation of ideas about cultural, religious, and gender rights) has generated in Western societies a multiplicity of culturally differentiated and often conflicting norms, a plurality of overlapping, intersecting, and interacting moral universes (conceptions of the good life and how to live it), which pose major challenges, not least in a legal context, in what for many is now a very confusing world.

At least since 2001, this crisis has been directly associated with Islam. For many people Islam *is* the problem, an icon of difference perceived (misconceived) as an aggressive existential threat, a view which events in the Middle East, Europe, and elsewhere have reinforced. Political Islamism ('jihad'), and the words and deeds of groups such as Al-Qaeda, the so-called Islamic State or Boko Haram make the threat seem very real, and the predominant view is that it must be met by tough security measures, accompanied by policies restricting practices which seemingly run counter to liberal-secular values. Thus it might seem that Huntington was right, that there is an apparent 'clash of civilisations'[10] involving two deeply entrenched, hostile 'sides', each asserting its own moral superiority, with protagonists engaging in what Edgar Morin has called 'la réduction d'autrui à son pire aspect'.[11]

In consequence, the context for any discussion of diversity and its governance is now very different from what it was at the time of *Parekh* or in earlier periods.

British multiculturalism, Jenkins (1967) to Parekh (2000)

Jenkins on 'integration'

Let me now go back in time, as it were, and reflect on the changing British (perhaps even 'English') response to 'other' cultures and faiths during the last fifty years.

Briefly, after 1945 there were three main groups of immigrants to Britain: refugees from Eastern Europe, workers and families from Southern Europe, and Empire/Commonwealth migrants from the Caribbean and South Asia. Despite some hostility, the first two rarely posed difficulties for the governance of diversity. The third did, evidenced by 'race riots' in the late 1950s in London and elsewhere directed against the African Caribbean population, and antipathy towards the newly arriving Asians. Racism of a traditional kind was a key issue in a period of rapid decolonisation.

There was a three-fold response: legislation controlling immigration and redefining British nationality; regulating 'race relations' through laws to combat discrimination; and a move towards a specific form of 'integration', devised by Home Office advisers under the Labour Home Secretary: the so-called 'Jenkins formula':

> Integration is perhaps a rather loose word. I do not regard it as meaning the loss, by immigrants, of their own national characteristics and culture. I do not think we need in this country a 'melting pot', which will turn everybody out in a common mould . . . I define integration, therefore, not as a flattening process of assimilation but as equal opportunity, coupled with cultural diversity, in an atmosphere of mutual tolerance.[12]

The key phrase was 'coupled with cultural diversity'. The future for immigrants was within a shared public sphere of norms and values, with equal opportunity in employment, housing, education, health and welfare, equality before the law, and protection from discrimination. Distinctive beliefs, values, practices, religion, language were for the private sphere; immigrants and minority ethnic groups would be 'Here but Different', governed through a 'weak' form of multiculturalism which influenced practice elsewhere.

Race and Law,[13] by Anthony Lester (one of Jenkins's advisors) and Geoffrey Bindman, a human rights specialist, illustrated very well the liberal discourse of the 1960s and 1970s on 'race' and 'colour', as it was usually called (occasionally the newish term 'ethnicity' begins to appear). Their book's principal focus was the legal history of anti-discrimination decisions, and the politics of the debates about possible legislation. It demonstrated the then contemporary importance of the 'equality' and 'anti-discrimination' agendas (see further below, in the section 'Scrutinising minorities, shifting agendas'), though, depressingly, much of what the book described in its introductory chapter concerning hostility towards other 'races' could equally well apply to the discourse around immigrants and refugees in 2015–2016.

The Swann Report

Although 'culture' and 'religion' as such did not figure prominently in the diversity discourse of the 1960s and 1970s, their significance emerged strongly in the 1980s and 1990s. An important landmark was the Swann Report of 1985 on mother tongue and bilingual education.[14] *Swann* specifically addressed issues raised by the existence of an 'ethnic' school sector, privately run and funded, which typically offered lessons in the language, culture, and religion of the migrants' society of origin, out of regular school hours and on informal premises. These so-called 'Saturday schools', also found in other countries with substantial immigrant populations, reflected the wishes of parents for their children's education, and some believed they offered a model for the mainstream state school system. Experiments with bilingual education, which became increasingly important in the USA in 1970s and 1980s, and the accompanying academic debates on the developmental significance of bilingualism, were also matters which *Swann* addressed.

The report came out against mainstreaming bilingual education, arguing that assisting children to achieve fluency in their mother tongue should be left to the ethnic school sector. Although the report recommended making mother tongue languages (e.g. Arabic, Mandarin, Urdu) available as subjects in the secondary school curriculum, it also reasserted the traditional viewpoint that 'the English language is a central unifying factor in being "British", and the key to participation on equal terms as a full member of the society'.[15] Historically, British schools, explicitly or implicitly, had always been concerned with maintaining a language (English) and mainstream British culture, and this remained the predominant view, as legislation such as the Education Reform Act of 1988 subsequently confirmed.[16]

Although in many schools the ethos in the 1980s was quite different from that which prevailed in the monocultural educational system of the 1960s, anti-racist, multicultural, and multilingual education made only modest headway, often in the face of strenuous opposition,[17] with *Swann* later castigated for encouraging practices in schools which privileged pupils' ethnicity. Indeed, while open to diversity in many ways, *Swann* itself had put down a marker: there were limits to the difference that was acceptable in the public domain of education.

The Parekh Report

If 'culture' assumed increasing importance in educational debates in the 1980s and 1990s, so did religion or 'faith' as it came to be called. If, in the case of minorities, especially with origins in South Asia, their otherness was in the 1950s–1960s represented principally through the language of 'race' or 'colour' (as in 'race relations'), and in the 1970s–1980s through that of 'ethnicity' and 'culture', during the 1990s the principal marker became 'faith'. 'British discourse on racialised minorities', says Peach, 'mutated from "colour" in the 1950s and 1960s, to "race" in the 1960s–1980s, "ethnicity" in the 1990s, and "religion" in the present period'.[18] While in the 1970s and 1980s, for instance, activists and some academics sought to articulate a unity of people of African Caribbean and South Asian origin around the all-embracing class-race term 'black', dominant discourse gradually shifted towards ethnic and cultural, and eventually religious, specificities.

Although Peach (cited above) allocates the discourse of 'ethnicity' to the 1990s, and, writing in 2005, that of 'religion' to 'the present period', the shift towards religion, or 'faith', or at the very least the demands for such a shift, was already apparent in the early 1990s, indeed in the case of claims for faith schools, from the mid-1980s. The Swann Report had considered, but rejected, a proposal to extend the so-called 'voluntary-aided' educational sector to include Muslim and other minority religious schools alongside the existing Christian and Jewish establishments. One response, notably in the city of Bradford, was an attempt by Muslim community leaders to acquire and reorganise several schools (primary, middle, and secondary) on a religious basis, and such claims continued to be pressed into the 1990s, culminating in legislation in 1997 to include Muslims and others within the voluntary-aided sector.[19] If 'multicultural education', as in the Swann Report, epitomised the 1980s, 'faith schools' came to symbolise the 2000s. The Rushdie Affair (1989), with the subsequent debates about the blasphemy law, and in due course Islamophobia, and proposals, initially controversial, to ask about religious affiliation in a future census (implemented in 2001), also pushed the faith agenda. Consequently, government policy during the 1990s–2000s came increasingly to emphasise religion as a mode of recognising and working with minorities, responding to their own efforts to define themselves in religious terms. This re-focusing on a communitarian, faith-based multiculturalism was eventually reinforced in measures such as the Racial and Religious Hatred Act, 2006, which confronted 'racism by [religious] proxy'.

Allen[20] interprets such developments as cultural racism, with religion the latest marker through which xenophobia is expressed. It is rather that the terms of engagement 'mutated', to use Peach's term, in a dialectical relationship between three sets of actors: the state, faith institutions, and believers. Tariq Modood[21] rightly emphasises the part played by the Rushdie Affair and contends that the shift was not solely or even principally top-down, but reflected the increasing tendency for minority ethnic groups to stress religious faith as a key element of public and private identity. And not just Muslims. In the mid-1990s a swathe of Hindu societies emerged among students in British post-secondary education institutions,[22] with, after 9/11, an accelerating rejection of 'Asian' as an all-embracing ethnic identity in favour of one specifying religion. Not that the 'old' language ever faded away; 'black' continued (and continues) to be asserted by some as a unifying identity, for example, and national origin still identifies both new immigrants and settled minorities in many contexts. There are multiple markers of difference, and the popular pejorative and insulting term 'Paki' might be said to conflate many of them, simultaneously signalling 'race', ethnicity, culture, national identity, as well as religion.

At the same time, there were also, through the 1990s, and increasingly through the 2000s, significant changes in the nature and composition of the immigrant population with arrivals from a wide range of countries and cultures, entering via different routes, with different legal and other statuses. Britain's increasingly 'super-diverse'[23] minority population fed resentment about immigration and multiculturalism. In the midst of all these shifts and developments, enter *Parekh*.

As the report is discussed at length elsewhere in this volume, I will limit myself to a few observations.

First, in classic fashion, *Parekh* pursued the long-standing equality and rights agenda with a substantial part of the report concerned with inequality in criminal justice, education, arts and the media, health and welfare, employment, immigration, political representation. This was familiar territory from the 1960s onwards, and the sad thing was that it needed to be said once again. There was, however, an additional focus on identity and difference: 'The fundamental need', said *Parekh*, 'both practical and theoretical, is to treat people both equally and with due respect for difference; to treasure the rights and freedoms of individuals; and to cherish belonging, cohesion and solidarity'.[24]

Second, if *Parekh* was significant because it saw inequality/difference as complementary and closely connected, it was its stance on difference which attracted attention and disapproval. As is well known, *Parekh* received a mixed reception, some of it very hostile, especially where its remarks on identity were concerned.[25] Indeed, at the launch, Jack Straw, the then Home Secretary, 'strongly part[ed] company with the commission . . . over their view of Britishness'.[26] 'The report's suggestion that the word "British" is racist has finally frightened even those ministers who thought they could never go wrong by appeasing such doctrines', said the *Daily Telegraph*.[27] The sophisticated arguments proposed in the report about the need to 'rethink the national story' were deemed incomprehensible and unacceptable.

Third, the report had not yet caught up with the shift towards a faith agenda; references to religion/faith, if not minimal, certainly do not have the prominence and weight attributed to 'race' and ethnicity. In the case of Islam, for example, there is no mention of Shari'a which ten years later was at the forefront of debate about the Muslim presence. The main thrust is on anti-Muslim racism and Islamophobia, described as 'one of the most serious forms of cultural hostility in Europe',[28] and the report recommended legislation on discrimination on religious grounds. Nonetheless, it did contain a proposal for establishing a commission on the role of religion in public life,[29] which many years later bore fruit in CORAB and its report in 2015.

Despite reservations of the kind expressed by the Home Secretary, the government accepted *Parekh*'s recommendations, and subsequently asserted it had implemented many of them. Indeed, a Runnymede Trust review of progress in

2004 said *Parekh* had been 'influential in shaping the latest phase of thinking on race equality',[30] and in a 'Foreword' to that review Bhikhu Parekh himself claimed that 'the large issues relating to British identity, pluralism, cohesion, and recommendations on the traditional narrative of the country's history are now being debated in public'.[31] Yet even taking into account the government's response, Runnymede observed that minorities in Britain still experienced economic and social disadvantages (and discrimination) in the areas covered by the report, while racism, casual and sometimes violent (and latterly Islamophobia), remained an everyday occurrence. Significantly, the Commission for Racial Equality's evaluation of its work over some thirty years, which identified ten areas for advancement, was entitled *A Lot Done, A Lot to Do*.[32]

Although these evaluations were relatively upbeat about the small, but nonetheless important, changes taking place, and subsequent research records 'a rather optimistic picture of ethnic minority integration in Britain',[33] other assessments have been more sceptical. Thus a later Runnymede report in 2014 concluded that '[e]thnic inequalities in education, employment, health and housing are widespread in England and Wales and persistent', and in some areas increasing.[34] For example, although a professional, managerial, and business minority middle class was emerging in many walks of life, progress was often painfully slow. And latterly minorities were disproportionately affected by recession and austerity, albeit unevenly across different ethnic groups.

Scrutinising minorities, shifting agendas

A plethora of reports

Parekh was, however, but one of many reports, consultations and commissions of inquiry that have sought to address the situation of minorities in Britain over the last fifty years, and increasingly in the last twenty years. From Roy Jenkins's essays and speeches in 1967 to the British government's *Counter-Extremism Strategy*[35] and the CORAB Report, both in 2015, there have been over a thousand interventions by various ministries (principally the Home Office), non-departmental public bodies, and NGOs pronouncing directly or indirectly on diversity and its governance. And this does not include a very large number of academic books and papers, nor media articles, TV programmes, novels, plays, and films.[36] *Parekh* must be situated among this plethora of interventions, and the various changing themes which they have addressed.

Modes of consulting and reporting

The UK has had a long-standing tradition of appointing commissions and similar bodies to report on issues of the day and make recommendations on policy. From the early nineteenth century, for example, there were influential reports on such matters as Irish and Welsh education, ecclesiastical courts, children's employment, and railway gauges. Important inquiries were often conducted by an authoritative 'Royal Commission', such as those on the press (1974–1977), or care for the elderly (1998). Much earlier, in 1903, the Royal Commission on Alien Immigration[37] had led to the 1905 Aliens Act, the first major legislation controlling immigration into the UK.

In recent years weighty Royal Commissions appear to have given way to ad hoc commissions or working parties established by central government, such as that on arranged and forced marriages in 2000.[38] These commissions are usually chaired by a prominent member of the British establishment, often a judge or similar, who is believed to have a 'safe pair of hands' (a metaphor from the game of cricket), along with a committee of the 'great and good' (experts in the field under review or otherwise prominent in public life). There are consultations,[39] inviting responses, oral and written, often to an official paper summarising a problem and possible solutions, collection of evidence, and finally a report. Though recommendations may be genuinely sought, sometimes the device is used to 'kick an issue into the long grass' (golfing metaphor), i.e. keep it out of sight, as with the Chilcot Inquiry into British involvement in the Iraq War, established in 2009, finally reporting in 2016. Besides such working parties, there are reports commissioned by central or local government with a single reviewer and investigative team. This method, which relies largely on interviews and documentary evidence, seems increasingly in favour.[40]

In addition to these interventions by central or local government, parliamentary bodies such as the House of Commons Home Affairs Committee have played an increasingly important role, inviting comments on issues such as female genital mutilation and Shari'a councils,[41] and interviewing selected individuals and organisations. There have also been numerous influential reports from non-departmental public bodies such the Commission for Racial Equality (now absorbed in the Equality and Human Rights Commission), and from working parties set up by non-governmental bodies (sometimes with government blessing). *Parekh* was one such, as was the Runnymede's earlier review of *Islamophobia* (1997);[42] the CORAB Report is a recent example. Indeed, NGOs and civil society institutions make numerous (sometimes influential) interventions on such matters, responding in depth to consultations and commissions. Some have become favoured professional 'consultees', submitting opinions, being interviewed by parliamentary committees, commenting to the press, and publishing their own in-house research.[43] Besides these are the myriad scholarly publications, newspaper articles, television documentaries, Internet blogs, feeding public debate.

Many of these governmental, non-governmental, and other interventions (including those by academics) have minimal impact, and are quickly forgotten by all but nerdish scholars. Such was the fate of *Our Shared Future* (2007) by the short-lived Commission on Integration and Cohesion.[44] To someone in government it had seemed a good idea at the time.

Changing narratives

Where does *Parekh* fit into this official and unofficial scrutiny of minorities? Running through all these reports are various 'narratives' or agendas, often illustrated in their titles. Indeed, significant changes in those agendas can often, though not always, be read off from the titles alone. Detailed analysis is beyond the scope of this essay, but briefly key themes have included:

- inequality-discrimination
- identity-difference
- integration-cohesion
- radicalisation-security
- Britishness and British values.

The first was influential in the early years, and perhaps reached its apogee in *Parekh*, though, as indicated earlier, *Parekh* also had a benign take on identity/difference. No sooner had *Parekh* reported, however, than the northern cities of England experienced a wave of disturbances (in 2001, albeit before 9/11) which were seen as deeply threatening. The response was a series of inquiries the most influential being the Cantle Report on *Community Cohesion* which recorded that 'many communities operate on the basis of a series of parallel lives',[45] a phrase (originating in Germany) that achieved wide currency and led to the view that Britain was 'sleepwalking to segregation', as Trevor Phillips, then Chairman of the Commission for Racial Equality, put it.[46] By 2005, however, the shocks of 9/11 and 7/7, the bombings on the London transport system, which raised the terrifying prospect of attacks by young Muslim suicide bombers, meant that integration-cohesion became increasingly bound up with radicalisation-security as the principal preoccupation of government.

From PREVENT to 'muscular liberalism'

One response was the PREVENT strategy initiated by the Labour government in 2005 when, in the light of the July 2005 attacks on the London transport system, it brought together 'a diverse range of people with different skills and knowledge' in seven working groups, under the rubric of 'Working Together to Prevent Extremism'.[47] The outcome was a series of recommendations aimed at engaging with young Muslim men and women to counter extremist ideas and enhance their quality of life (education, jobs), and building institutions within the British Muslim community with which government could hold dialogues.

While the PREVENT strategy sought to enrol Muslims in the fight against extremism, and sought appropriate community channels for that purpose, it was also concluded (following the Cantle Report) that minorities could not be permitted to enclose themselves in separate communities,[48] and this fed growing misgivings about the extent and nature of diversity. Indeed, by the mid-2000s, the overriding narrative became 'diversity has increased, is increasing, and ought to be diminished', with multiculturalism held responsible for all manner of terrible things: 'parallel lives', ghettoisation, radicalisation of young people, unfair access to resources, and so on, issues readily taken up by right-wing populist, anti-immigrant parties.

These concerns influenced reviews of the PREVENT strategy undertaken by the Coalition government which came into office in May 2010, and informed a widely reported speech by the Prime Minister, David Cameron, at a security conference in Munich in February 2011.[49] He began by talking about terrorism, and Britain's posture regarding military matters, but soon moved on to Islam and the threat posed by 'young men who follow a completely perverse, warped interpretation of Islam'. He was, however, careful to distinguish that Islam from other forms, and added: 'the ideology of extremism is the problem; Islam emphatically is not'. This then led him to propose that these young radicals 'find it hard to identify with Britain, because we have allowed the weakening of our collective identity':

> Under the doctrine of state multiculturalism, we have encouraged different cultures to live separate lives, apart from each other and apart from the mainstream. We've failed to provide a vision of society to which they feel they want to belong. We've even tolerated these segregated communities behaving in ways that run completely counter to our values.

'Frankly', he went on:

> we need a lot less of the passive tolerance of recent years and a much more active, muscular liberalism. A passively tolerant society says to its citizens, as long as you obey the law we will just leave you alone. It stands neutral between different values. But I believe a genuinely liberal country does much more; it believes in certain values and actively promotes them. Freedom of speech, freedom of worship, democracy, the rule of law, equal rights regardless of race, sex or sexuality. It says to its citizens, this is what defines us as a society: to belong here is to believe in these things. Now, each of us in our own countries, I believe, must be unambiguous and hard-nosed about this defence of our liberty.

He repeated these sentiments, especially his objections to 'passive tolerance', even more forcefully in 2015 and 2016.[50] In the media and across the Internet the overriding message was 'multiculturalism has failed'. At last, some said, Britain (like France, Germany, and the Netherlands) had turned its back on multiculturalism for which no one had voted and which was undermining British culture and society.

'Britishness' and 'British values'

Although narratives changed, one seemingly ever-present theme in recent years has been that of 'Britishness' and 'British Values', though what these signalled in *Parekh* and in later developments was somewhat different.

Between the 1960s and the early 1990s there was in fact very little public discussion of the nature and significance of 'Britishness'; there had never been in

Britain anything like the extended debate about national identity (and language) that had been pervasive in France since the Revolution. There were exceptions: the Conservative politician, Enoch Powell, in the late 1960s, challenged the right of immigrants born in the Caribbean or South Asia to aspire to a British, or English, identity, and in the mid-1980s the committee which produced the Swann Report (of which Bhikhu Parekh was a member) defended itself against the accusation that their recommendations sought to 'undermine an ill-defined and nebulous concept of "true Britishness"',[51] adding,

> The identity of our society represents an amalgam of all the various forces which have been and indeed are still at work within it and the many influences which have impinged upon it from outside. Thus to seek to represent 'being British' as something long established and immutable fails to acknowledge that the concept is in fact dynamic and ever changing, adapting and absorbing new ideas and influences.

This comment perhaps anticipates what was said later in *Parekh,* but public references to Britishness were more usually flippant ('KNICKERS TO THE FRENCH; The perfumed halls of Galeries Lafayette compete with the grey flannel Britishness of Marks & Spencer'[52]), or downbeat ('Queen Salutes "Britishness" of German-Built Luxury Liner'.[53]) When the *Evening Standard* headlined 'Britishness Still Counts',[54] it was referring to motor cars. Yet change was in the air, and indeed in academic circles had been for some time, notably with Paul Gilroy's *There Ain't No Black in the Union Jack* (1987)[55] and Linda Colley's highly influential and much quoted *Britons: Forging the Nation* (1992).[56] Certainly Gilroy and Colley influenced *Parekh,* as did a paper on ethnicity (1992)[57] by Stuart Hall, himself a member of the commission. Between Gilroy and Colley (and Hall), however, came the Rushdie Affair, which elicited a strong, at times racist, strain of anti-Islamic feeling, reflecting the sense that the Muslim reaction represented a rejection of liberal, democratic, *British* values.

As noted above, Parekh himself had suggested that despite the immediate rejection by the Home Secretary and the media of the report's observations concerning Britishness, it had stimulated subsequent public debate about identity in the UK. Indeed, a search of headlines in the mainstream press from the mid-1980s onwards found that of 582 references to 'Britishness' or 'British values', only forty-nine occurred between 1990 and 1999 while in 2000 alone there were twenty-three, mostly concerning *Parekh.*[58] Thereafter earnest accounts of 'Britishness' have rarely been absent from the media or speeches by political and religious leaders; over three-quarters of the 582 references occurred after 2005. But there are differences. Whereas *Parekh* (like *Swann*) engaged with what Britishness signified in term of identity, and sought a legitimate place within in it for Britons whose roots were in Asia, Africa, or the Caribbean (bringing 'black' into the Union Jack, as it were), later accounts were more concerned with 'values', British values and the threat to these from immigration and, increasingly after 7/7, Islam (or 'jihad'). Consequently, the inculcation of British values in schools, to draw young people away from Islamist extremism, became a frequent motif in political speeches and media articles, notably, and most recently, in the *Casey Review*, published in December 2016, which *inter alia* recommended the introduction of an 'Oath of Integration with British Values and Society', that intending migrant settlers should take on arrival.[59]

Speaking in 2006, the then Prime Minister, Tony Blair, had developed this theme by emphasising that in the current context integration is 'not about culture or lifestyle':

> It is about integrating at the point of shared, common unifying British values. It isn't about what defines us as people, but as citizens, the rights and duties that go with being a member of our society. Christians, Jews, Muslims, Hindus, Sikhs and other faiths have a perfect right to their own identity and religion, to practice their faith and to conform to their culture. This is what multicultural, multifaith Britain is about. That is what is legitimately distinctive. But when it comes to our essential values – belief in democracy, the rule of law, tolerance, equal treatment for all, respect for this country and its shared heritage – then that is where we come together, it is what we hold in common; it is what gives us the right to call ourselves British.[60]

'Our tolerance', he added, 'is part of what makes Britain'. There were similar interventions by David Blunkett (as Home Secretary)[61] and Gordon Brown (then Chancellor of the Exchequer).[62]

In fact there are four senses in which the idea of 'Britishness' has been deployed:[63]

- As a certain way of life (warm beer, cricket, old ladies cycling to church), echoing George Orwell's essay 'The Lion and the Unicorn',[64] themes taken up in 1993 by the then Prime Minister, John Major, in a speech to the Conservative Group for Europe.[65] Tony Blair's 'Cool Britannia' tried to put an updated slant on this.
- As a political and ethnic (indeed 'racial') identity, e.g. as discussed in *Parekh*.
- As adherence to certain values which guide personal conduct such as 'decency', 'discipline', 'respect', and 'tolerance' (sometimes equated with the values of conservative middle England, as interpreted by the United Kingdom Independence Party).
- As adherence to the civic values of a liberal, democratic society.[66]

The latter are referenced in Tony Blair's speech, cited above, and in numerous publications by the later Coalition and Conservative governments.[67] Indeed, they are summarised in the official definition of 'extremism' as 'vocal or active opposition to fundamental British values, including democracy, the rule of law, individual liberty and mutual respect and tolerance of different faiths and beliefs'.[68] From 2005 they became a mantra of Labour, Coalition, and Conservative governments alike.

These narratives and agendas overlap and intersect in complex ways, with one segueing into another: the ethnic into the civic and vice versa, the personal into the political, as perhaps with 'tolerance'. But the predominant sense after 2005 was of Britishness as the epitome of liberal democratic values, and not infrequently as guardian of the Judeo-Christian tradition. This shifting focus was partly a response to the profound changes in the ways of life of the descendants of earlier waves of immigrants, now born and brought up and apparently settled in Britain – how they saw themselves and were seen by others – and the challenges this seemingly represented. The growing (re)attachment to faith, increasing visibility of the religious infrastructure, claims for the recognition of Islamic principles, reinforced the view that Muslims were leading separate, parallel (and transnational) lives in families thought to maintain values at odds with contemporary, liberal principles, for example in relations between men and women. Muslims and their families (often believed to be a breeding ground for extremism), came under constant public scrutiny by politicians and religious leaders, by feminists and human rights activists, and in the media, and the object of frequent policy initiatives, with what Isaiah Berlin[69] called 'freedom to', wear the veil, build mosques, pray in the street, arrange marriages, threatened or eroded, sometimes criminalised.

The prioritising of 'faith' was thus double-edged, especially when it meant 'Islam'. As Cesari[70] and others have argued, events during the 2000s combined to constitute Islam as an 'existential threat'. There was a 'great fear' of Islam (the literal meaning of Islamophobia), accompanied in some quarters by the belief that Muslims were seeking to impose through violence or subversion a strict interpretation of Islamic norms and values incompatible with the liberal-secular (sometimes 'Judeo-Christian') principles and practices espoused by Western democracies. 'Fear', 'threat', 'panic' increasingly characterised the multitude of reports and legislative measures and other interventions by successive Labour, Coalition, and Conservative governments, with the focus increasingly on Islam or at any rate 'Islamic extremism'. In retrospect, then, *Parekh* appeared at a critical moment just before major changes in British policy towards minorities, especially Muslims. 'Passive tolerance' was abandoned, with *Jenkins* (and *Parekh*) giving way to 'muscular liberalism', and an emphasis on 'British' civic values, principally as against those imagined as Islamic.

While the boundary between acceptable and unacceptable once shifted towards greater accommodation of difference, it now shifted in the opposite direction, with increasing lack of sympathy with 'otherness', and a growing tendency to criminalise it. This is not to deny that there should be limits, though it is necessary to clarify what exactly is at stake, understand the underlying social and cultural dynamics, avoid exceptionalising, and question whether criminalisation is the answer. But, with respect to Islam, there have been calls by some politicians and in the media for limits which are tantamount to demanding Muslims cease to be Muslim if they are to remain in Britain.

Put crudely, the shift from *Parekh* to 'muscular liberalism' and the abandonment of 'passive tolerance' can be summarised as follows. Whereas *Parekh* advocated a regime that acknowledged that people of non-European origin were part of the British polity, and should become so in the full sense, as equals, free from discrimination (bringing black into the Union Jack), and that

British society must make appropriate changes to enable this to happen, the regime which emerged after 2005 appeared to say, principally to Muslims, but to immigrants in general: 'You are not part of "us", and perhaps never will be, unless *you* change fundamentally and reject your old ways'. *Parekh* could not foresee, let alone address such developments.

Grounds for optimism?

Where does this leave the relatively benign approach to diversity which previously characterised British multiculturalism?

Although things look bleak, one must not forget the multiplicity of practical accommodations that have been negotiated, and the ground-level 'conviviality' which observers have identified,[71] and which hopefully remains intact. Multiculturalism and integration have not been the complete failures they are sometimes portrayed; the values which *Parekh* represented have not entirely disappeared. There are therefore some grounds for optimism.

Legal pragmatism?

First, 'legal pragmatism'. Within legal and political circles there is much discussion about what to do about ethnic, cultural, and religious diversity and whether, and if so how, difference might be accommodated. Consider, for example, the following *obiter dicta*:

- 'We live in a very diverse society and the justice system has got to be able to cope with that' (Lord Woolf, former Lord Chief Justice).[72]
- 'All [religions] are entitled to equal respect whether in times of peace or, as at present, amidst the clash of arms' (Judge Munby).[73]
- In the context of a plural society the courts 'must pay appropriate regard' to different values and practices (Lady Justice Arden).[74]
- 'It is not the case that for a Muslim to lead his or her life in accordance with [Shari'a] principles will be in conflict with the requirements of the law in this country' (Lord Nicholas Phillips, former Lord Chief Justice and President of the Supreme Court).[75]
- 'Judges have to show, and have to be seen to show, respect to everybody equally, and that requires an understanding of different cultural and social habits. It is necessary to have some understanding as to how people from different cultural, social, religious or other backgrounds think and behave and how they expect others to behave' (Lord Justice Neuberger, President of the Supreme Court of the United Kingdom).[76]

Certainly, religion in the courts raises many difficult questions, with judges confronted with matters of doctrinal interpretation or conflicts of rights. Sometimes, albeit by no means always, in responding to cases involving other cultures, or where there is a conflict between cultural and other rights or claims, the courts have fostered quotidian accommodation. Without going into detail, there are various approaches:

- Avoid determining the validity of a religious belief, by accepting the individual's subjective interpretation of doctrine. (I call that a 'Pirandellian perspective' – *Così è (se vi pare).*)[77]
- Sidestep the religious issue, e.g. by treating a contract as a contract, whatever the religious basis.
- Seek an accommodation between religious and other claims, but if this is impossible, balance the salutary and deleterious effect of allowing one claim but not the other.[78]

In legal circles, then, there seems at least a willingness to dialogue, to accept the validity of a genuinely held belief, and perhaps find ways of allowing a practice (albeit in modified form) even if not in strict accordance with the law.

A negotiated order?

Second, although there is much opposition to the recognition of Islamic norms and practices, various initiatives have brought Muslims and others together to seek ways in which English and Islamic law can reach a mutually acceptable form of accommodation through dialogue and negotiation.

Pragmatically dealing with complex issues with no easy solution and negotiating boundaries has historically characterised British multiculturalism,[79] and this has necessarily involved dialogue. In the mid-1980s, for example, there were seminars involving Muslim scholars and legal practitioners comparing English and Islamic law concerning marriage and divorce, identifying commonalities and divergences. Despite philosophical differences, there was often agreement on practicalities. A similar perspective has been apparent at recent 'round table' meetings to discuss why relatively few Muslim marriages are formally and legally registered, often to women's disadvantage, and how registration might be encouraged. Those engaged in such conversations are among the 'critical friends' of Islamic institutions, who repudiate the simplistic representation of Shari'a councils, for example, while engaging with the problems posed by the ways they work.

Dialogue and negotiation are foregrounded in Ayelet Shachar's 'transformative accommodation',[80] and John Eekelaar and Maleiha Malik's advocacy of 'cultural voluntarism'.[81] Such (intercultural) dialoguing is not always easy, or indeed possible, entailing as it does difficult processes of boundary negotiation in which the limits of what may be tolerated are under scrutiny, differences identified, and perhaps reconciled, with incremental change all that can be expected. Dialogue is a major theme of the CORAB Report, and in many respects CORAB, whose tenor is often against the grain of contemporary political and popular sentiment, takes us back to *Parekh* while addressing the dilemmas of the current context, which of course *Parekh* could not.[82]

Fractal cities

Third, long-term changes in British society.

By the mid-2000s, the arrival of new immigrants from many different countries and cultures, speaking different languages, and entering via different routes, with different legal and other statuses, made London 'super-diverse'. Such cities are highly heterogeneous and fragmented with different things happening at the same time in different places, and at different times in the same place. They are what Ed Soja calls 'Fractal Cities',[83] 'fragmented and polarized', with many different trajectories apparent: 'parallel lives', yes, but integration too, with varying degrees of cultural diversity, the scene of creative new hybridities, where actors are in dialogue, responding, joking, playing, crossing, mixing, and engaging in negotiations through which new orders emerge, though certainly not always harmoniously.[84]

Two publications with similar names illustrate contrasting readings of contemporary London. One, *Londonistan*,[85] portrays the UK capital as a haven for Islamic fundamentalists. The other, *Londonstani*,[86] an ironically entitled novel, builds on the linguistic and cultural 'mixity' found on the ground. In fact, London is both *Londonistan* and *Londonstani*, but optimistically I would emphasise the latter.

Concluding observations on commissions

The *Religious and Cultural Diversity* project sought to compare the work of four commissions and their reports across national space and across time. It is clear, however, that while the reports may be using similar language/concepts/terminology, the meaning they attribute to the various terms is clearly not always the same ('integration', 'accommodation', 'intercultural' are some examples), and that needs to be teased out. Moreover, the meaning of a concept such as 'Britishness' (in the case of *Parekh*), may shift over time within the same national space. Indeed, this may be a consequence of the reports themselves whose promotion of a certain understanding of a concept may be received positively or negatively in the public square. What, for instance, was *Parekh*'s influence on the British public understanding of multiculturalism? Did the hostile response feed into the growing backlash during the 2000s?

Second, commissions are not operating in a social or political void. Their deliberations and recommendations are taking place against a background of complex and interacting social, cultural, religious, and political change, local and global, not least regarding what is happening to the internal dynamics of migrant families. The relationship between the work of the commissions, such long-term changes, and more immediate events should be foregrounded.

Third, and related to that, I would emphasis the politics of it all – not political theory, not political philosophy, but *politics*, the power relations revealed by the internal workings of the commissions on the one hand, and their relations with the world in which they operate. The question of power is, as always, central.

This means triangulating the work of a commission (comparatively and over time) with respect to its discourse, the social realities 'on the ground' (perhaps the

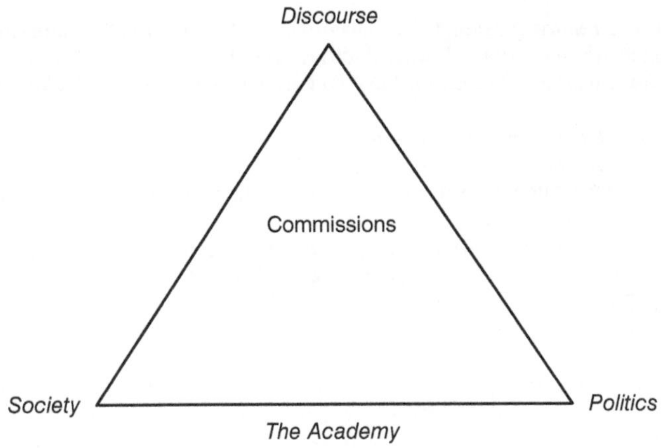

Figure 5.1 Triangulating the debate on commissions

principal anthropological contribution), and the powers that be (see Figure 5.1). There is also the question of how the work of the commissions interacts with the work of academics, whether they are directly or indirectly involved (as members), and beyond that of what is happening more generally to academic, principally socio-scientific, knowledge in the public square. (If any attention is actually paid to what academics say.)

A final reflection. The organisers of the conference where this essay was first presented urged contributors to bring forward innovative suggestions as to how best to deal with our current predicaments. As an anthropologist I had difficulties in responding to this. We're good at identifying how things are, less happy prescribing remedies, and predicting what's next seems impossible – anthropologists are poor futurologists.

British multiculturalism has historically taken shape through what has been at times a painful process of dialogue and negotiation. Whether it is possible to maintain that tradition in an era which lacks sympathy with 'otherness', and looks to criminalise it, or where opponents assert views which are non-negotiable, and mindsets, on both sides, are apparently unyielding, has to be in doubt. 'Other' beliefs and practices seem likely to remain in serious contention for the foreseeable future; Islam(ism) will continue to be seen as an existential threat to be met with tough security measures and restricting policies and practices thought to menace liberal-secular values.

Nonetheless, I naively believe that the tradition of British multiculturalism as a 'negotiated order' is still influential, and that a return to the religious and ideological wars of previous centuries is not unavoidable. Strategy is clear; if operationalisation problematic. This may be a comforting illusion, but we must, I believe, remain optimistic.

Notes

1 Parekh Report, *The Future of Multi-Ethnic Britain* (Runnymede Trust Commission on the Future of Multi-Ethnic Britain, Profile 2000) (*Parekh*).
2 Grillo R, *Muslim Families, Politics and the Law: A Legal Industry in Multicultural Britain* (Ashgate 2015).
3 CORAB, *Living with Difference: Community, Diversity and the Common Good. Report of the Commission on Religion and Belief in British Public Life* (Woolf Institute 2015) (CORAB Report). On CORAB, see Chapters 2 and 4 in this volume, by Tariq Modood and Maleiha Malik, respectively.
4 Sartori G, *Pluralismo, multiculturalismo e estranei*, 2nd edition (Rizzoli 2002).
5 E.g. Putnam R, 'E Pluribus Unum: Diversity and Community in the Twenty-first Century' (2007) 30 Scandinavian Political Studies 137.
6 Grillo R, 'Cultural Essentialism and Cultural Anxiety' (2003) 3 Anthropological Theory 157.
7 See, for example, www.lemonde.fr/idees/chronique/2010/05/01/ce-cher-pays-d-enracinement-ils-mangent-notre-pain_1345444_3232.html, accessed 30 August 2016.
8 Giddens A, *Modernity and Self-identity: Self and Society in the Late Modern Age* (Polity Press 1991).
9 Habermas J, 'Notes on a Post-secular Society', www.signandsight.com/features/1714.html/, accessed 11 January 2016.
10 Huntington SP, 'The Clash of Civilizations' (1993) 72 Foreign Affairs 2.
11 Morin E, 'Edgar Morin contre la "rhinocérite"'. *Le Point* 1886, 6 November 2008.
12 Jenkins R, *Essays and Speeches* (Collins 1967) 267.
13 Lester AP and Bindman G, *Race and Law* (Longman 1972).
14 Swann Report, *Education for All. Report of the Committee of Inquiry into the Education of Children from Ethnic Minority Groups. Cmnd. 9453* (HMSO 1985) (Swann Report). The earlier Rampton Report was a continuation of the agenda on 'race': *West Indian Children in Our Schools: Interim Report of the Committee of Inquiry into the Education of Children from Ethnic Minority Groups. Cmnd. 8273* (HMSO 1981).
15 Swann Report (n 14) 38.
16 Donald J and Rattansi A, 'Introduction' in Donald J and Rattansi A (eds), *'Race', Culture and Difference* (Sage 1992).
17 See Halstead M, *Education, Justice and Cultural Diversity: An Examination of the Honeyford Affair, 1984–85* (Falmer Press 1988).
18 Peach C, 'Muslims in the UK' in Abbas T (ed) *Muslim Britain: Communities Under Pressure* (Zed Books 2005) 18.
19 See discussion in Grillo RD, *Pluralism and the Politics of Difference: State, Culture, and Ethnicity in Comparative Perspective* (Clarendon Press 1998) and in Halstead (n 17).
20 Allen C, 'From Race to Religion: The New Face of Discrimination' in Abbas T (ed) *Muslim Britain: Communities Under Pressure* (Zed Books 2005).
21 Modood T, 'Muslims and the Politics of Difference' (2003) 74 Political Quarterly 10.
22 Raj DS, '"Who the Hell Do You Think You Are?" Promoting Religious Identity among Young Hindus in Britain' (2000) 23 Ethnic and Racial Studies 53.
23 Vertovec S, 'Superdiversity and Its Implications' (2007) 30 Ethnic and Racial Studies 1024.
24 *Parekh* (n 1) 10.
25 Anne-Marie Fortier has a good account and analysis of the response to the Parekh Report in Fortier A-M, *Multicultural Horizons: Diversity and the Limits of the Civil Nation* (Routledge 2008).
26 Available at www.lgcplus.com/future-of-multi-ethnic-britain-launch-home-secretarys-speech/1362303.article, accessed 7 January 2016.
27 'Turning Point at Runnymede', *Daily Telegraph* (London, 13 October, 2000).

28 *Parekh* (n 1) 6. This builds on the earlier Runnymede inquiry: Runnymede Trust, *Islamophobia: A Challenge to Us All* (Runnymede Trust 1997).
29 *Parekh* (n 1) 311.
30 Runnymede Trust, *Realising the Vision: Progress and Further Challenges* (Runnymede Trust 2004) 2.
31 The continuing influence of *Parekh* can be seen in later reports including Mayor of London, *Delivering Shared Heritage: The Mayor's Commission on African and Asian Heritage* (Greater London Authority 2005); Ajegbo K, *Diversity and Citizenship Curriculum Review* (Department for Education 2007); Maylor U and Read B, *Diversity and Citizenship in the Curriculum: Research Review* (Institute for Policy Studies in Education, London Metropolitan University 2007).
32 Commission for Racial Equality, *A Lot Done, A Lot to Do: Our Vision for an Integrated Britain* (CRE 2007).
33 Heath A and Demireva N, 'Has Multiculturalism Failed in Britain?' (2014) 37 Ethnic and Racial Studies 161, 217. (Incidentally, the article does not mention *Parekh*.)
34 Finney N and Lymperopoulou K, *Local Ethnic Inequalities: Ethnic Differences in Education, Employment, Health and Housing in Districts of England and Wales, 2001–2011* (University of Manchester/Runnymede Trust 2014) 4.
35 Home Office Counter Extremism Directorate, *Counter-Extremism Strategy, Cm 9145* (Home Office 2015).
36 They are far too many to list here. They are included in a file (*Commissions, Inquiries, Reports etc. on Ethnic and Religious Minorities in Britain, 1900–2016, Draft December 2016*) which I have placed on the Internet at www.academia.edu/28147725/Commissions_Inquiries_and_Reports_on_Minorities_in_the_UK_1903_2016_REVISED, and www.researchgate.net/publication/311576130_Commissions_Inquiries_Reports_Consultations_and_Other_Interventions_on_Ethnic_and_Religious_Minorities_in_Britain_c_1900-end_2016_Draft_December_2016.
37 Royal Commission on Alien Immigration, *Report of the Royal Commission on Alien Immigration* (British Parliamentary Papers IX, HMSO 1903).
38 Home Office, *A Choice by Right: The Report of the Working Party on Forced Marriages* (Home Office Communications Directorate 2000), discussed in Grillo *Muslim Families* (n 2).
39 A flow chart on how to organise a consultation may be found in *Working Together: Co-operation between Government and Faith Communities* (Home Office 2004) 15.
40 Examples include Jay A, *Independent Inquiry into Child Sexual Exploitation in Rotherham, 1997–2013* (Rotherham Metropolitan Borough Council 2014), Clarke P, *Report into Allegations Concerning Birmingham Schools Arising from the 'Trojan Horse' Letter (HC 576)* (UK Government 2014), Casey L, *The Casey Review: A Review into Opportunity and Integration* (Department for Communities and Local Government 2016) (*Casey Review*), and the inquiry into the application of Shari'a law, instituted by the then Home Secretary, Theresa May, in 2016; www.gov.uk/government/news/independent-review-into-sharia-law-launched, accessed 26 May 2016.
41 Home Affairs Committee, *Female Genital Mutilation: The Case for a National Action Plan* (House of Commons Home Affairs Committee 2014); www.parliament.uk/business/committees/committees-a-z/commons-select/home-affairs-committee/inquiries/parliament-2015/inquiry6/, accessed 30 June 2016.
42 Runnymede Trust, *Islamophobia* (n 28).
43 See Grillo *Muslim Families* (n 2) *passim*.
44 Commission on Integration and Cohesion, *Our Shared Future* (Commission on Integration and Cohesion 2007).
45 Cantle Report, *Community Cohesion* (Home Office 2001) Section 2.1.
46 Phillips Trevor, 'After 7/7: Sleepwalking to Segregation', originally 2005, now available at www.humanities.manchester.ac.uk/socialchange/research/social-change/summer-workshops/documents/sleepwalking.pdf, accessed 25 April 2012.

47 PREVENT Working Groups, *Preventing Extremism Together* (Home Office 2005) 2.
48 McGhee D, *Intolerant Britain? Hate, Citizenship and Difference* (Open University Press 2005) 64.
49 Cameron D, 'PM's Speech at Munich Security Conference, 6 February 2011', www.gov.uk/government/speeches/pms-speech-at-munich-security-conference, accessed 15 February 2011.
50 Prime Minister's Office Press Release, 13 May 2015, www.gov.uk/government/news/counter-extremism-bill-national-security-council-meeting, accessed 7 January 2016; Prime Minister's Office Press Release, 18 January 2016, www.gov.uk/government/news/passive-tolerance-of-separate-communities-must-end-says-pm, accessed 25 January 2016.
51 Swann Report (n 14) 6.
52 Glaister Dan, 'KNICKERS TO THE FRENCH', *The Guardian* (London, 15 December 1992) 12.
53 *Associated Press* (London, 6 April 1995).
54 Morton Ian, 'Britishness Still Counts', *Evening Standard* (London, 29 September 1994) 69.
55 Gilroy P, *There Ain't No Black in the Union Jack: The Cultural Politics of Race and Nation* (Hutchinson 1987).
56 Colley L, *Britons: Forging the Nation 1707–1837* (Pimlico 1992).
57 Hall S, 'New Ethnicities' in Donald J and Rattansi A (eds), *'Race', Culture and Difference* (Sage 1992).
58 Data obtained via www.nexis.com/.
59 *Casey Review* (n 40) 168.
60 Blair T, 'Our Nation's Future: Multiculturalism and Integration', www.number-10.gov.uk/output/Page10563asp, accessed 8 December 2006.
61 Blunkett D, 'Integration with Diversity: Globalisation and the Renewal of Democracy and Civil Society' in Griffith P and Leonard M (eds), *Reclaiming Britishness* (Foreign Policy Centre 2002); Blunkett D, *New Challenges for Race Equality and Community Cohesion in the 21st Century* (Home Office 2004).
62 Brown G, *On Britishness* (British Council Annual Lecture, 7 July 2004); Brown G, *The Future of Britishness* (2006). Full text available at www.theguardian.com/politics/2007/feb/27/immigrationpolicy.race, accessed 14 January 2016. There were other speeches by Brown on the same theme.
63 The Commission for Racial Equality identified eight 'dimensions' of Britishness (Geography, National Symbols, People, Values and Attitudes, Cultural Habits and Behaviour, Citizenship, Language, Achievements) in *Citizenship and Belonging: What Is Britishness?* (CRE 2005).
64 In Orwell G, Critical Essays (Secker & Warburg 1946). See also Samuel R, 'Introduction: The "Little Platoons"' in Samuel R (ed) *Patriotism: The Making and Unmaking of British National Identity, Vol 2: Minorities and Outsiders* (Routledge 1989).
65 www.johnmajor.co.uk/page1086.html, accessed 7 January 2016.
66 e.g. Cameron D, 'British values aren't optional', *Mail on Sunday* (London, 15 June 2014).
67 E.g. Communities and Local Government, *Creating the Conditions for Integration* (Communities and Local Government 2012) 4; *Casey Review* (n 40) *passim*.
68 HM Government, *Tackling Extremism in the UK* (HM Government 2013) 1.
69 Berlin I, *Liberty* (Oxford University Press 2002).
70 Cesari J, *Why the West Fears Islam: An Exploration of Muslims in Liberal Democracies* (Palgrave/Macmillan 2013).
71 E.g. Wessendorf S, *Commonplace Diversity: Social Relations in a Super-Diverse Context* (Palgrave/Macmillan 2014).
72 Originally quoted in www.ligali.org/newsletter/news_july2004.htm, accessed 1 January 2015.

73 www.judiciary.gov.uk/wp-content/uploads/JCO/Documents/Speeches/law-morality-religion-munby-2013.pdf, accessed 1 January 2015.
74 [2007] EWCA Civ 399 para. 46.
75 Phillips N, 'Equality before the Law', Speech by Lord Nicholas Phillips, Lord Chief Justice, East London Muslim Centre, 3 July 2008, 8, http://innertemplelibrary.wordpress.com/2008/07/04/equality-before-the-law-speech-by-lord-phillips-of-worth-matravers/, accessed 13 August 2008.
76 Supreme Court speech 'Fairness in the Courts: The Best We Can Do', 10 April 2015, www.supremecourt.uk/docs/speech-150410.pdf, accessed 25 January 2016.
77 Pirandello L, *Così è, se vi pare!* (Dutton 1917); translation *Right You Are! (If You Think So)* by A. Livingston (Dutton 1922).
78 See the Canadian Supreme Court's decision weighing a witness's right to wear a face veil, against the right of a defendant to examine the witness's demeanour when presenting evidence (R. *v.* N.S., 2012 SCC 72).
79 See Grillo R, 'British Multiculturalism: A Negotiated Order' in Académie royale de Belgique (ed) *Les minorités: un défi pour les États Actes du colloque international (22 et 23 mai 2011)* (Académie royale de Belgique 2012).
80 Shachar A, *Multicultural Jurisdictions: Cultural Differences and Women's Rights* (Cambridge University Press 2001).
81 Eekelaar J, 'From Multiculturalism to Cultural Voluntarism: A Family-based Approach' (2010) 81 The Political Quarterly 344; Malik M, *Minority Legal Orders in the UK* (The British Academy 2012).
82 There was some overlap of membership between CORAB and Parekh, and interestingly the media response has been similar. CORAB perhaps puts rather less emphasis on equality issues.
83 Soja E, *Postmetropolis: Critical Studies of Cities and Regions* (Blackwell 2000) 155.
84 See *inter alia* Back L, 'The Fact of Hybridity: Youth, Ethnicity and Racism' in Goldberg DT and Solomos J (eds), *A Companion to Racial and Ethnic Studies* (Blackwell 2002).
85 Phillips Melanie, *Londonistan: How Britain Is Creating a Terror State Within* (Gibson Square Books 2006).
86 Malkani G, *Londonstani* (Fourth Estate 2006).

References

Ajegbo, K. 2007. *Diversity and Citizenship Curriculum Review*. London: Department for Education.

Allen, C. 2005. 'From Race to Religion: The New Face of Discrimination'. In *Muslim Britain: Communities Under Pressure*, edited by T. Abbas, 49–65. London: Zed Books.

Back, L. 2002. 'The Fact of Hybridity: Youth, Ethnicity and Racism'. In *A Companion to Racial and Ethnic Studies*, edited by D.T. Goldberg and J. Solomos, 439–454. Oxford: Blackwell.

Berlin, Isaiah. 2002. *Liberty*. Oxford: Oxford University Press.

Blair, T. 2006. 'Our Nation's Future: Multiculturalism and Integration'. www.number-10.gov.uk/output/Page10563.asp.

Blunkett, D. 2002. 'Integration with Diversity: Globalisation and the Renewal of Democracy and Civil Society'. In *Reclaiming Britishness*, edited by P. Griffith and M. Leonard, 65–77. London: Foreign Policy Centre.

Blunkett, David. 2004. *New Challenges for Race Equality and Community Cohesion in the 21st Century*. London: Home Office.

Brown, G. 2004. *On Britishness*. British Council Annual Lecture, 7 July 2004.

Brown, Gordon. 2006. *The Future of Britishness*. www.theguardian.com/politics/2007/feb/27/immigrationpolicy.race.

Cameron, David. 2011. 'PM's Speech at Munich Security Conference, 6 February 2011'. www.gov.uk/government/speeches/pms-speech-at-munich-security-conference.
Cantle Report. 2001. *Community Cohesion*. London: Home Office.
Casey, L. 2016. *The Casey Review: A Review into Opportunity and Integration*. London: Department for Communities and Local Government.
Cesari, J. 2013. *Why the West Fears Islam: An Exploration of Muslims in Liberal Democracies*. New York: Palgrave/Macmillan.
Clarke, Peter. 2014. *Report into Allegations Concerning Birmingham Schools Arising from the 'Trojan Horse' Letter (HC 576)*. London: UK Government.
Colley, L. 1992. *Britons: Forging the Nation 1707–1837*. London: Pimlico.
Commission for Racial Equality. 2007. *A Lot Done, A Lot to Do: Our Vision for an Integrated Britain*. London: CRE.
Commission for Racial Equality/ETHNOS Research and Consultancy. 2005. *Citizenship and Belonging: What Is Britishness?* London: CRE.
Commission on Integration and Cohesion. 2007. *Our Shared Future*. London: Commission on Integration and Cohesion.
Communities and Local Government. 2012. *Creating the Conditions for Integration*. London: Communities and Local Government.
CORAB. 2015. *Living with Difference: Community, Diversity and the Common Good. Report of the Commission on Religion and Belief in British Public Life*. Cambridge: Woolf Institute.
Donald, J., and A. Rattansi. 1992. 'Introduction'. In *'Race', Culture and Difference*, edited by J. Donald and A. Rattansi, 1–8. London: Sage.
Eekelaar, John. 2010. 'From Multiculturalism to Cultural Voluntarism: A Family-based Approach'. *The Political Quarterly* 81 (3): 344–355.
Finney, N., and K. Lymperopoulou. 2014. *Local Ethnic Inequalities: Ethnic Differences in Education, Employment, Health and Housing in Districts of England and Wales, 2001–2011*. London: University of Manchester/Runnymede Trust.
Fortier, Anne-Marie. 2008. *Multicultural Horizons: Diversity and the Limits of the Civil Nation*. London: Routledge.
Giddens, A. 1991. *Modernity and Self-identity: Self and Society in the Late Modern Age*. Cambridge: Polity Press.
Gilroy, P. 1987. *There Ain't No Black in the Union Jack: The Cultural Politics of Race and Nation*. London: Hutchinson.
Grillo, R. 2003. 'Cultural Essentialism and Cultural Anxiety'. *Anthropological Theory* 3 (2): 157–173.
Grillo, R. 2012. 'British Multiculturalism: A Negotiated Order'. In *Les minorités: un défi pour les États. Actes du colloque international (22 et 23 mai 2011)*, edited by Académie royale de Belgique, 173–203. Brussels: Académie royale de Belgique.
Grillo, R. 2015. *Muslim Families, Politics and the Law: A Legal Industry in Multicultural Britain*. Aldershot: Ashgate.
Habermas, J. 2008. 'Notes on a Post-secular Society'. www.signandsight.com/features/1714.html.
Halstead, M. 1988. *Education, Justice and Cultural Diversity: An Examination of the Honeyford Affair, 1984–85*. London: Falmer Press.
Heath, A., and N. Demireva. 2014. 'Has Multiculturalism Failed in Britain?' *Ethnic and Racial Studies* 37 (1): 161–180.
HM Government. 2013. *Tackling Extremism in the UK*. London: HM Government.

Home Affairs Committee. 2015. *Female Genital Mutilation: Follow-up. Sixteenth Report of Session 2014–15*. London: House of Commons Home Affairs Committee.

Home Office. 2000. *A Choice by Right: The Report of the Working Party on Forced Marriages*. London: Home Office Communications Directorate.

Home Office. 2004. *Working Together: Co-operation between Government and Faith Communities*. London: Home Office.

Home Office Counter Extremism Directorate. 2015. *Counter-Extremism Strategy, Cm 9145*. London: Home Office.

Huntington, Samuel P. 1993. 'The Clash of Civilizations'. *Foreign Affairs* 72 (3): 22–49.

Jay, Alexis. 2014. *Independent Inquiry into Child Sexual Exploitation in Rotherham, 1997– 2013*. Rotherham: Rotherham Metropolitan Borough Council.

Jenkins, Roy. 1967. *Essays and Speeches*. London: Collins.

Lester, Anthony Paul, and Geoffrey Bindman. 1972. *Race and Law*. London: Longman.

McGhee, D. 2005. *Intolerant Britain? Hate, Citizenship and Difference*. Milton Keynes: Open University Press.

Malik, Maleiha. 2012. *Minority Legal Orders in the UK*. London: The British Academy.

Malkani, G. 2006. *Londonstani*. London: Fourth Estate.

Maylor, Uvanney, and Barbara Read. 2007. *Diversity and Citizenship in the Curriculum: Research Review*. London: Institute for Policy Studies in Education, London Metropolitan University.

Mayor of London. 2005. *Delivering Shared Heritage: The Mayor's Commission on African and Asian Heritage*. London: Greater London Authority.

Modood, T. 2003. 'Muslims and the Politics of Difference'. *Political Quarterly* 74 (1, Supplement 1): 100–115.

Morin, Edgar. 2008. 'Edgar Morin contre la "rhinocérite"'. *Le Point* 1886, 6 November 2008.

Orwell, George. 1946. *Critical Essays*. London: Secker & Warburg.

Parekh Report. 2000. *The Future of Multi-Ethnic Britain: Report of the Commission on the Future of Multi-Ethnic Britain, Runnymede Trust. Commission on the Future of Multi-Ethnic Britain*. London: Profile.

Peach, C. 2005. 'Muslims in the UK'. In *Muslim Britain: Communities Under Pressure*, edited by T. Abbas, 18–30. London: Zed Books.

Phillips, Melanie. 2006. *Londonistan: How Britain Is Creating a Terror State Within*. London: Gibson Square Books.

Phillips, T. 2005. 'After 7/7: Sleepwalking to Segregation'. www.humanities.manchester.ac.uk/socialchange/research/social-change/summer-workshops/documents/sleepwalking.pdf.

PREVENT Working Groups. 2005. *Preventing Extremism Together*. London: Home Office.

Putnam, Robert. 2007. 'E Pluribus Unum: Diversity and Community in the Twenty-first Century. The 2006 Johan Skytte Prize Lecture'. *Scandinavian Political Studies* 30 (2): 137–174.

Raj, D.S. 2000. '"Who the Hell Do You Think You Are?" Promoting Religious Identity among Young Hindus in Britain'. *Ethnic and Racial Studies* 23 (3): 535–558.

Rampton Report. 1981. *West Indian Children in Our Schools: Interim Report of the Committee of Inquiry into the Education of Children from Ethnic Minority Groups. Chairman A. Rampton. Cmnd. 8273*. London: HMSO.

Royal Commission on Alien Immigration. 1903. *Report of the Royal Commission on Alien Immigration, British Parliamentary Papers IX*. London: HMSO.

Runnymede Trust. 1997. *Islamophobia: A Challenge to Us All*. London: Runnymede Trust.
Runnymede Trust. 2004. *Realising the Vision: Progress and Further Challenges*. London: Runnymede Trust.
Samuel, R. 1989. 'Introduction: The "Little Platoons"'. In *Patriotism: The Making and Unmaking of British National Identity. Vol. 2: Minorities and Outsiders*, edited by R. Samuel, ix–xxxix. London: Routledge.
Sartori, G. 2002. *Pluralismo, multiculturalismo e estranei*, 2nd edition. Milan: Rizzoli.
Shachar, Ayelet. 2001. *Multicultural Jurisdictions: Cultural Differences and Women's Rights*. Cambridge and New York: Cambridge University Press.
Soja, Edward. 2000. *Postmetropolis: Critical Studies of Cities and Regions*. Oxford: Blackwell.
Swann Report. 1985. *Education for All. Report of the Committee of Inquiry into the Education of Children from Ethnic Minority Groups. Cmnd. 9453*. London: HMSO.
Vertovec, Steven. 2007. 'Superdiversity and Its Implications'. *Ethnic and Racial Studies* 30 (6): 1024–1054.
Wessendorf, S. 2014. *Commonplace Diversity: Social Relations in a Super-Diverse Context*. Basingstoke: Palgrave/Macmillan.

Part II

Laïcité facing religious diversity

France's Stasi Commission and its legacy

6 *Laïcité* is a most liberal legal frame

Reflections on the work of the Stasi Commission

Patrick Weil[1]

Introduction

The proposition that French *laïcité* is liberal might be shocking to some, so in this essay I want to elaborate on that. In matters of religion, one has always to distinguish – and many French won't – between legal regime and social beliefs. For instance, in the United States it is constitutional and legal to engage in blasphemy, to burn the Bible or the Quran or a cross if it is not deemed hate action. But if you suggest any doubt about the existence of God the possibility of being elected president of the United States is zero. So there is a big gap between what is legal and constitutional and what the social belief of a majority of a people is, or what they expect from their compatriots to trust them.

In France, too, the law and jurisprudence as such should be considered as different and distinct from the majoritarian social belief. Considering this distinction, French jurisprudence is not very far from US jurisprudence. France, like the United States, does not have established churches, and the same kinds of questions have been posed to courts for the last century and a half. But the social beliefs of the people are quite different. In the last survey done in Europe, a large majority of the French – two-thirds – are non-believers, either atheists or agnostics. In this regard the French are an exception in Europe, together with the Czechs and the Swedish.[2]

But, *laïcité* is first and foremost a legal regime that was established in 1905 mainly as a compromise with the Catholic Church – with the majority of French bishops approving the compromise – after more than a century of conflict, since the French Revolution. Then, the compromise was condemned by the Vatican, where a conservative pope had just been elected. Conflict was resumed until it was settled mainly in two stages after the First and Second World Wars. The spirit of compromise, present at the origins, still prevails and the 1905 law has become a principle of the Republic and *de facto* more important than the Constitution. Two French public figures, such as Caroline Fourest and Tariq Ramadan, both like the law of 1905, and all the French like it just as all Americans like the First Amendment. But it permits different interpretations that can be in conflict and which may end up in court. The main principle of this legal regime is individual freedom of conscience. This is the first and the most important principle of *laïcité* and the two following principles were in fact proclaimed to support the principle of freedom of conscience. The separation of state and church was born in the context of the previous domination or intrusion of the Catholic Church in public affairs: after the fall of the Empire in 1871, the Church was against the Republic and favored the monarchy; for instance, in the Dreyfus Affair, the Church favored the condemnation of Dreyfus. Second, separation and neutrality means that the state does not take a position on the existence of God.[3] It means also the absence of pressure on the 'consciences' of the citizens. This can have practical effects: French civil servants have to remain neutral and they cannot show their belonging to a given faith. The third principle is the respect for all faith and spiritual choices and implies the possibility for believers to organize themselves in churches and religious communities.

So that is the legal regime. The reason why I think it is liberal is because the concern is not only to guarantee individual freedom of religion; it is also to guarantee individual freedom *from* religion. Freedom of conscience not only means that one is free to believe, it also respects non-believers and the possibility of moving from believing to not believing and vice versa.[4]

Many people in France and outside France confuse *laïcité* with a regime of freedom with the state being antireligious. *Laïcité* is a legal regime which in the first place is concerned with the protection of the freedom of conscience including of a believer. People who mobilize against the influence of religion in public affairs have historically fought for and brought into law this regime of freedom. Some of them have continued to fight against the influence of religion, not at the state level but in the 'minds' of their compatriots. They should find a word for that struggle other than '*laïcité*', which should be reserved for the legal regime of freedom of conscience associated with the separation of state and church. Instead of speaking of *laïcité* they should talk of fighting to free the conscience from religious dogma.

How can freedom of conscience be guaranteed in practice? By ensuring that people can circulate and experiment with the four types of spaces that organize the city:[5]

1) the sacred space, for example a mosque or a church, where the rule of faith is followed;
2) the private sphere, in the home where an individual creates his or her own rules – if you want people to take off their shoes when they enter your home or not to smoke, then that is your rule, and you can impose it on people who do enter your home;
3) the state's space, where the French *laïcité* regime imposes neutrality and the USA imposes non-establishment;
4) finally, the remaining space, where the freedom of expression of all spirituality prevails, which is the public space.

In the proceedings of the Stasi Commission in France, of which I was a member, there was a question on the categorization of the school as a type of space. Is a public school a state space because it has civil servants, i.e. teachers, who have the obligation of neutrality? Or is it a public space because it has children who are not under an obligation of neutrality. We faced a 'conflict of borders'. Sovereignty doesn't concern only states. Religions never accept the sovereignty of states. They always claim 'sovereignty' and extension of their sphere of power. Look at the United States, in particular recent religious jurisprudence in the *Hobby Lobby* case,[6] where by discretionary power the Christian owner of the company demanded exemption from the general law that stated employers will pay for their employees' benefits including reimbursement for contraception or abortion. So faith's claim to 'sovereignty' never ends. Thus truces are found in all democratic societies, creating the kind of borders such as the ones traced above. However, we consequently find recurrent conflicts of interpretations, and numerous cases in courts, related in fact to borders of sovereignty.

But these borders are the sites where the limits of sovereignty are challenged and changed. The challenge to wearing the headscarf would center in France on students, in Germany, on teachers.

The French 'affaire de foulard' and the Stasi Commission

In France, the public school has always been a border zone, a *zone frontière*. In 1989, at a high school in Creil, a few young students wanted to wear the headscarf. The minister of education went to the French Administrative Supreme Court, the Conseil d'état, for advice and received, basically, the following advice: the headscarf can be worn as long as it is a sign of personal identity, but if it is a tool for putting pressure on others it is forbidden. For each case, the school board was to decide. And that's where the problems began because it was extremely complicated for school boards to make this kind of decision. A few years later, at the beginning of this millennium, tensions arose in some schools. The policy of the ministry was to avoid attracting media attention, telling the school board not to intervene or draw attention to the issue. This led to teachers complaining to their local MPs and MPs from all parties deciding to establish a parliamentary committee with the aim to investigate the issue of religious symbols in public schools.

At this point President Jacques Chirac became wary of what recommendations might emerge from a committee created by the National Assembly, and composed of only MPs, and decided to create a presidential commission. I was asked to be part of it. I presume this was because I had written a report for the previous left-wing government, proposing to reform immigration and citizenship laws, that was quite well received.[7] The result was the passing of two laws that are still, in 2017, in the law books.[8] Chirac thought the composition of a presidential commission should reflect the diversity of French society. I was not only a left-wing academic and policy adviser, I was also perceived as a non-religious Jew. The commission included another Jew, who was a religious woman, as well as two Muslims – a woman and a man – Protestants, atheists, Catholics and others. The composition of the commission aimed to ensure that every citizen felt 'represented'. The commission decided that we wanted public hearings. Any time the commission met for public hearings, the hearings were televised on the parliament channel. Some hearings were, however, kept out of the public eye, for example when witnesses did not want to express themselves in public; and of course sometimes we met for our own internal discussions.

But we – that is the commission members – accepted that our discussions and the non-public hearings would be recorded. All these records are now in the French national archives and after thirty years, i.e. in 2033, they will be made public. All our votes, and all the discussion that led to the votes have been recorded with a view to be available for the next generation of researchers. The way we worked is worth mentioning here, with a few words of clarification. With a different composition, the commission would most probably have agreed upon different kinds of recommendations. Bernard Stasi, the chair, was a former minister, a center-right politician. The commission also contained another – left-wing – politician, and academics René Rémond, Alain Touraine, Mohamed Arkoun, Jean Baubérot, Gilles Kepel and myself. We could have produced a socio-historico-political analysis of the issues at stake. But this was not the route the commission took. We had to propose workable solutions. For that reason, I think, the role of the person who became the *de facto* vice-president of the commission and turned out to be undoubtedly the most important member was the former president of the French Administrative Supreme Court Marceau Long. Marceau Long headed the Conseil d'état when it issued the abovementioned 1998 opinion. In terms of knowledge of jurisprudence and of the law, he was of course the most reliable colleague in the commission. And not only that. The rapporteurs, the people who wrote the draft on the report, were, in the main, members of the Conseil d'état; Long's authority over them was obvious. Consequently I felt, as I have previously written, that because we adopted a very legalistic approach, i.e. we wanted to resolve the problems by proposing solutions that could be endorsed by parliament and then accepted by the courts, we were in some ways acting like a court or how the Conseil d'état acts in its advisory role towards the government.

Such legalistic approach explains in part why we didn't talk at all about identity. I don't recall that we even mentioned the word 'identity'. It was only later, that Nicholas Sarkozy, when he became president, launched a public and official discussion on French identity. After the report was published, we were even criticized for not having heard more Muslim women wearing the headscarf, and I understand, having heard the Canadian example, that the impact of hearing a diversity of opinion on the way people wear a scarf would have been good both for the image of our commission and for providing citizens who were following the hearings on TV with an understanding of how diverse in its meaning a headscarf can be. The issue for us was first that we had very limited time to discuss and write the report.[9] Our main purpose was to find solutions to concrete 'problems' that had been raised, which was not easy. So once we all agreed that wearing the scarf could have different meanings, we did not want to spend hours hearing people telling us that the scarf has different meanings. This was not an issue between the members of the commission tasked with writing the report.

We wanted to hear people that could help us find the solutions we were looking for. Moreover we had different issues to deal with, the main issue of course being the wearing of the scarf in public schools. I would like to offer some explanations. The proposal to ban conspicuous religious signs in public schools was not done on the basis of a judgment on our part about what a headscarf is, nor on what *the* meaning of a religious sign entails. That was not the case, because we accepted completely that it can have different meanings. And the majority of us believed it is not the role of the state, and of a general law, to define an official meaning. The issue was that in a rising number of high schools, groups of males were going after Muslim female students who did not wear a scarf, telling them: Look, wearing the scarf is authorized in public school – your 'sister' wears one, because she has the right to wear it. So if you don't wear it, it's obviously your choice, because you have the possibility of wearing it. And if you don't wear it, it's because you're a bad Muslim, you're a whore, and so on, and we're going to pick you up on it. These students were harassed, sometimes aggressively, either in the courtyard or sometimes in the classroom. That was the extent of the problem we had to face. And after envisaging many different options we decided on the ban, which would preserve the liberty of conscience of the majority of Muslim girls who were not wearing the scarf. But it was putting at stake those who voluntarily wanted to wear it. The practical solution was the following: in the French system, some 20 percent of students attend private schools under contract with the state. Ninety-five percent of their expenses are financed by public taxes. The majority of schools on such contract are Catholic schools, but they have an obligation under their contract to welcome pupils of any faith. We knew this. What has happened since the introduction of the 2004 law[10] is that most Muslim girls who want to wear the scarf now attend Catholic schools until they can attend a Muslim school that is under contract. Their number, for now, is very low.

This might suggest inconsistency in the French system. The fact is, many French appreciate it. And so we have this dual system which is the product of conflicts between the Catholic Church and the state and the truce that was signed in 1959. However, the truce was temporary and sovereignty wars resumed with the status quo only returning after big demonstrations in 1983–1984 and again in 1993–1994. Thus, when the first headscarf incident occurred in 1989, in a college in Creil, the guns were – so to speak – still smoking after the first big demonstration. The issue of Muslim headscarves has therefore to be understood against the backdrop of conflicts and compromises that preexisted and were not necessarily related to the issue of Muslims. In this frame the most religious students attended private schools under contract and not public schools.

At that time we thought the pressure on Muslim girls to wear a headscarf came not only from local groups of male students. Years later we received confirmation that a campaign had been initiated by the Muslim Brotherhood. A youth leader of this organization admitted in his published memoirs:[11] the organization were campaigning and targeting high school after high school in order to 'encourage' and pressure Muslim girls to wear headscarves.

Impact of the Stasi Commission report

What about the impact of our report? It has proven to have had some positive effect, at least in the public schools where conflicts over the headscarf have almost disappeared. But the report was not only about public schools. It was about the adaptation of the law of 1905 to the new religious landscape of twenty-first-century France. And a closer examination of the impact of the report in the days following its issue reveals that the main debate did not revolve around the proposal of a headscarf ban, but about the proposal to institute an official holiday recognizing minority faiths in the calendar. This was a personal idea that I submitted to all the religious authorities that attended the public hearings. They approved of it, and my colleagues in the commission approved of it as well, unanimously. President Chirac had also said to the chair of our commission that he was in favor of it, but the prime minister opposed it. Some socialists who are very anti-religious didn't like it either and it was not proposed to parliament. The rationale was not simply the recognition of the new religious diversity prevailing in France. It was also that today, if you are a Muslim or a Jew or a Buddhist, you can go to your boss and say I am Muslim, I am a Jew, I am a Buddhist, and I don't want to work on Eid, on Yom Kippur, etc. But under French tradition a sort of privacy of faith exists, which is not legal but is a social custom; it is a tradition not to express your faith publicly. If you want to express it, you can; if you don't want to express it, you are not obliged to. It is therefore considered not fair that a non-Christian should have to approach her employer and be forced to inform the employer of her faith in order to obtain the right to a holiday. If an option in the official calendar offered a day of break for any religious holiday then, of course, the presumption would be that someone who wanted her day at Eid is a Muslim. But she could be an atheist who finds that Eid falls on a date that year that creates an opportunity for a long weekend. And the following year, she might choose to take a break on a Jewish holiday. So, in addition to being a sign of respect for all minority faiths in the official calendar of the French Republic, it would favor a greater respect for privacy. What is interesting is that this proposal was returned in the report drafted by the parliamentary commission that proposed the ban on the full veil.[12]

The issue of religious holidays returned because the social reality is that Eid and Ramadan have become such large social events in France that at some point the French legislator will have to adapt the official holidays regime to the diverse religious landscape of the country. Since our report was passed, French Muslims have access to chaplains within the army, the prisons and the hospitals. Space in cemeteries is provided and the construction of mosques or the opening of places of worship are progressing. It is not always easy and perfect, but progress has been made on the ground.

Conclusion

By way of conclusion I want to state that in recent years two trends have been observed. First are the conflicts of interpretations which lead to litigation in the courts. And courts are making decisions on different issues. Sometimes they reject the government's position: after a woman who was driving with a full veil was issued a fine, a traffic court decided to revoke the fine.[13] In a case of a marriage annulled because the woman was not a virgin, an appeals court stated divorce may be granted for any reason, including the former one, but the annulment of a marriage because the woman is not a virgin would create a cause of annulment in the jurisprudence and would put at risk all future marriages of non-virgins. This would be a terrible regression *vis-à-vis* women's marriage rights and more. So courts have functioned and continue to function as legal regulators of conflicts of interpretation and of power.

Second is the reaction in the sphere of politics, especially reaction to the work of commissions. The political reaction to the work done by the Stasi Commission was exactly the opposite of the academic reaction. The academic reaction has been vehemently opposed to the report for being so antiliberal.[14] The political reaction was that the commission did not go far enough, with claims such as: 'The only place where a ban is in effect is public schools. That's nothing. Look at these scarves, they are everywhere now in the companies, in the street etc. Let's do something'. And what did those who made such claims do? They found the burka to forbid, the full-face veil to attack. They wanted to find an occasion, a pretext, an object on which they could pass a general ban. They were absolutely frustrated by the fact that the ban of conspicuous religious signs only applied in public schools, which was seen as authorizing the headscarf everywhere else (except in the civil service). They hated the fact that the ban was so limited, and they found something else to fight against. Ultimately, they passed this general ban, and got the support of the European Court of Human Rights, but that is another story.[15]

Notes

1 Patrick Weil is a senior research fellow at the French National Research Center in the University of Paris I, Pantheon-Sorbonne. He was a member of the 2003 Presidential Commission on secularism, established by Jacques Chirac (i.e. the Stasi Commission, named after its chairman Bernard Stasi).

2 See www.pewforum.org/2012/12/18/global-religious-landscape-exec/.

3 See J. Rivéro, "De l'idéologie à la règle de droit: la notion de laïcité dans la jurisprudence administrative," in *La Laïcité* (Université d'Aix-Marseille, Centre de sciences politiques, 1960), p. 266.

4 Article 31 of the law of 1905 even specifies criminal penalties for any person 'who, by assault, violence or threats against an individual or by making him afraid of losing his job or being exposed to damage to his person, his family or his wealth' prevents another person from practicing or contributing to a religious organization. The same holds for any person forcing another to participate in or contribute to any religious organization.

5 M. Hénaff and T.B. Strong, "Introduction: The Conditions of Public Space: Vision, Speech and Theatricality," in *Public Space and Democracy*, ed. M. Hénaff and T.B. Strong (University of Minnesota Press, 2001), pp. 2–8.

6 *Burwell v. Hobby Lobby*, 573 U.S. ___ (2014).

7 *Rapports au Premier Ministre sur les législations de la nationalité et de l'immigration* (La documentation française, 1997).

8 La loi n° 98-170 du 16 mars 1998 relative à la nationalité and Law no 98-349 du 11 mai 1998 relative à l'entrée et au séjour des étrangers en France et au droit d'asile.

9 Commission de réflexion sur l'application du principe de laïcite dans la republique, Rapport au President de la Republique, available at http://lesrapports.ladocumentation-francaise.fr/BRP/034000725/0000.pdf.

10 Art. 1 of the Law of March 15, 2004 (No. 2004-228) regulating, in application of the principle of *laïcité*, the wearing of signs or clothing which conspicuously manifest religious affiliations in public elementary, middle and high schools, JORF (Official Organ of the French Republic) number 65 of March 17, 2004, p. 5190.

11 Farid Abdelkrim, *Pourquoi j'ai cessé d'être islamiste, itinéraire au cœur de l'islam en France* (Editions les points sur les i, 2015).

12 Rapport d'information fait en application de l'article 145 du règlement au nom de la mission d'information sur la pratique du port du voile intégral sur le territoire national (Commission Gerin), l'Assemblée nationale le 26 janvier 2010, p. 141, availabe at www.assemblee-nationale.fr/13/rap-info/i2262.asp.

13 www.elle.fr/Societe/News/Niqab-au-volant-a-Nantes-le-PV-annule-1445507#.

14 See, among others, J.R. Bowen, *Why the French Don't Like Headscarves: Islam, the State, and Public Space* (Princeton University Press, 2008); J.W. Scott, *The Politics of the Veil* (Princeton University Press, 2007). See M. Nussbaum, "Veiled Threats?," *NY Times Opinionator*, July 11, 2010, available at http://opinionator.blogs.nytimes.com/2010/07/11/veiled-threats; and especially M. Nussbaum, "Beyond the Veil: A Response," *NY Times Opinionator*, July 15, 2010, available at http://opinionator.blogs.nytimes.com/2010/07/15/beyond-the-veil-a-response.

15 Law n° 2010-1192 to Forbid the Concealment of One's Face in Public was promulgated on October 11, 2010 and published in the Journal Officiel on October 12, 2010. On this topic see Patrick Weil, "Headscarf versus Burqa: Two French Bans with Different Meanings," in *Constitutional Secularism in an Age of Religious Revival*, ed. Susanna Mancini and Michel Rosenfeld (Oxford University Press, 2014), pp. 196–215.

7 Retrospective on the French Stasi Commission

Jean Baubérot and John R. Bowen

John R. Bowen: In 2015, at the request of Marie-Claire Foblets, Jean Baubérot and I set out to write an essay on the work of the Stasi Commission in France. We decided to try a novel method of working: we would engage in a dialogue on the subject, treating a number of aspects of the commission's work, roughly in chronological order.

We decided on this approach for a number of reasons. We have known each other for over ten years, and take a great deal of interest in each other's work. I find Jean's combined perspective of historian and sociologist to be particularly appealing in the French intellectual context, where the famous *pensée unique* often flattens out discussions of current issues. Jean draws on his historical work to give him an independent basis for resisting these often arid discussions. I believe that he appreciates the somewhat different distance (American, anthropologist) that I try to maintain in my own contributions to debates in France on religion, law, and politics. We thought a dialogue would bring out these different but compatible perspectives.

We also looked forward to working together because we have become friends. For many months, several years ago, we had frequent lunches together as part of our joint effort to "catalogue" the workings of the Paris Maison des Sciences de l'Homme, at the request of its Administer, Michel Wieviorka. We exchanged much about our lives and projects as well as about the work at hand. We thus saw personal and intellectual pleasure in working together again.

We realized that we had both closely followed the work of the Stasi Commission, but from two very different vantage points. Jean was a member of the commission and thus party to the subtle and shifting discussions that took place in the hallways and over lunches as well as in formal settings. He was the only member to abstain from the final report, as he explains in what follows here. I followed the sessions through the new Senate streaming service—a sort of French C-Span. My interest came from my project of writing an analysis of the entire set of debates and discussions about Islamic headscarves, going strong during a period when I was often in Paris. I tried to understand and explain—and not simply denounce or mock—what seemed to much of the rest of the world to be a rather silly fixation on a dress code. So, Jean and I each found ourselves, in very different ways, in complex roles that combined insider and external perspectives.

Our narrative strategy was to play to this feature of our respective involvements. One of us would initiate an exchange from his perspective, and the other would respond. We met over lunch to map out the process, and then sent our writings to each other over email. We hope that the result is interesting. We think that it preserves a certain sympathetic resonance between us that also allows for moments of dissonance, as we express different inferences about what we perceived was going on during this period.

Introduction

Jean Baubérot: On June 20, 2003, a little after 8 pm, when I was at a restaurant with friends, I received a telephone call from, I was told, the "Office of the President of the Republic." The contact person asked me if I would accept a seat on a commission to be established under the direction of the Médiateur de la République, Mr. Bernard Stasi, and would be charged with reflecting on the "principle of *laïcité*."[1]

Even though I found this proposal agreeable, it still surprised me somewhat. In fact, the previous month I had dined at the Elysée Palace with a dozen other invitees and some members of President Chirac's entourage, including Blandine Kriegel, who was his advisor on societal issues. We discussed ideas for the celebration of the next centenary of the 1905 law on the separation of Church and State. In particular, we suggested a grand event on December 5, 2005, the anniversary of the law's promulgation, with a speech given by the President of the Republic. For me, this initiative was thus a logical continuation of the evening at the Elysée. Additionally, Bernard Stasi (1930–2011), who had become a naturalized French citizen at the age of 18 and was a political personality of the center right, was known for his courageous stance in favor of immigrants. He had notably written a book, *L'immigration, une chance pour la France*[2] (*Immigration, an Opportunity for France*), which had been met with opposition from his own camp . . . and with admiration by intellectuals. I thus responded to the proposal in the affirmative.

A few days later, I received a telephone call from Bernard Stasi. He said he wanted to talk with me to get to know me better before the commission began its work. During the course of the conversation he could not help but tell me that it is the "Elysée" that directly chose the members of the commission (including myself!), apart from three or four people that he was able to add. He appeared slightly aggrieved. I explained that, in my opinion, the creation of the commission is a logical continuation of the reflection launched by the Elysée in view of the centenary of the 1905 law. The Constitution defines the French Republic as "laïque" ("secular") and at the same time as "indivisible" and "democratic and social," and it seems important to utilize this opportunity to clarify what this "laïque" characteristic signifies. He agreed.

In the two phone conversations, there was never any mention of the wearing of the headscarf or even of Islam in general. Based on the process that went from the dinner at the Elysée to the set-up of a commission, I had not made any connection between the commission and the report "Pour une nouvelle laïcité" ("*For a new* laïcité") which had just been written by François Baroin, the vice-president of the UMP group in the National Assembly. Baroin recommended, among other things, to return to the 1989 Council of State opinion and to "ban the wearing of the veil in schools." But, he added, considering the sensitive nature of the subject, it would be good to defuse it by setting up "an ad hoc commission."[3] Although I had been interviewed by François Baroin as part of his work, I associated this recommendation with the creation of the *Mission d'information parlementaire sur les signes religieux à l'école* (Parliamentary Information Mission on Religious Signs at School), chaired by Jean-Louis Debré, which had began its work on June 4.

I thought that Jacques Chirac did not necessarily agree with Baroin's proposal. In the Creil case of 1989[4] he had appeared rather favorable to the toleration of the headscarf. On the other hand, the appointment of Bernard Stasi as president of the commission did not seem to be in line with the Baroin report, which considered a segment of immigrants to pose problems for *laïcité*. Perhaps the president was looking for a solution "from above" by setting up another commission that would deal with the broader question of secularism. Certainly, since 1989, there had been the attacks of September 11, 2001. But there was also, in France, the shock of having the right-wing leader, Jean-Marie Le Pen, make it to the second round of the presidential election of May 2002. Was this not the time for a show of national unity against the far-right? The prohibition of religious symbols at school seemed to be a measure that would profoundly divide French society. Contrary to what some would later claim, the commission's hearings showed just how divided public opinion was.

Opinion of the Conseil d'Etat (Council of State)

> In schools, the wearing by pupils of signs by which they intend to manifest their belonging to a religion is not in itself incompatible with the principle of *laïcité*, insofar as it constitutes the exercise of freedom of expression and manifestation of religious beliefs, but this freedom should not allow pupils to display signs of religious affiliation which, by their nature, by the conditions under which they are worn, individually or collectively, or by their ostentatious or assertive nature, would constitute an act of pressure, provocation, proselytism or propaganda, infringe the dignity or the freedom of pupils or other members of the educational community, compromise their health or safety, disrupt the course of teaching activities and the educational role of teachers and, finally, would disturb the order in the establishment or the normal operation of the public service.[5]

This analysis was not accurate, as subsequent events have shown. Even if it is not possible to prove, it seems plausible that Jacques Chirac, through the Stasi commission, wanted to take personal ownership of the future law. It is, however, not entirely impossible that he had not yet taken a decision in June–July 2003, and did so subsequently. In any case, for me, but also for many other members of the commission at the beginning of the commission, the "die was not cast."

Why the Stasi Commission?

John R. Bowen: I saw another worry behind the formation of the commission. By June 2003, when you received that telephone call, sporadic worries about France's failure to integrate Muslim immigrants—and more seriously, their children and grandchildren—had developed into a loud and increasingly shared shouting for the state to "do something" about the problem. Many were particularly shocked when the October 2001 friendly soccer match between France and Algeria, attended by the French Prime Minister, was marked by young, French-born boys of Algerian descent waving the Algerian flag, booing the Marseillaise, and then ending the match by storming the field. Coming so shortly after the attacks of 9/11, the disarray gave the lie to the myth of unity-through-football that had swirled around the 1998 World Cup victory of France's "white, black, Arab" team.

If the perceived problem was the failure of integration, why did President Chirac appoint a commission to reflect on French secularity, *laïcité*? That topic pointed towards a second sequence of events, one that began in 1989, a year of two notable geo-political shifts. The fall of the Berlin Wall signaled the demise of a bipolar world. The Ayatollah Khomeini's fatwa against the writer Salman Rushdie, and the formation of the Islamic Salvation Front in Algeria, signaled the rise of a new perceived enemy, the Islamist at our doorstep. That fall, intellectuals and journalists together sounded alarm bells at the arrival of a public school of three young Muslim girls wearing headscarves. Appeals to ban such attire warned of the impending arrival of Islamist forces. Prominent intellectuals called on the state to not give in to Islamism—compromise would be "the Munich of the Republican School" said some, recalling the compromises with Hitler that led to the Second World War.[6]

In the ensuing years, a number of special media exposés linked the two narratives: Islamic scarves signaled and indeed caused national disunity. Often these exposés focused on public schools, which in France hold a special place as the battleground between the Republic and the Church. The presence of scarves in public schools weakens national unity, so the narratives went, by performing communalism (*communautarisme*), and by inducing pressure on other Muslim girls to also don scarves. Because there was little or no evidence for such causal links, much of the writing in this vein concerned other sorts of offenses against *laïcité*: pupils breaking the fast on school grounds, or challenging teachers on matters of history, or calling each other by racial or religious slurs on the playgrounds. Over the course of 2003, any "headscarf affair" received extensive media attention.

So for me, the announcement of the Stasi Commission clearly responded to both these concerns: failures of integration and scarves in schools. What I and others sitting on the outside looked for was how the new commission was going to work those highly-mediatized anxieties into a task and report that sounded so placid and cool-headed: to reflect on *laïcité*.

Jean Baubérot: Of course, you're right that cases involving the headscarf were regularly played out in the media. But, for my part, I attached much more importance to the issue of failures at integration, and the appointment of Bernard Stasi at the head of this commission seemed to me to really take it in that direction. It was also the opinion of the philosopher Mohammed Arkoun, another member of the commission, who considered the "headscarf issue" to be "derisory" and was convinced that the commission would play a pacifying role, by including Islam in a republican dynamic of integration. Arkoun was a very dear friend and we both tried in the early 1990s to have the political authorities create an Institute for Islamic Studies within French higher education.[7] Based on wishful thinking, we were almost convinced that the Stasi Commission could become a sort of anti-Debré Commission, demonstrating that one should not focus on the headscarf issue! The appointment to the commission of two "heavyweights" of the French intellectual world, namely the sociologist Alain Touraine of the School of Higher Studies in Social Sciences, well known for his work on cultural diversity, and the political scientist René Rémond, President of the National Foundation of Political Science, a committed Catholic and proven supporter of what the French media called "laïcité ouverte," seemed to bolster this approach. I considered both to be more liberal than myself in matters of secularism.

Jacques Chirac's letter setting out the terms of reference asked the commission to carry out "an in-depth and serene reflection . . . while avoiding prejudices and confusions that too often obscure the debate" in order to "favor a peaceful implementation of the principle of *laïcité*." This principle was defined as "the neutrality of the public service, respect for pluralism, religious freedom, freedom of expression, but also the strengthening of cohesion and fraternity among citizens, equality of opportunities, eradication of discrimination, gender equality and the dignity of women."[8] Naturally, this letter can be interpreted in several ways and today I do not read it as I read it then. Perhaps I was also influenced by the process that led to the law of 1905. Initially, the situation was very conflictual between the Republic and Catholicism. What was called the "laïcité intégrale" was on the agenda. Gradually, things evolved towards a liberal solution. Article 1 of the law proclaims that "the Republic guarantees the freedom of conscience and guarantees the free exercise of worship," which had not been the starting point. On the commission, in any case, I then pleaded for the same, *mutatis mutandis*, to be done in the case of Islam, indicative of a liberalism similar to that of the period 1905–1908 with regard to Catholicism.[9]

Why these nominations to the commission?

John R. Bowen: I agree with you that the problem of integration was the real problem facing France. But what is at issue with integration? I think of integration as achieved when all citizens have access to the major opportunities a society offers: education, employment, security, and the acknowledgment by others of one's rights as a citizen, equal to all other citizens.[10] This conception of integration is not, I think, alien to France. It is, rather, an elaboration of the value "equality," one that includes social and cultural rights as well as political rights.

But then, why would the commission focus on *laïcité*, rather than, say, issues of equal employment opportunity? President Chirac's definition of *laïcité* includes equal employment opportunity and ending discrimination, but neither the choice of commissioners nor the content of the hearings suggested that these issues really were part of the commission's charge. Nor does anyone in France ever use the term *laïcité* to include those other issues. The commissioners were nearly all experts on some combination of religion, immigration, education, and *laïcité*, with the addition of local-level leaders. Clearly some dimensions of *laïcité*—those having to do with immigration (i.e. Muslims) and education—were behind the choice of commissioners. Even before the commission started work, one could have inferred that the focus would be on something like "problems with *laïcité* having to do with Muslims, and particularly in schools." Headscarves were destined for discussion.

But if that were the agenda, then the selection of commissioners also foretold a curious angle on the subject. The two members who had written about Islam were Gilles Kepel and Mohammed Arkoun. Both were noted scholars: Kepel had written about Islamic movements in Egypt and in Paris, and he had (and has) garnered the most attention for his focus on issues of violence and jihad. Arkoun was a philosopher and a proponent of Islamic reform, but he had little or no contact with ordinary Muslims in France. If the views, experiences, and interests of young Muslims living in France were to be central to the commission's work, then is it not surprising that no one on the commission had any knowledge of the topic?

***Jean Baubérot*:** It should not be forgotten that the commission drafted a set of 26 proposals. Even if, afterwards, those which concerned the wearing of religious symbols were very clearly in the forefront, we did not discuss this topic until the end of our work. The impression I had at first was widely shared. Thus, the journalist Béatrice Gray affirmed: Jacques Chirac "broadly widened a debate which had polarized around the question of the headscarf," by creating a commission "to redefine *laïcité*."[11]

To complete your remarks, I will say that in terms of principles, there was a very broad consensus in France that integration implies equal rights on both the political and social area. On the other hand, there is strong dissensus on "cultural rights." This very expression appears to many as incompatible with the idea of "republican universalism" and the "indivisible" character of the French Republic. Significantly, most of the time the stereotyped expression of the "one and indivisible (French) Republic" is used, whereas "one" is not in the Constitution. It was at the time of the French Revolution, in the Jacobin Constitution of 1793, that the wording "one and indivisible" was used. Against this very dominant stance, Alain Touraine stood as the great defender of cultural rights; in 1989 he had signed the petition opposing that of the philosophers who denounced a "Munich of the Republican school." This second petition warned of the risk of a "Vichy of integration."[12]

Touraine's choice seemed to point to an opening towards recognizing cultural rights. Admittedly, the commission also included Régis Debray, one of the "republican philosophers" who supported the first petition. But it would have been abnormal for this stance to have no representative.[13] Debray is more intelligent and open to dialogue than, for example, Elisabeth Badinter. To have chosen him rather than her seemed a good thing. As for Gilles Kepel, unless I am mistaken, he had not really engaged in debates on the need (or not) to exclude young girls wearing headscarves from school.

Be careful not to perform a retrospective analysis. The composition of the commission appeared balanced. The different conceptions of secularism were represented, and there were also several people whose position I, personally, did not know at all, or even whether they even had one. I do not think it was possible, when the commission was set up, to really plan what would happen next. In fact, the ability to withstand insidious pressures, the strength of character, the skill to maneuver . . . each and every one of these elements proved to be at least as important as a person's starting positions.

How did the commission do its work?

***Jean Baubérot*:** Indeed, we can compare the work of the commission to a funnel. In the first instance it tackled very broad problems. Little by little, it turned to a proposed law on the wearing of religious symbols in public schools, a subject that it had treated for a long time only indirectly or allusively.

Thus, in early July, the work of the commission began with two general statements on *laïcité* in France; That of Henri Pena-Ruiz, who like Debray embodied the "republican philosopher" stance, and mine. The discussion that followed did not focus on a specific orientation. We were at the stage of questions then rather than of affirmations. During the summer, I worked with Jacqueline Costa-Lascoux, who was very involved in educational issues, to develop a grid summarizing the main themes at stake in the field of *laïcité*. Several members of the commission had requested one and we had received the mandate to create it. However, this grid was quickly forgotten because, as of the beginning of September, priority was given to the hearings focusing on general elections.

Specifically, the commission met in the Senate on Tuesdays and Fridays, from 9 am to 1 pm and conducted hearings, broadcast live by the Public-Sénat television channel. Soon, it turned out that these were not enough and other sessions, including closed-door sessions with public servants, were added and took place in different ministries. These sessions were followed by a meal where the morning hearing, and other possible topics, were discussed.

In July, I had acquiesced in the idea of the meal, which avoided an extra meeting in the afternoon. However, I gradually realized that while these meals created a very friendly atmosphere, they were not an ideal setting for in-depth discussions. Speaking while eating favors short interventions, "first-impressions" which are immediately understandable by everyone. The dogmatic claims were difficult to challenge directly and repeatedly, without becoming marginalized in the group that was being constituted. These meals played an important role in producing a certain overall conformity and placing significant limits on opposition to the staff, even when the latter made some contestable decisions.

***John R. Bowen*:** You are right to caution against a view from later on that could be said to be too "retrospectivist." But I had the same impressions when I followed the commission's working method—from a distance, without your knowledge from the inside. My concerns were not about the breadth of conceptions of *laïcité* represented on the commission—as you say, Alain Touraine and you had very different ideas from those held by Régis Debray and Henri Pena-Ruiz. They were rather about the combination of two features of the commission. First, that no one had any real sense of everyday life as faced by the young women and men constantly discussed—and there apparently had been no plan to include their perspectives until the last minute. (Nor were the sociologists who were directly involved in field studies with those women and men even interviewed!) Second, that "laïcité" did not really describe the practical agenda that unrolled during the commission's work.

Let me expand on that last point—again, as seen from the exterior, and you can correct me on my misunderstandings. Had "reflecting on *laïcité*" described the commission's focus, then one would have expected the commissioners to focus on the place of religion in public life (the schools, politics), on whether changes in the 1905 law were required to adequately fund the building of mosques, churches, and temples, the ways in which *laïcité* is taught in schools and to future teachers, the curricula followed in private religious schools that receive state subsidies and are supposed to follow the national curriculum, the appropriateness of the high degree of state involvement in regulating religion, most recently regarding the CFCM, and other, related questions about *laïcité*.

These issues were not, however, those on which the commissioners focused. Yes, representatives of several religions were asked for their opinions about the 1905 law, and an intriguing suggestion was made to give Jews and Muslims each a state holiday, but most of the testimony turned on problems of the poor suburbs: violence against women, challenges to authority in the schools, poverty, and, central to most of the testimony, the *voile*. In each case, topics were deemed relevant only insofar as they involved the public actions of Muslims and the question of the headscarves. So, when unemployment was introduced as a problem afflicting many residents in the poor suburbs, the hearings did not then pursue the causes of and best solutions for unemployment, but stopped at a general plea for doing something about the problem. For problems of discrimination, the commission stated that it was regrettable and something ought to be done, as it contributed to tensions, but that was not the focus of the report either and little new light was shed. The major specialists on those questions did not testify; the topics were thrown into the mix, one might speculate, to "cover the bases."

Jean, you can best speak to what was happening "behind the scenes," but from the outside it appeared that whoever was setting the agenda was not mainly concerned with the problems listed above, but with a different question, which could be stated as approximately "problems in and around poor schools that are linked to the rise of Islam."

***Jean Baubérot*:** I fully understand why and how an observer, and especially an expert on Islam like you, could have a different vision of the work of the commission. What I am trying to convey is what I called the funnel process, which is probably hardly noticeable from the outside. This process allowed for a narrowing of the content, through the gradual transformation of shared questions and even perplexities within the commission into what I think was in a way the imposition of a decision. The latter expression would, of course, be rejected by the other members who ultimately voted for the draft law. Yet it fits my analysis if we take the term "imposition" to mean something quite different from a physical constraint. Things happened in such a way that even those who had reservations considered that the draft law should be adopted.

In the drafting of the report, lengthy discussions on the most appropriate wording caused much delay. I insisted, on several occasions, that we not quibble too much about the commas, and Schwartz did the same. The first part of the report describes recent historiographical changes, in contrast to the "republican" legend. The liberal model prevailed in 1905 over the "combative and anticlerical" model. The expression "pacte laïque," which I had launched in 1990 and which had been strongly contested by the "republican philosophers," was adopted.[14] The second part indicates that "*laïcité à la française* [is] a legal principle applied with empiricism." It is the third and the fourth parts, however, which begin by "taking better account of all spiritual convictions" and by "promoting *laïcité* and the fight against discrimination," which best illustrate my funnel metaphor. They were drafted more quickly, given the delays that we were working against.

The commission did not operate in a confrontational manner. At times there was agreement between specialists in *laïcité* like Pena-Ruiz and myself to rectify simplifications. Debray was one of those who argued that we should solicit sooner and in greater numbers the opinions of young Muslim women. The rector of the Paris Academy, Maurice Quenet, was furious after hearing a female school principal who was very hostile to the headscarf, and whom he assumed had made certain erroneous statements. And so on. Moreover, although some of the staff's decisions provoked some internal criticism, the conviviality that prevailed and the practical reasons invoked (rightly or wrongly) prevented them from turning into real opponents. Indeed, at the management level there was a division of roles. The president, Bernard Stasi, delighted everyone with his intelligence, his brilliance and his humor. He formed many connections. Rémy Schwartz had most of the power, which is not unusual for the rapporteur of a commission. However, on two or three occasions towards the end of the work, Stasi demonstrated some annoyance at the way Schwartz was actually leading the effort.

You correctly note that specialists who should have been heard were not heard. This was the result of a choice of the commission when it organized its work. It was decided that no author would be interviewed. I understand that such a decision may seem surprising, and that its consequences may be considered unacceptable. It was based on two reasons. First, we assumed that we had all read all the important works and that we would refer to them along the way, or at least take into account the analyses of their authors. After that, given the number of books and authors on the many subjects that were the subject of our work, we were never going to finish unless we made such a drastic selection that we would be severely criticized. Interviewing some would mean displeasing all the others!

At the time, this option did not appear absurd to me. Many interviews of actors in politics and civil society seemed unavoidable and we could not multiply infinitely the number of people heard. To learn, at the beginning of the summer, that we would be part of such a demanding commission, gives rise to a haunting question: "How am I going to manage, during the autumn, an already overloaded schedule with this extra work?" The few exceptions to this rule on the part of the staff provoked the most hostile reactions from members of the commission.

Of course, the decision was ultimately counterproductive: the immediate pseudo-knowledge of the "field" took precedence, in the end, over the necessary distance and the objectivation process, even though the hearings of leaders of associations were going in very different directions. Above all, the forcefulness of some members could not be rebalanced, at least not entirely. It was difficult to oppose, on a recurring basis, the work of a colleague to the more or less peremptory statements of a member of the commission. On decisive points there was a quasi-monopoly of speech in cases where, in principle, there should have been a tandem. The very determined people overshadowed those who were more amenable. Kepel imposed himself over Arkoun, sometimes acting as if he spoke on behalf of both. And, of course, other perspectives on Islam and Muslims were denied any attention. Schwartz has positioned himself as *the* specialist in legal issues. Marceau Long, vice-president of the Council of State at the time of the 1989 opinion, is a gentle personality who could not counter the rapporteur.

Conversely, Touraine curiously revealed himself as rather indecisive. His defense of cultural diversity had little effect. On the other hand, I asked that a comparative diagnosis be made of the state of *laïcité* in France, before deciding whether or not to take a tougher stance; the option selected would then apply to all religions. But the staff evaded this request. René Rémond, who initially publicly expressed his opposition to a law prohibiting religious symbols,[15] changed his mind when he understood that no changes would be made to anything that could affect Catholicism (chaplaincy, Debré law on public funding of private schools and its successor provisions, Alsace-Moselle, etc.).

Each of these reasons may seem a little anecdotal, but taken together they explain in part how the imbalance gradually emerged and how a dominant group spirit developed. That being said, it must be remembered that the commission also worked on the proposal for a Jewish holiday and a Muslim holiday to be incorporated into the calendar of public holidays. On the other hand, with regard to the fight against discrimination, it made several proposals, including that of "eliminating discrimination induced by public policies" (ten concrete measures were provided) and of supporting the project that would lead to the creation of the HALDE (High Authority against Discrimination and Equality). It also proposed religious education at school, and the development of studies on Islam in higher education.

What was the impact of the commission's report?

John R. Bowen: Your remarks on how the commission worked underscore the crucial role of the person who sets the agenda, in this case the rapporteur Rémy Schwartz, who had made clear in previous writings that he thought the

State Council's jurisprudence to be insufficient to deal with the problem, and that Muslim girls did face pressure to wear headscarves in schools, "in violation of their individual freedom."[16]

Now for the aftermath: despite the breadth of the commission's recommendations, only two received much immediate attention. One was soon to become the law forbidding the display of ostentatious signs (*signes ostensibles*) of one's religious affiliation in public schools. The other, which was proposed within the commission by Patrick Weil, was that Aïd al-Kebir and Kippur become national holidays. The idea was a new one, and I think a good one, but it was immediately quashed by President Chirac.

Media coverage of "problems" in schools and hospitals rose sharply as the commission was doing its work, and when Chirac pre-empted the commission by a week by calling the headscarves "aggressive," he merely added to the bandwagon of politicians either supporting a new law or seeing it as inevitable and thus requiring their support. Despite street demonstrations against a new law, the National Assembly and the Senate passed it and the president signed it, to become known as the "Law of March 2004."[17] It was set to go into effect that fall. Official Muslim organizations divided in their response, with the major confederation of French Islamic organizations, the UOIF, urging schoolgirls to test the law, and Paris Mosque directing Muslims to follow it. But then two French reporters were taken hostage in Iraq by a group demanding that France repeal the law or they would execute the hostages. The action immediately brought unanimity in France against the terrorists, and a backing away from calls to show up at schools in scarves.

In the end, little of note occurred with the return to school that September. Official estimates were that a few dozen girls were expelled for their scarves, several hundred took them off, and about one hundred left school altogether. The national press moved on to uncover new "problems," such as Muslim boys pulling out sandwiches at the end of the fasting period. Some politicians began to expand the law's reach well beyond its text, as when here and there a mayor would refuse entry to city hall to a woman in a headscarf, or a minister banned parents on school outings from scarf-wearing. I also see the law as one step toward more restrictions on Muslims, with the 2010 law against full face veils, the denial of citizenship to women who wore such clothing, and the extension of the right to impose headscarf bans to private organizations that contracted with the government.

On a different track were the efforts to justify the law to a skeptical English-language audience. Even if much of the popular support (and parliamentary arguments) for the law centered on a perceived threat from Islamists and on assigning an anti-woman meaning to the scarf itself, the commission had refrained from making such claims, and based its recommendations on an argument about *laïcité* and the specific mission of the public schools. The major English-language justifications of the law were narrower still, arguing that the law was passed to protect non-scarf-wearing girls from pressure to wear the veils, based on evidence of such pressure. I was involved in a mock trial of France, where the lawyer for French was none other than Rémy Schwartz! He stated that young Muslim women had pleaded with the commission to "Save us! Save us!" Although no public record of such pleas is available, it became easy for such insiders to make such claims.[18]

Jean Baubérot: I agree with your analysis of the consequences of the law. So I would rather concentrate on the way in which the votes were obtained. You are absolutely right that there were two main proposals immediately identified by the media. The first, that of the addition of two holidays, a Jewish and an Islamic, had been adopted unanimously.[19] The commission had worked it out in great detail. Thus, in order to not reduce the total number of school days, the summer holidays, which are longer in France than elsewhere, were shortened by two days. The reception of this proposal showed bad faith; both the left and the right failed to take into account the whole proposal. You're right, it was immediately dismissed (and not only by Chirac). As a result, it had a perverse effect: some members of the commission would have had more difficulty voting for the prohibition of conspicuous signs without this first proposal. It would introduce, it was said, a great novelty: France would become the only "non-Muslim" country to have a holiday that relates to Islam. Was it meant to be window dressing?

On the part of Patrick Weil, the author of the proposal, certainly not. On the part of the staff, I'm a little skeptical. Let's say that its president, Bernard Stasi, did not make an effort to take it into consideration. But, in general, Stasi appeared to me to be out of sorts during the post-commission period. So, he did not write a long article in *Le Monde* to report on his work,[20] as one might expect. On the other hand, he who had been so charming throughout the course of the commission, easily lost his temper at the round tables. He sometimes reacted angrily to critics instead of trying to convince. I found him to be ill at ease. I'm not sure that he completely accepted the second, and main, proposal.

The second proposal, banning conspicuous political and religious clothing and signs (the law would in the end retain only religious symbols), had been adopted in a less consensual manner than that concerning the additional public holidays. The scope of the measure was debated. Should we ban all political and religious signs in public schools? No, because one can give a political or religious meaning to any sign. So it was necessary to establish a restrictive list of the so-called "conspicuous" and prohibited signs; "discreet signs" would still be permitted (examples were given: such as pendants with stars of David, small crosses or small Korans). Should it also prohibit conspicuous commercial signs, clothing with trade marks? It would have meant a return to the school uniform of the beginning of the century; a majority voted against it. Should it be limited to students? Yes, the ban should not include parents of students when they come to school, nor students at university. Wearing the "veil" (since that is what it was essentially about[21]) was generally allowed. A limitation was made to this freedom for pupils (all considered minors) at public schools, with a "proportionate sanction." The specific mission of the secular school was thus prioritized. In addition, the law would not apply to private institutions, even those under contract, even though the latter receive more than 80 percent of their financing from public funds. Thus "freedom of conscience" would be respected.

For my part, I tried to counter this proposal with two suggestions. The first, which I knew in advance had no chance of being adopted, was to give the force of law to the opinion of the Council of State. I had prepared a short text with the help of a jurist. It was mostly a matter of principle and I hardly insisted on it. My second proposal, which was more realistic in view of the state of mind of the commission, was to limit the prohibition to "conspicuous religious dress," stating that the bandana was not included in these forms of "dress." Thus, young girls could cover their hair without anyone knowing whether they were doing it out of religious conviction or as a fashion statement. It seemed to me that many people were in favor of this proposal. But Rémy Schwartz refused to put it to a vote and no one protested.

So on Tuesday, December 9, in the late morning,[22] we voted only on the staff's proposal. Nobody voted against it. There were three abstentions: Alain Touraine, Ghislaine Hudson, who was the school principal on the commission, and me. For my part, I did not vote "no" because I did not deny that there are extremists who claim to act in the name of Islam. There had already been attacks in France during the Algerian civil war, and in the context of post-9/11, I did not want to risk becoming hostage to or the standard bearer for the issue for anybody. Given the French tradition, where we are not accustomed to minority positions, the refusal to accept the proposal seemed sufficient to me.

Probably a little disappointed by the result, the rapporteur said we could still change our mind by the end of the afternoon. Knowing the reservations of some others, I responded: "Good, then we will be six or seven who abstain." But without provoking a single reaction, Schwartz specified that he spoke only of the abstainers: "for the others, it is finished!" In the afternoon, Alain Touraine rallied first. Ghislaine Hudson was in a delicate position: the union of principals, her union, was clearly in favor of a law prohibiting religious symbols. She agreed to change her vote at around 5 pm. I was, therefore, the only one who did not vote for the proposed law.

You have indicated the main consequences of the law. Indeed, when it came into effect, at the start of the 2004 school year, the taking of two hostages in Iraq and the demand made by the terrorists to abolish the law changed the situation. As far as I am concerned, I remember how difficult it was for me to convey a dialectical message in the media ("I am both against the law and against giving in to the blackmail of terrorists"). It is true that the effect on young girls who wore headscarves and were excluded from school was minimized.[23] And, from the beginning, some campaigned to extend the law beyond its scope. I will say, however, that the commission repeatedly referred to the future Anti-Discrimination Authority, of which Stasi hoped to become the president. It counted on this institution to ensure that the scope of the law would not be exceeded. Chaired by Louis Schweitzer,[24] the HALDE effectively played this role during the presidential term of Jacques Chirac.

Subsequently, Nicolas Sarkozy first normalized, then abolished the HALDE, placing it under the Defender of Rights,[25] which significantly changed things. This led to the law of October 11, 2010 prohibiting the concealment of the face in public spaces (i.e. the ban on the burqa and the niqab), in the name of public order (and without invoking *laïcité*). Several members of the commission

(Patrick Weil, Rémy Schwartz himself, . . .) opposed this new law. More broadly, we are witnessing the imposition of a "new *nouvelle laïcité*."[26] This seems to me to be characterized by a twofold movement: first, a diminution of the separation between the political power and the religious authorities and, second, a shift in the neutrality requirement of the state (which must be an impartial arbitrator) to individuals, from the state sphere to the public space. This is not in the spirit of the 1905 law, but is in a certain tradition of mistrust of civil society. However, this "new *laïcité*" is not the one promoted by the Observatoire de laïcité, set up at the end of 2012. Within the framework of existing laws, the Observatory tries to preserve the spirit of the 1905 law on the separation of Church and state.[27]

Looking back, would you have preferred a different procedure? Do you wish you had (re-)acted differently?

***Jean Baubérot*:** Although some of my friends in Quebec have criticized the Bouchard-Taylor Commission, in comparison to the Stasi Commission, it worked better.[28] First at the level of resources: the Stasi Commission had no premises and, apparently,[29] for the most part, its modest expenses were taken from the budget of the Mediator of the Republic. Bouchard-Taylor had a budget of Can $5 million (they spent Can $3.7 million). In addition to the fact that they went into the field themselves, this allowed them to conduct investigations, including checking media reports. These investigations have shown that the reported facts were often severely distorted. In total, 13 academic papers were written, on which the commissioners were able to base their report. The time allotted to the two commissions was also very different. After organizing its work in early July 2003, the Stasi Commission met for three months from the beginning of September. The Bouchard-Taylor Commission lasted a little over a year.

The Stasi Commission was originally titled "Independent Commission for Reflection on the Application of the Principle of Laïcité in the Republic." By an almost Freudian slip, the term "independent" disappeared from the title of the final document! The difference in the relationship of the two commissions to the political sphere is marked by what happened immediately after the submission of their respective reports. Jacques Chirac relied on the Stasi Commission to announce the introduction of a bill on the prohibition of conspicuous religious symbols in public schools. Conversely, there was, in the National Assembly of Quebec, a unanimous vote in favor of maintaining the crucifix that hangs above the seat of its president (it is a political symbol), while the Bouchard-Taylor Commission asked for its transfer "to a place that can enhance the significance of its heritage." Overall, however, the Bouchard-Taylor Commission made it possible not to toughen the "laïcité québécoise,"[30] and this is what the Quebec Liberal government wanted. The two examples, by virtue of their differences, show that (as is logical) politics always has the last word.

Should I have proceeded otherwise? Of course, knowing what happened, if it were to be done over, I would have adopted a more vigilant position from the outset. Thus, I would have tried to oppose the decision not to include specialists in the hearings. In particular, because the televised broadcast of the debates was widely followed by journalists. But such an attitude of principled mistrust would have been misunderstood, and I would have been immediately marginalized. Should I have been lobbying for abstentions? All the members of the commission knew the reasons for my position and I felt that everyone should speak in conscience.

I continue to think that the 2004 law is much more ambivalent than that of 2010. There has never been as much talk of "cultural diversity" in France as in the years between 2004 and 2008.[31] Conceived as a limitation of a freedom that remained the general rule, the commission's proposed law could be the price to pay for the acceptance of this diversity.[32] I think that was Alain Touraine's analysis in the end. But, since 1989, the "Islamic headscarf" has been globally "demonized" by those opposing its toleration in public school. The process that led to the 2004 law strengthened and formalized this "demonization," inducing a chain of events. The question is whether a victory of Ségolène Royal, the PS candidate for the presidency of the Republic in 2008, would have changed the situation. To a certain extent, it seems to me that it would have. But we will never know.

In the end, what mattered was that I publicly disagreed by my vote, especially since at the end of the day I was the only one to do so, which I had not foreseen at all. This allowed for a minimum of debate to be maintained and prevented the law from being presented as self-evident to any reasonable person.[33] Thus, some rumors claimed that it was the closed nature of hearings that had shown "the scale" of the "Islamist threat" and prompted the members of the commission to propose the law. Several people told me that they would have believed this rumor if the vote had been unanimous. But, besides the rumor being incorrect, it established a very contestable continuum between the wearing of the headscarf itself and extremism. I rejected this continuum and for me the main problem was drawing the line in the right place between inclusion and exclusion.

I thought, and I still think, that it was the Council of State which, in 1989, drew the appropriate boundary. And this not only on a political and strategic level, but also on a pedagogical level. Its opinion could have taught students both the consistency of democratic freedoms and the fact that they contained limits, in the name of respect for other freedoms. Recent events, including the Daesh terrorist attacks, have only strengthened my conviction in this regard. A *laïcité* unifying all those who oppose violent extremism is the best defense against Daesh, isolating it, removing its attractiveness to young people trying to find their way in life. Among those fighting for this *laïcité*, there is Latifa Ibn Ziaten, a woman who wears a headscarf. Her son Imad was the victim of the terrorist Mohamed Merah in 2012. In collaboration with the national educational system, she meets thousands of students each year to tell them: "practicing one's faith in peace, respecting republican values and the beliefs of others" is "possible."[34] Decidedly, the boundary is currently not in the right place!

***John R. Bowen*:** As the question concerns you most directly, let me merely add one more indication that the line you mentioned has not been clearly drawn in France. The Stasi Commission was careful to limit its opinion about the Islamic scarf to public schools, based on the specific mission and character of these schools. The report refrained from assigning a general meaning to the act of wearing a headscarf. Even if one disagrees with the law of 2004, the self-restraint of the commission on that particular question I think you and I both find admirable. But it had little or no effect. Indeed, as I write these words, today's issue of *Le Monde* features a declaration by Prime Minister Manuel Valls that the headscarf represents "the subjection of women."[35] In order to validate and support the work of persons such as Mme. Ibn Ziaten, the frontier, as you say, between the acceptable and the unacceptable, must be moved.

Notes

1 For another reflection on my participation in the Stasi Commission, see J. Baubérot, "L'acteur et le sociologue. La Commission Stasi," in D. Naudier and M. Simonet (eds), *Des sociologues sans qualités? Pratiques de recherche et engagements*, Paris, La Découverte, 2011, pp. 101–116; see also J. Baubérot, *Une si vive révolte*, Paris, L'Atelier, 2014 (on the Stasi Commission, see pp. 201–208).

2 Paris, Robert Laffont, 1984.

3 J. Baubérot, *Les sept laïcités françaises*, Paris, éd. de la Maison des sciences de l'homme, 2015, pp. 107ff.

4 The refusal by three pupils, in the autumn of 1989, to obey the internal regulations of the school in Creil (Oise), which prescribed that they take off their headscarf in class, had become, to the surprise of many, a national controversy. It was in this matter that the Council of State issued its well-known opinion.

5 Council of State (France), opinion of November 27, 1989, n° 346.893.

6 On these sequences of events, and my own analysis of the headscarf "affairs" and the work of the Stasi Commission in more detail, see John R. Bowen, *Why the French Don't Like Headscarves: Islam, the State, and Public Space*, Princeton NJ, Princeton University Press, 2007, pp. 81–127.

7 On this (failed) attempt, see J. Baubérot, *Une si vive*, p. 158 and "Mon ami Mohammed Arkoun," jeanbauberotlaicite.blogspirit.com, Note of September 25, 2010.

8 Commission presided over by Bernard Stasi, *Laïcité et République*, Paris, La Documentation française, 2004, pp. 6ff.

9 *Stricto sensu*, one should speak of the laws (plural) on the separation, because three laws adopted in 1907 and 1908, in the face of the intransigence of Pope Pius X, supplemented the liberal provisions of the law of 1905. See M. Larkin, *Church and State after the Dreyfus Affair: The Separation Issue in France*, London, Macmillan, 1974 (published in French as *L'Eglise et l'Etat en France. 1905: la crise de la séparation*, Toulouse, Privat, 2004) and *Religion, Politics and Preferment in France since 1890: La belle époque and its legacy*, Cambridge, Cambridge University Press, 1995 (2nd ed. 2002).

10 For a recent overview of these issues, see Richard Alba and Nancy Foner, *Strangers No More: Immigration and the Challenges of Integration in North America and Western Europe*, Princeton, NJ, Princeton University Press, 2015.

11 Béatrice Grey, *Le Monde*, July 2, 2003.

12 See John R. Bowen, *Why the French*, p. 85; Jean Baubérot, *Les sept laïcités*, p. 92.

13 There were two: besides Debray, Henri Pena-Ruiz was also a member of the commission.

14 Commission presided over by Bernard Stasi, *Laïcité et République*, p. 28.

15 This violated a rule established by the commission: in order to prevent polemics among the members via the media, each member had a duty to reserve his or her opinion until the publication of the report.

16 Rémy Schwartz, "Le jurisprudence de la loi de 1905," in *La laïcité: Archives de philosophie du droit*, 48: 85–94. Paris: Dalloz, 2005, p. 93.

17 On the 2003–2004 period of media attention, parliamentary debates, street demonstrations, and the law's effects, see John R. Bowen, *Why the French*, pp. 98–152.

18 On this Public Broadcasting Service program, where the judge was Cheri Blair, see http://pennstatelaw.psu.edu/WOT/episodeI/episodeI.html.

19 This unanimity seemed to me to be due not to a general assent but to the fact that those who had reservations were aware that this proposition was "the price to pay" for obtaining a favorable vote on the interdiction of conspicuous symbols.

20 He only replied to the critique that the pope had directed at the results.

21 Even if the wearing of the kippa was also prohibited (but a difference being that Jews did have certain private schools under contract). To be sure, and to make the law one of a general nature, a prohibition should cover an (imaginary) "grand cross."

22 The report was presented to President Chirac on December 11.

23 Cf. the testimonies of young girls in I. Chouder, M. Latrèche, and P. Tévanian, *Les filles voilées parlent*, Paris, La fabrique éditions, 2008.

24 Until then the president of the Renault automobile group.

25 The successor to the Médiateur de la République, a position held by Bernard Stasi in 2003.

26 Cf. S. Hennette Vauchez and V. Valentin, *L'affaire Baby Loup ou la nouvelle laïcité*, Paris, Lextendo éditions, 2014.

27 And it publishes brochures providing details on the scope of *laïcité* in France.

28 Cf. J. Baubérot, *Une laïcité interculturelle. Le Québec, avenir de la France?*, La Tour d'Aigues, L'Aube, 2008. For a comparison of the two commissions, see pp. 213–218. One can find, in the rest of this work (dedicated to the "reasonable accommodation crisis" in Quebec between 2006 and 2008), elements that explain and analyze the successes and failures of the Bouchard-Taylor Commission.

29 Since this information was never provided to the members of the commission.

30 On this *laïcité*, cf. M. Milot, *Laïcité dans le nouveau monde. Le cas du Québec*, Turnhout, Brepols, 2002; "Laïcité au Canada: liberté de conscience et exigence d'égalité," *Archives de Sciences sociales des religions*, April–June 2009, 146, pp. 81–98.

31 For one example: M. Wieviorka, *La Diversité, rapport à la Ministre de l'Enseignement supérieur et de la recherche*, Paris, Robert Laffont, 2008.

32 Two years after the vote of the law of March 15, 2004, I wrote that "two readings [are] possible of this law: one, pessimist, where it appears as an indication of the French tensions, and which can produce a 'catho-laïque' 'communautarisme majoritaire' (analogous to the United States, with the meeting of the neo-conservatives and some Protestant fundamentalists); another, more optimistic, where, paradoxically, this law has been a sort of republican orgasm, leading to a relaxation and making it possible to engage in a policy more favorable to cultural diversity." And I concluded: "The future will decide" (*L'intégrisme républicain contre la laïcité*, La Tour d'Aigues, L'Aube, 2006, p. 126s.). The "Sarkozy years" have, indeed, decided!

33 Significantly, the media spoke of the "Sages" of the Stasi Commission.

34 L. Ibn Ziaten, *Dis-nous Latifa, c'est quoi la tolérance?* Paris, L'Atelier, 2016, back cover.

35 Bastien Bonnefous, "Islamisme: Valls provoque la polémique au sein du PS," *Le Monde*, April 7, 2016.

8 Theoretical perspectives on cultural and religious diversity in two national reports

Karel J. Leyva

In some liberal democracies, when reflecting on ways to manage social cohesion, the issue of cultural and religious diversity emerges as being of utmost importance. Spirited debates occur in pluricultural societies on subjects such as reasonable accommodations, various *laïcité* regimes, or the relevance of proposed normative responses within contemporary political theories. These reflect not only the tensions arising from social relationships between national majorities and ethnic, cultural and religious minorities, but also the concerns of intellectuals, private foundations and governments faced with resolving these tensions. Several commissions have been created to advise on the general direction that should be taken by their respective societies in regard to diversity, in contexts as diverse as the United Kingdom, France, Belgium, Quebec and Australia, just to name a few.[1]

In this essay, I explore the theoretical perspectives of the discourse on cultural and religious diversity in the reports of two of these commissions, established consecutively in France (2003) and Belgium (2004). I will begin by analysing the report of the *Commission de réflexion sur l'application du principe de laïcité dans la République* (Commission for Reflection on the Application of the Principle of Laïcité in the Republic of France),[2] using the analytical framework of two ideal-typical republican positions which developed in the 1990s.[3] From this perspective, I hope to be able to show that while the report may deserve some of the criticisms it received, especially in regard to its proposal to ban the wearing of religious symbols, it certainly cannot be reduced to that single aspect. Next, I will describe the report of the *Commission du dialogue interculturel* (Commission for Intercultural Dialogue),[4] focusing on its underlying recognition theory. I will try to explain how the hypothesis that the theoretical influences of the report stem from communitarian philosophy is tenuous, and that an attentive reading of the way specific aspects of the report are worded leads us to look elsewhere for its possible philosophical influences. This is not a trivial issue: 'communitarian' is the same term that was used several years later to label the report of the *Assises de l'interculturalité* (Round Tables on Interculturalism),[5] which was basically a continuation of the *Commission du dialogue interculturel* (Commission for Intercultural Dialogue) [CID]. In both cases, the goal of the commissions was to offer recommendations, taking into account cultural and religious diversity as a legitimate aspect of Belgian society. And in both cases, the term 'communitarian' was used for the purpose of discrediting the reports produced by these commissions. While I believe that this description of the two reports does not do them justice, I will only elaborate on the CID, specifically because it was the starting point of the work conducted by the steering committee of the Assises, even though the latter is worthy of a similar analysis.

The Stasi report

On 11 December 2003, the *Commission de réflexion sur l'application du principe de laïcité dans la République* submitted its report to Jacques Chirac, then president of France. In an effort to balance national unity with respect for diversity, the document articulates 26 recommendations to adapt French *laïcité* to the new challenges presented by cultural and religious diversity. While the Stasi Commission has been the subject of numerous academic works, it has rarely been examined from a purely republican viewpoint. The attention of researchers has been focused primarily on its recommendation to ban religious symbols in public schools and the resulting legislation, and their response has generally been critical toward the commission and its report.

The Stasi report was harshly criticized for several reasons, the first being that it did not provide any proof to back up its assertion of the existence of an Islamist threat to public order. Second, it did not engage in a serious discussion of the rights that were in jeopardy. Third, it contains conceptual inconsistencies, particularly when it recommends restrictions to deal with the supposed pressures to which young Muslim girls are subject. Finally, it did not show that the ban of the veil in schools could have positive results. The recommendation in question is based on questionable motives, be they legal or public.[6] Furthermore, it has been pointed out that the report involves a resurgence of the idea of assimilation as the main goal of French governmental policy, as well as an undermining of the movement in favour of the right to be different. This assimilation policy interferes with the integration of groups of Muslim immigrants.[7] It is apparent that the position of the Stasi report on the veil issue is a new legal demarcation of freedom of religious expression, indicating a sharp break with the previous rationale of the French government.[8] In fact, according to some, France had been leaning toward multiculturalism, particularly with the establishment of the *Conseil français du culte musulman* (French Council of the Muslim Faith) and the creation of Muslim secondary schools under state contract. However, the Stasi report was apparently distancing itself from the inclusive *laïcité* that characterized the previous French model, moving toward an exclusive *laïcité*.[9] This viewpoint is again based on the recommendation to ban religious symbols in schools.[10]

From a similar perspective, some studies focus on the use of the concept of *laïcité* by the political elite as a key component of national identity, when addressing the issue of cultural diversity, especially in the debates surrounding the adoption of the French law in 2004. It has been argued that an analysis of the arguments used to justify this law, as much by the Stasi report as by the political elite, shows that the issue of the Islamic headscarf was approached through a symbolism influenced mainly by a communitarian version of republicanism. More importantly, the rhetoric of the Stasi Commission seems to reveal that, despite the fact that the concept of community was used pejoratively in the report, the communitarian dimension of French republicanism predominates. And while the report contains liberal elements (for example, the requirement of freedom of conscience, legal equality of religious choices and neutrality of public authority), its arguments for a prohibition of religious symbols in schools includes a 'republican communitarian' dimension. This is because *laïcité* is presented as a cultural value, originating from France's history and expressing the idea of well-being upheld by the state. In the report, the primary function of the principle of *laïcité* would therefore be to reaffirm national unity and prevent social disintegration. Republican communitarianism, according to which the political community is based on collective national identity, connects the threat of communitarian leanings to the fear of multiculturalism.[11]

Others believe that the opinions of the commission on the meaning of Muslim religious symbols are culturalist and stereotypical;[12] they are based neither on reliable literature about Islam nor on the perspective of young girls who wear the veil. In fact, these viewpoints are based solely on the testimony of public agents (principals, teachers, medical staff, etc.) who feel threatened by accommodation requests. As for the rationale asserting the intrinsic incompatibility that exists between religious expression and state neutrality, it is inadequate to justify banning religious symbols. On one hand, Muslims are treated severely (their requests being labelled as disturbing, as a hindrance to public service function). On the other hand, the French state systematically violates the principle of neutrality when it maintains, for example, the concordatory status of Alsace-Moselle, a situation in which *laïcité* is presented as a principle 'applied empirically'. As well, even though the Stasi report emphasizes that the secular state has made reasonable accommodations in favour of Christians and Jews, it applies an inverse interpretation of this idea when it asks Muslims to limit the public affirmation of their identity. Now, in reality, the responsibility of accommodations should fall on institutions and not on individuals; it does not involve limiting identity, but rather affirming it as much as possible, as long as the request does not cause undue hardship. On the contrary, the French interpretation considers existing laws and rules to be a priori legitimate, without analysing their impact on minorities. Muslims are therefore subject to a duty of confidentiality, 'proof of their "reasonable" consent to the demands of republican living together'.[13] So, the Stasi report suffers from a typical tendency of French republicanism, which involves judging French society by enshrined ideals and judging minorities based on the interpretation of their practices. Basically, the Stasi report 'promotes a conservative republicanism, that assimilates liberal, democratic and universal values with French ethnocultural norms, and judges the practices of minorities according to criteria that do not distinguish sufficiently between the two levels'.[14]

This brief overview of the analyses of the report shows that the great majority of the works studied in relation to the Stasi Commission have focused mainly on its proposal to ban religious symbols in public schools, and on the resulting legislation. Among the aspects most frequently criticized are the prevalence of the ideal of assimilation in the report, its attack on religious freedom, the use of the principle of *laïcité* to justify the ban, and the lack of evidence that would allow for an assessment of the danger confronting the Republic. Some studies present the report as being communitarian, incompatible with multiculturalism; it is characterized by conservative republicanism. I would like to show how reading the report in light of the responses given by contemporary French republicanism to cultural and religious diversity reveals other aspects which are often overlooked, and which account for its composite character.

The republican approach of the Stasi report

The Stasi report draws a clear distinction between historically entrenched religions in France and 'new religions' that have been established there throughout the past few decades. The first category of religions, fundamental riches of French society, do not represent a threat to *laïcité*: they have been able to adapt to *laïcité*, although not without a struggle, particularly where the Catholic Church is concerned. As for the second category, they must find a place in society through integration while avoiding community withdrawal; these require 'the application of the principle of revised *laïcité*'.[15] Therefore, the central issue of the report is how to reconcile societal unity and respect for diversity in the face of this new reality in which Islam holds a place of great significance. It is concerning Islam, and more specifically, the communitarian leanings of certain Islamic groups, that the Republic must act urgently, because 'community withdrawal' undermines the foundation of the social pact: no individual can 'favour allegiance to a particular group over affiliation with the Republic'.[16] This serious concern about community withdrawal exists throughout the report, which affirms that *laïcité* must allow 'community affiliations' to be eclipsed. It fears that 'community feeling' could drift toward a 'fixed communitarianism' which would threaten social cohesion, and that individuals would overlap 'in a mosaic of closed-off communities'.[17] Some religious demands threaten the stability of the Republic. *Laïcité* appears here as weak, threatened by demands that tend to allow 'community beliefs' to take precedence, as well as by 'communitarian groups' that take advantage of social unease in neighbourhoods open to the development of 'communitarian logic' in order to mobilize supporters.[18] The commission is aware that the difficulties encountered were 'still minor'; however, it emphasizes that 'they are real, strong and serve as warning signals of dysfunction, especially since the recent rapid diffusion of these phenomena is so troubling'.[19]

At school, where students must confront 'pressures and manipulation by politico-religious activists', finding a solution 'is a social emergency'.[20] Public order is disturbed precisely in the area where students should be protected from the 'frenzy of the world'.[21] The state must act to prevent 'their minds from being plagued by the violence and frenzy of society' and by 'worldly passions'.[22] In fact, at school, just 'wearing conspicuous religious signs – a large cross, a kippa or a veil – is enough to disturb the tranquillity of academic life'.[23] However, problems are not limited only to the wearing of the veil: requests for time off, interruptions of courses for religious reasons, and the refusal of girls to submit to identity checks by a male screener . . . these all disturb academic tranquillity. Even though it concerns an 'activist minority', such attitudes seriously jeopardize the principles regulating public service. Furthermore, teachers protest when they see mothers of students wearing a headscarf during school outings. Staff members feel bewildered, distressed and unable to fulfil their mission; morale is affected and they feel unhappy, uneasy and miserable. They have the 'sense' that the rules are not clear, and that the hierarchy does not provide adequate support. They feel like victims of an 'ongoing' guerrilla war 'against *laïcité*'.[24] Alone in these situations, 'they question the official statistics that minimize the difficulties experienced on the ground';[25] they recognize the tension that arises from demands related to identity and religion, the establishment of clans and the 'communitarian clusters in playgrounds'.[26] For this reason, they request a ban of religious symbols so 'school principals are not left alone to deal with the issue of deciding if a symbol is ostentatious or not'.[27]

After hearing, on one hand, representatives of major religions and leaders of human rights associations resist a ban of religious symbols and, on the other hand, 'practically the entirety of principals and numerous teachers' request the ban, the commission believed that the issue did not arise from 'freedom of conscience, but [from] public order'. On one side, the ordinary course of teaching could no longer be ensured; on the other side, 'pressure' was being exerted on minor girls, requiring them to wear a religious symbol. In such circumstances, the Republic could not remain indifferent to the 'cry of distress' from these girls. 'Schools must remain places of freedom and emancipation'.[28] So it was by shifting the issue of protection of fundamental rights to the issue of security, order and rational autonomy that the Stasi report justified its proposal to ban religious symbols in public schools.

Now, most of the elements presented so far are characteristic of the discourse present in a French republican movement that took shape mainly after the events that occurred in 1989 in France, when three young girls were expelled from a public school in Creil for refusing to remove their veil. Generally speaking, within this movement, which is commonly called traditional republicanism,[29] critical viewpoints are conveyed regarding certain expressions of diversity often associated with demands of cultural and religious particularisms, and multiculturalism is rejected.[30] This republicanism asserts that not only does multiculturalism belong to the Anglo-Saxon world,[31] but it also promotes legal inequality, fosters community withdrawal and places cultures above politics and groups above individuals.[32] In certain cases, it goes so far as to consider that recognizing specific rights of cultural and religious minorities is tantamount to a desire that France embark on 'the American path of minority exuberance'. Furthermore, this republicanism feels that promoting positive discrimination measures designed to reduce cultural, social and economic inequalities represents an attack against republican principles,[33] and it 'disguises a return to ideology that is fundamentally reactionary'.[34] Multiculturalists are therefore considered 'reactionaries in the literal meaning of the term'[35] and their 'ideology' is a major force in the development of the French extreme right.[36] Some authors fear that under a multicultural policy, France would gradually split into ethnic, racial or religious communities that are 'adversarial, hostile, separate, demanding, to the point of obtaining their own rules and laws'.[37] And some authors also assert that multiculturalism is based on 'unconditional respect' of the right to difference, preferring to refer to multicultural society as 'multi-community society'.[38] Diversity in itself does not present 'any unsolvable problem to republican tradition in its French version, distinguished by the regulating value of assimilation' and by the principle of *laïcité*:

> it is institutional multiculturalism or normative multi-communitarianism, as a political horizon or project, that constitutes a threat or a challenge to the French-style republican tradition, particularly since it directly attacks the principle of laïcité and the juridico-political norm of the equality of all citizens before the law.[39]

This movement therefore sees reasonable accommodations, differentiated rights and multicultural policies as threats against republican principles and as a danger to France's future, which justifies their attachment to a combative perception of *laïcité*, whose corollary is a consensus for the ban of religious symbols.[40] This view is more obvious in the work of Guy Coq, who feels that the 'generous feeling' which has resulted in the 'willingness to welcome individuals by lauding a multicultural society' ignores the reality that 'social cohesion and societal integration of all members are cultural phenomena and have cultural conditions'. This is why at least a minimum of cultural coherence proves to be necessary if we do not want society to break 'into withdrawn humans, sometimes enemies, striving to establish actual rival sub-societies in the appropriation of common spaces'.[41] Coq does not define what this minimum includes, but his viewpoint is based on the following idea: since religion is no longer able to ensure social connection,[42] as Catholicism did previously, French social cohesion has no alternative but to be based on attachment to French culture and national history, of which *laïcité* is an expression. Here is his reasoning for not accepting differentiated rights: allowing the headscarf in schools involves endangering the social connection represented by *laïcité*, the bonding agent that substitutes for religion. This author feels that, in the context of the 'rise in fundamentalism', all attempts to introduce the veil 'into classrooms take on a very specific meaning: an obvious calling into question of the definition of school as being secular'.[43] Wearing religious symbols at school, or as Coq states, 'branding schools with their symbols', jeopardizes the consensus gained by *laïcité*. On one hand, this would be giving in to the demands of traditional Muslims[44] and, on the other hand, it would mean that this right must be given to fundamentalist groups of all other religions. In this regard, the very wearing of the headscarf is identified with communitarian withdrawal[45] and fundamentalism: the scarf being both a sign of religious belonging referring to a culture other than French culture, and an act of protest by groups or individuals who manipulate young girls, forcing them to wear the veil. As well, headscarf wearing also represents the 'main risk of proselytism' in schools,[46] because 'the scarf in itself, by virtue of its very presence, can exert insidious pressure on young Muslims who are daily treated like "bad Muslims" by veiled youths'.[47] This perspective is in contrast to the view of the Council of State, which, during the veil issue of 1989, established that the wearing of the headscarf did not in itself represent an act of proselytism or pressure on young Muslims. As did other republicans,[48] Coq criticized the decision of Minister Jospin to rely on this viewpoint in order to choose a negotiation path rather than institute a ban. If for some,[49] Jospin betrayed 'the mission of schools', Coq felt that the minister committed a 'political error' with 'very adverse consequences', allowing Islamist minorities, who manipulate headscarf wearers, to discover that *laïcité* has a soft underbelly, that basically they can easily take action in a school to undermine secular vision, and trigger a serious crisis in terms of the principles of this much-hated secular society. As well, it also afforded the opportunity to try to destabilize the huge majority of young Muslims who had taken the step of integration, but could easily be disrupted by this deliberate conflict.[50]

Coq therefore believes that wearing 'ostentatious signs of one's affiliation' is unacceptable in schools[51] and that opposition to this action is appropriate, 'regardless of who initiates it: students or teachers'.[52] The adjective '*ostentatoire*' (ostentatious) deserves further attention as concerning its definition. It is this particular word that was used in the circular issued by Minister François Bayrou in 1994, which proposed a ban of 'ostentatious symbols, which in themselves constitute elements of proselytism or discrimination'.[53] However, the document does not actually define the word; it merely contrasts it with 'discreet symbols'. As for the Stasi Commission, several years later it used the word '*ostensible*' (conspicuous) in its recommendation to ban religious symbols in public schools. Now, in Coq's view, there seems to be no difference between 'conspicuous' and 'ostentatious', because he calls an 'ostentatious display of affiliation' that which 'a glance at the person cannot miss'. This display would be undesirable, because

> it invites us to first see a member of a particular community, whereas *laïcité* requires us to see a unique person and a member identical to all others of the academic institution, and of the global society for which it is the school's mission to symbolize constant unity and cohesion.[54]

Coq has 'the deep conviction that the veil in class represents an obstacle to the integration of many people'.[55] It must therefore be banned, and even members of religions who do not pose a threat to the stability of the Republic should also conform to this law. 'I acknowledge that this rule of discretion must also apply to other religions in the classroom. So I would not be shocked if kippas and ostentatious crosses were also banned by a circular or law'.[56]

Traditional republicanism often contests rationales that use human rights as a basis for authorizing the wearing of religious symbols in schools. In certain cases, since veiled girls are perceived as objects of systematic manipulation and the headscarf is interpreted as a symbol of feminine subservience, the freedom of conscience argument falls apart, quite simply because 'allowing the Islamic headscarf does not equate to welcoming free beings'.[57] Respect for difference is not seen as a valid argument, either because the headscarf, perceived as being opposed to the principle of rational autonomy and to human rights, is not honourable,[58] or because 'the type of difference demanded upsets the precarious balance of the secular space in the classroom'.[59] As well, 'the absoluteness' of individual rights undermines the principles underlying the right to educate: 'any political demagogy on new rights is almost certainly damaging to secular spaces'.[60] Finally, some authors such as Finkielkraut, maintain that this 'return to human rights against secular schools' would not only be philosophically ludicrous and morally reprehensible, it would also be a 'political crime' that would cost dearly afterwards. In the name of tolerance, individual freedom and democracy, communitarian affiliations would display themselves 'violently where formerly it would have been permissible to put a stop to them'. For this author, 'we no longer know how to differentiate between human rights and tribal rights'.[61]

I would like to point out that the Stasi report echoes several of the arguments presented so far, particularly by articulating the following ideas: (a) the very wearing of the headscarf at school is enough to disturb public order; (b) *laïcité* is weakened and threatened by religious demands; (c) even though the problems encountered are minor, they represent a serious threat for the future; (d) there is therefore a social emergency requiring urgent measures; (e) one of these measures includes the ban of religious symbols; and (f) this ban will protect young girls from communitarian pressures that threaten their autonomy. It also resembles traditional republicanism in regard to the value placed on the nation as a principle of social integration, where each individual must be able to recognize himself or herself, as well as the particular fact that it does not give statistics that would allow an assessment of the actual danger to which the Republic is exposed; its argumentation is based on generalizations. It also reveals itself to be very close to this republicanism in its strong vision of citizenship, in that it calls for affiliations to be eclipsed and emphasizes citizens' obligations.[62] It holds that in a secular society, people must be able to distance themselves from their own tradition.[63] It affirms that the strength of 'French cultural identity can stimulate the melting pot of integration',[64] that *laïcité* affects national identity[65] and that it can be the catalyst for the integration of everyone into society.[66] In the same way, it opposes everything that undermines not only confidence in the Republic, but also 'identification with the nation'.[67]

That being said, the report differs from traditional republicanism in some aspects. First of all, even though the ban of religious symbols seems to be the corollary of combative *laïcité*, the commission advocates open and dynamic *laïcité*. It feels that 'the time for combative *laïcité* is now over, making room for peaceful *laïcité*'.[68] To meet the new challenges of cultural and religious diversity, '*laïcité* should not be on the defensive; it can stop acting like a fortress under siege'.[69] Then, while sharing the fear that traditional republicans have of communitarian withdrawal, the report speaks out against the illusion of a 'disembodied republican pact'[70] and accepts the challenge to forge unity in respect for diversity. In a quest for balance, this report therefore avoids polarization. Finally, unlike traditional republicanism, which questions the legitimacy of arguments based on human rights to defend the right of wearing the headscarf, the commission takes pains to justify the ban of the veil by virtue of conditions established by the European Convention on Human Rights and Fundamental Freedoms.[71]

Other elements, which allow us to observe a distancing of the commission from traditional republicanism, seem to bring it nearer to some aspects of another movement which played a major role in diversity debates in France: the multiculturalist movement. This movement attempts to deal with cultural and religious diversity by way of the social inclusion of individuals and groups belonging to cultures which are distinct from the dominant culture, by showing support for cultural rights and public recognition of difference. Its advocates accept the legitimacy of cultural and religious demands as being part of the active exercise of citizenship, and approve the establishment of accommodations in order to treat individuals equally. They therefore extol multiculturalism as a viable integration model for contemporary France, being the complete opposite of traditional republicanism, which they often simply call 'French republicanism'[72] and condemn its abstract universalism.[73] They harshly criticize the common usage of *laïcité* in France, as it often becomes a defensive battle, 'frequently suspected of being thrust into the spotlight in an effort to belittle Islam, resist its presence or oppose it.[74] They also censure the lack of sociological foundations of this movement[75] and rely on the results of sociological research for their position on the veil issue, pointing out that the voices of the girls in question have never been heard. In fact, it is precisely because the voices of Muslim girls have never been considered in the debate on the wearing of the veil that Françoise Gaspard and Farhad Khosrokhavar (1995) attempted to understand this phenomenon. Rather than using abstract principles or suppositions, they proceeded on the basis of empirical data obtained from a field study conducted in the suburbs, especially in the neighbourhoods considered 'intense'. This study allowed them to draw conclusions opposite from those maintained by traditional republicans, showing that girls wearing the veil are not necessarily manipulated and that the situation often involves a choice allowing for their integration. Within this movement, some authors openly promote French-style multiculturalism,[76] which can recognize, over and above the diversity of society and of its groups, the need for individuals to establish their identities on the basis of multiple affiliations. This multiculturalism sees itself also as capable of: promoting the visibility of difference in public spaces (including media spaces); organizing encounters between groups in order to prevent withdrawal based on differences; and prompting a new dynamic for relations between the state and society which follow the path of positive discrimination. The last of these becomes necessary in order to correct existing inequalities; it is a discrimination that 'does not target groups', but rather promotes differentiated action by the state. It is a question of finding a middle ground, a 'moderate multiculturalism', that rejects both communitarian confinements and liberal individualism.[77]

I believe I have shown above how the underlying logic of the Stasi report intersects with traditional republicanism, particularly concerning the veil issue.

Now, I will argue here that it is moving toward multicultural goals, in that it rejects the idea that religious choices should be privatized 'and denied any social dimension or capacity of public expression'.[78] Free spiritual or religious expression in public spaces is presented as legitimate, even as essential to the democratic debate: 'representatives of different spiritual options are entitled to act as such in public debate, like any component of society'.[79] In this way, the report considers the French government's trend to set aside formal equality and to promote real equality, while in practice adopting reasonable accommodations, as 'progress'. The commission applauds such measures as: government authorities taking into account needs related to religious holidays like Eid al-Adha and Yom Kippur; municipal facilities being made available to community organizations; and an annual calendar of religious holidays being distributed to all government offices, the latter approving time off for these holidays. The commission also endorses the fact that government offices take into consideration religious dietary restrictions: more than ever before, school, hospital and prison cafeterias offer diverse menus. Finally, the report commends the fact that municipalities are making it easier to build places of worship and that ritual sacrifice is becoming more acceptable.[80]

In addition, the commission is not content to simply list these observations. It encourages the French government to 'continue to make improvements'.[81] It criticizes the problems that still exist in relation to taking vacation days during important Muslim and Jewish holidays, and protests the fact that school exams are planned during these holidays, 'depriving those who have permission to be absent from the possibility of participating in them'.[82] As a result, the commission considers that, without affecting the Catholic calendar, it would behove the Republic to recognize the most holy days of the two other main monotheistic religions that exist in France, as Buddhists plan their main annual holiday on a Sunday in May. So in schools, most of the students would not work on the days of Yom Kippur and Eid al-Adha. Students would have to offset these two extra holidays.[83] Incidentally, this is not, however, to 'question the historical place occupied by Christian culture and denominations in society',[84] but rather to ensure full freedom of worship of all religious choices, within the framework of French *laïcité*.

The report also condemns the fact that there are not enough Muslim chaplains in hospitals and prisons, and that there are none in schools or in the army. Pointing out that 'some measures prefer specific religions that do not have the same resources as others', the report proposes that a general Muslim chaplain be appointed 'under the same conditions as general chaplains of other religions',[85] and that chaplains be recruited for prisons and the army.[86] In particular, it laments the fact that in hospitals, the grooming of the deceased is not always ensured 'in respect of religious rules, even if these are compatible with public order requirements and service constraints', and that it is sometimes impossible to bury the dead 'in accordance with religious traditions and in compliance with the laws of the Republic'.[87] Thus, tackling the way *laïcité* is used by public authorities to refuse the orientation of graves in cemeteries, the commission proposes 'that the Ministry of the Interior call for respect of religious convictions, particularly when cemetery plots expire'. In conjunction with religious leaders, the reclaiming of plots must be done under conditions that respect denominational requirements, adapting the layout of ossuaries'.[88] Religious convictions must also be respected in relation to food, 'pork substitutes and Fish Friday must be offered in the context of mass catering (schools, penitentiaries, hospitals, businesses)', provided that the consideration of these religious requirements is compatible with the effective operation of the service, 'according to the principle that Quebecers call "reasonable accommodation"'.[89]

The report also speaks out in favour of state financing for community and religious buildings, which is unthinkable for traditional republicans. The condition established for such financing harmonizes perfectly with the goals of multicultural republicanism, which rightfully requests the organization of encounters between groups to prevent withdrawal based on differences.[90] In its call to clarify the criteria for allocating assistance for community buildings, the commission says 'yes to funding those who foster interaction, meetings, openness to the city; no to funding those organizations that refuse to dialogue with the rest of society.[91] Stating that 'social intermingling, learning to live together and respect for cultural and spiritual differences in a secular framework' cannot be assured by schools alone, the commission recommends the implementation of a civil service[92] and the prioritizing of common sports facilities that promote social intermingling.[93] Not only is this intercultural dimension found in moderate multiculturalism; it is the central focus of the Commission for Intercultural Dialogue.

The Commission for Intercultural Dialogue

La Commission du dialogue interculturel (The Commission for Intercultural Dialogue) [CID] was established by the Belgian federal government in February 2004, two months after the Stasi report was submitted in France. Among the reasons behind this initiative was a desire to counteract the rise of anti-semitism, racism and xenophobia in Belgium, but also to engage in discussions to halt a line of thinking that advocated for a ban on wearing the veil in schools, based on the Stasi report.[94] Thus, some authors presented the CID as 'a Belgian-style Stasi Commission', which nevertheless arrived at different conclusions and recommendations since it was guided by a separate philosophy.[95] In this section, I will present some features of the CID report that allow us to better grasp its underlying recognition theory. I will also discuss the concept expressed during the publication of the report that it included 'echoes of a particular communitarian philosophy imported from America'.[96] I will attempt to show how an attentive reading of the report leads us to look elsewhere for its possible philosophical influences.

The pluralist perspective of the CID

The basic policy option that underpins the report is the recognition of the cultural groups that make up Belgian society. This recognition relies to a certain extent on: the role that culture plays in individual personalities; the consequences of a lack of cultural recognition for members of culturally deprived populations; and the political and social implications of assimilationist ideology.

Interpreting this in the same way as E.B. Tylor, as a complex set of knowledge, representations, images and values that individuals acquire as members of a society, the CID emphasizes the essential role of culture in the making of who we are: culture produces the normative, intellectual and imaginary framework that enables individuals to think and act. Recognition of cultural diversity therefore involves an understanding of the extraordinarily complex character of culture, that only exists through unique cultures and is personified in specific individuals. Furthermore, it would be erroneous to try to delineate cultural identities, because they result from a mixture of factors that obviously includes the personal history of the individual. Thus, part of the social diagnosis proposed by CID is based on the effects of stigmatization suffered by foreign youths who live under the strain of two different cultures, that of their parents and that of Belgian society. This causes them to assert themselves in their difference, to the point that some take deviant paths such as delinquency or violence. The commission maintains that in order for dialogue to be possible, the country must distance itself from assimilationist logic and support these youths in the affirmation of their plural identity. It must also work more on the cultural policies implemented in Belgium, which have proven to be inadequate. These policies find themselves under pressure, on one hand, from intolerant discourses, and on the other hand, from those who have been influenced by assimilation ideology and fear that cultural recognition policies will lead to communitarian withdrawal. Believing that, on the contrary, identity withdrawal is a consequence of assimilationist logic, the report requests that public authorities recognize cultural minorities in their distinctiveness; it is by cultural recognition that cultural minorities will be able to integrate into the community of citizens and open up to each other. The challenge facing Belgium is therefore how to promote cultural pluralism and to join it to the political, philosophical and community pluralism already established. This would be done by transforming the cultural diversity resulting from immigration into active plurality, that is to say allowing communities of foreign origin to fully enjoy citizenship, being accepted by Belgians of European origin. It must be understood that cultural rights do not, however, mean confining members of cultural groups in their distinctiveness, but involve respecting the freedom of choosing one's own identity. This type of recognition helps to create the conditions necessary for dialogue and increased interaction.

It is important to notice that recognition of cultural rights and plural identities is not only for the purpose of avoiding the tainting of identity: it mainly targets social integration of all individuals living on Belgian soil, regardless of origin, on the basis of interculturality as a political model, which model can only be maintained by citizen participation. We see here the creation of a project that, by attacking issues related to the challenges of cultural and religious diversity in contemporary Belgium, is careful to emphasize the social, political and economic conditions which allow the project to be carried out. How, in fact, can we encourage intercultural encounters if members of cultural minorities do not interact as peers with the rest of society? And how is such interaction possible if we do not deal with inequalities of access to employment, housing and citizenship? Therefore, the major consistent policy of the CID involves taking into account the set of factors that hinder intercultural relations, but in a complex manner, namely, by avoiding culturalist reductionism. Basically, in addition to the cultural dimension, the underlying recognition theory of the report contains an economic and participative dimension that makes active citizenship the axis around which the proposed intercultural project is formulated.

Being conscious that the cultural element evokes specific reactions, the report emphasizes that it is inseparable from the economic element, if only because the populations belonging to minority cultural groups are often formed as a result of economic immigration.[97] The precarious socio-economic situation of populations deriving from immigration reinforces the feeling of cultural non-recognition experienced by youths who are concentrated in certain neighbourhoods, schools and professions: hence, their resentment toward Belgian society. This concentration is not the result of the community withdrawal so feared by the Stasi Commission and traditional republicans. It involves social and economic conditions, the latter being connected to the racist perception to which immigrants fall victim.

> Racism can be defined as an ideology that attributes superiority to one group in relation to another. In the past, this ideology basically emphasized biological differences. Today, it generally resists cultures. 'Modern' racists believe that cultures should not be mixed, and that they must do everything possible to protect their own culture. They do not openly claim that people belonging to another culture are 'inferior', but rather 'different'. The use of this idea, along with its hidden reasoning, activates the same mechanism as does the 'race theory', namely that there are barriers between 'us and them', which justifies inequalities of rights and opportunities.[98]

In this way, the envisaged intercultural model sees cultural racism as a major factor in hindering the access of culturally different populations to social participation: racism is a barrier to dialogue. In order to counteract racist discriminations, the CID proposes restrictive measures, striving for more systematic legal recourse, as well as preventive measures, whose goal is to develop a positive image of subjects who are discriminated against, combat stereotypes and establish intercultural encounters. For example, a transformation within companies is necessary to eliminate the refusal to hire immigrants, and differences in salaries and positions, in addition to restructuring the employee selection process, particularly where disqualification of migrated individuals is allowed. Furthermore, positive action measures must be implemented regarding employment, education, housing and political representation, leaving behind formal equality which is blind to differences, in order to ensure real equality. Possible forms of positive action policies include the establishment of incentives by way of tax benefits and financial advantages for institutions and businesses. In the housing sector, racism and discrimination can be resisted by aiming for social and cultural mixing by increasing the number of adapted social housing units, informing owners and tenants of their rights and duties, and trying to take into account housing requests of culturally underprivileged populations.

This political strategy of social decompartmentalization, of elimination of the barriers hindering intercultural dialogue, is found in the way that CID handles the issue of religion in public spaces. In fact, in the same way that the report promotes a plural social space, where diverse cultural communities interact, dialogue and converse, religious plurality must also be developed through dialogue. So, just as the struggle against cultural racism involves implementing measures that promote mixing and emphasize diversity, the struggle against religious fundamentalism also requires the introduction of tools that foster openness, dialogue and diversity, instead of measures of exclusion and control. Finally, while the stigmatization of difference has serious consequences for cultural identities, the CID also points out the 'symbolic and psychological' importance of religious holidays for individuals, as it can be 'painful for many individuals and families to not be able to participate in major symbolic holidays of their culture'. It therefore believes 'that we must recognize the basic right of each individual to be able to enjoy the holidays they consider most important', and it 'recommends that public authorities study the possibility of allowing a choice of vacation days'.[99]

The line of argument in the report advocating cultural and religious recognition is based as much on the consequences of assimilationist logic for individual identities as on its adverse effects for social integration, and as much on the need to get individuals from non-European cultures to dialogue as on the need to allow them to participate as equals in social life. Yet, it is equally based on respect for the founding principles of liberal modernity: recognizing cultural rights is recognizing the freedom to belong to a group or not, to publicly express one's convictions without fearing rejection or stigmatization; it is also recognizing that all individuals have the same value without distinction of race, origin or gender. In short, recognition of cultural minorities 'relies on and must rely on a universalist perspective of humanity and citizenship'.[100] In this way, the report's discourse on recognition of identity establishes its discourse on the intangibility of democratic principles, reminding us that in order to be acceptable, any practice must be fully compatible with the set of values around which Belgian democracy is formed. It is therefore this combination of recognition of identity and implementation of positive action policies promoting economic and social integration of cultural minorities, of the expression of respect for these minorities with the preeminence of Belgian democratic culture, which is at the core of the recommendations proposed by CID to the federal government.

In light of what has been explained so far, an interpretation stating that the discourse of the report 'on recognition is overblown . . . because it contains echoes of a particular communitarian philosophy imported from America' seems misguided. This interpretation is based in part on the following ideas: (a) 'demands are made from the perspective of complaints and in the name of "cultural communities"'; (b) 'individuals are immersed in communities', these being 'fixed, and discourse and practices cement them even more'; and (c) 'the concept of universal citizenship makes room for a joining of communities, who will possibly communicate with each other, but are primarily linked by the marketplace in which they all desire to participate'.[101]

First, it must be noted that while the report attaches great importance to culture in the construction of identity, it recognizes that it is based on 'other elements such as gender, social class, ideology, and era'.[102] The consideration given by the report to cultural demands is mainly due to the observation of inequalities arising from existing differences between the majority culture and minority cultures, particularly those of immigrants. So it is not in the name of 'cultural communities' that the report goes about defending cultural rights, but in the name of the principle of equality of access to citizenship. In other words, even though its condemnation of the consequences of stigmatization of cultural identities takes into account the psychological factor (tainted identities, etc.), it chiefly targets the possibility of enabling all individuals to participate socially and politically in the public space of Belgian society. This goal is only considered feasible if we provide 'several tools to members of certain minority groups, for the purpose of lifting them out of their disadvantaged position in relation to employment, housing, education and political representation'.[103] This being understood, it proves difficult to establish that the discourse of the report anchors individuals in their communities. This is not only because the discourse motivating the report is that of citizen equality, but because far from wanting to enclose members within cultural groups with their particularities, the CID reinforces the right of individuals to ensure their freedom to choose their affiliation. It is actually the language of rights to equality and freedom that drives the report. Contrary to the perspective that universal citizenship is displaced by a joining of communities, the underlying recognition theory of the report explicitly includes a universalist vision of humanity and citizenship.[104]

The goal of socially and politically integrating cultural minorities by way of cultural recognition is also common to several contemporary philosophical movements, and can hardly be interpreted as being an exclusively communitarian element. For example, civic republicanism emphasizes the participation of minorities in social and political life, just as CID does. Based on the principle that participation in some structures and contexts is more difficult for certain types of citizens than it is for others, civic republicanism makes recognition of uniqueness a condition for equality. Members of cultural and religious minorities, like other citizens, share a wide range of interdependent relationships supported by political institutions to which they belong, having been called upon to participate in political and social life as equal citizens. When necessary, cultural and religious minorities need access to particular resources, representations, special rights and exceptions which place them on the same level as other citizens.[105] It is the same for those who, like J. Tully, desire to establish 'a form of constitutional democracy that adopts flexible criteria of public recognition of citizens'.[106] In this environment, citizens must also be free to contest, modify and negotiate the norms governing their participation, according to their developing identities. Some movements within liberalism advocate for the recognition of identities, subscribing to the idea that 'there are compelling interests related to cultural membership and cultural identity, which are fully consistent with liberal principles of freedom and equality'. They justify the adoption of measures to promote and encourage cultural and material prosperity of cultural groups, in respect of their identity. Kymlicka labels this common position 'liberal culturalist', feeling that it has become the dominant stance among liberals who are interested in diversity issues.[107]

It could be objected that these movements may be inspired, at least partially, by the communitarian movement. This is certainly plausible, but then what would be the relevance of finding echoes of communitarian philosophy in the underlying recognition theory of the report, when many non-communitarian theories also promote cultural recognition as a legitimate measure for our democratic societies? In fact, a liberal reading of the report would seem more appropriate, since it endorses openly liberal values, as is noticeable in its defence of state neutrality regarding the idea of well-being,[108] in its request for diversity management based on individual rights and liberties[109] and in the emphasis it makes regarding the ability of individuals to choose their affiliation, to 'change cultural heritage, by choosing several, by seeing themselves only in a part of the whole, or by adopting a new one'.[110] To put it succinctly, isolating a single aspect of the report, namely its position statement in favour of respecting cultural groups and their recognition, runs the risk of reductionism. On the contrary, it seems that the most promising path to take in order to explore the possible philosophical influences of the report is to focus on the way that all of these aspects are handled and articulated. This approach allows us to see that the combination of identity recognition and the policy of economic inclusion aimed at citizen participation, that characterizes the report, does not find its strongest advocates among communitarians. In fact, it seems very close to the view upheld by Nancy Fraser, among others. Quite obviously, there are differences between the theoretical choice of the report and that of Fraser (beginning with the fact that the CID does not endorse socialism as a model of fair distribution), but their respective positions share the basic insight that justice requires identity recognition as much as economic distribution. So in her article 'Social Justice in the Age of Identity Politics: Redistribution, Recognition, and Participation', whose title well conveys the perspective of the report, Fraser writes:

> It is my general thesis that justice today requires both redistribution and recognition, as neither alone is sufficient . . . Theoretically, the task is to devise a 'bivalent' conception of justice that can accommodate both defensible claims for social equality and defensible claims for the recognition of difference. Practically, the task is to devise a programmatic political orientation that integrates the best of the politics of redistribution with the best of the politics of recognition.[111]

Fraser makes the concept of 'participation parity' the normative linchpin of her theoretical framework and states, as does the CID, that social provisions for each member of society can interact on an equal level with other measures that prove to be necessary. People must not only have access to material resources that ensure their economic independence (objective condition), but also receive equal respect and obtain equal opportunities in their search for social standing ('intersubjective condition'). This condition confronts the system in which certain categories of people, and the qualities commonly associated with them, are devalued:

> One should say, rather, that it is unjust that some individuals and groups are denied the status of full partners in social interaction simply as a consequence of institutionalized patterns of interpretation and evaluation in whose construction they have not equally participated and that disparage their distinctive characteristics or the distinctive characteristics assigned to them.[112]

Is it not rather echoes of this perspective that are found in the CID's report? In my opinion, the Commission for Intercultural Dialogue represents the most significant effort made so far by a national commission aiming to transform both objective and intersubjective conditions, with the goal of causing members of underprivileged populations to participate as equal citizens within societies having a strong cultural and religious diversity. By addressing the various obstacles that hinder full citizen participation of all citizens in Belgian society, the CID has provided a normative response that is both philosophically consistent and morally convincing.

Conclusion

The analysis of the reports studied shows, first, it is very difficult to reduce them to simple social diagnostics, as they contain a philosophical dimension which establishes their positions regarding the adaptation of political models of managing cultural diversity. I believe I have shown that because the Stasi report, among others, sacrifices freedom of conscience in the name of public order and protection of autonomy, it has been interpreted as advocating a conservative republicanism. However, another reading is possible, when it is examined in light of the responses of contemporary French republicanism to diversity. While it is true that the debate about wearing the veil essentially reflects the ideas of a particular militant republicanism, the report also shares some of the ambitions of French multicultural republicanism. In both reports certain aspects have been isolated and used as a viewpoint for their overall interpretation. Now, neither the veil issue in the French case nor the appreciation of cultural belonging in the Belgian case are sufficient for a sweeping classification of the reports.

I will conclude by pointing out that, in spite of the substantive differences in the reports studied, they do have remarkable similarities. Both promote a pluralist perspective, requesting institutional changes that take into account the needs of individuals to connect to the particularities of their culture, and propose measures to ensure access of all citizens to real equality, since formal equality is inadequate. They both support involvement in a social dynamic that encourages interaction, encounters and dialogue between cultures. Moreover, the two commissions formulate similar recommendations, such as: the consideration of religious holidays;[113] the teaching of certain heritage languages in schools;[114] the integration of the history of peoples, migrations and cultures in academic programmes;[115] the creation of a charter ('of *laïcité*' in France, 'of citizenship' in Belgium) which defines the rights and obligations of each person, but which has no legal force in either case. Finally, the two commissions express their approval of religious education in public schools,[116] believing that it fosters understanding of different cultures and religious traditions.

The Stasi report and the CID report alike suggest to their respective governments that they must move forward regarding the consideration of cultural and religious diversity. However, the French government has actually chosen to take a step backward by choosing to apply a liberticidal proposal, while the Belgian federal government has simply stood still and allowed the CID report to be forgotten.

Notes

1 This research was conducted as part of the international project *Cultural and Religious Diversity in Four National Contexts: Comparative Study of the Identity Dynamic and Regulation of Religion (Quebec, France, Belgium, Britain)* directed by Solange Lefebvre and financed by the Canadian Social Sciences and Humanities Research Council (SSHRC). The author would also like to thank SSHRC for its financial support by way of a Joseph-Armand Bombardier scholarship.

2 Stasi, Bernard, *Rapport de la Commission de réflexion sur l'application du principe de laïcité dans la République* (Paris: La documentation française, 2003).

3 I will limit my remarks to two of the three positions identified by Jeremy Jennings in his study 'Citizenship, Republicanism and Multiculturalism in Contemporary France', *British Journal of Political Science*, vol. 30, no. 4 (2000), 575–597. In that work, Jennings introduces a classification that distinguishes between traditional republicanism, modern republicanism and multicultural republicanism. For an analysis of the Stasi report that takes into account modern republicanism, see my article 'Multiculturalisme et laïcité en France. Les trois républicanismes du rapport Stasi', *Dialogue. Canadian Philosophical Review* 54, no. 4 (2015): 647–684.

4 Delruelle, Edouard, and Rik Torfs, *Rapport de la Commission du dialogue interculturel. Rapport final et livre des auditions* (Brussels: Ministère de l'Égalité des Chances, de l'Intégration sociale et de l'Interculturalité, 2005).

5 Foblets, Marie-Claire and Christine Kulakowski, *Les Assises de l'Interculturalité* (Brussels: Ministère de l'Emploi et de l'Égalité des Chances, 2010).

6 Leane, Geoffrey W.G., 'Rights of Ethnic Minorities in Liberal Democracies: Has France Gone Too Far in Banning Muslim Women from Wearing the Burka?', *Human Rights Quarterly* 33, no. 4 (2011): 1032–1061.

7 Freedman, Jane, 'Secularism as a Barrier to Integration? The French Dilemma', *International Migration* 42, no. 3 (2004): 5–27.

8 Thomas, Elaine R., 'Keeping Identity at a Distance: Explaining France's New Legal Restrictions on the Islamic Headscarf', *Ethnic and Racial Studies* 29, no. 2 (2006): 237–259.

9 According to Timo Behr, 'The report of the Stasi Commission and the law of 2004 seemingly brought to a conclusion any attempts to move towards a more multicultural model of society'. Behr, Timo, *France, Germany and Europe's Middle East Dilemma: The Impact of National Foreign Policy Traditions on Europe's Middle East Policy* (Baltimore, MD: The Johns Hopkins University Press, 2009), 174.

10 Akan, Murat, 'Laïcité and Multiculturalism: The Stasi Report in Context', *The British Journal of Sociology* 60, no. 2 (2009): 253.

11 Heine, Sophie, 'The Hijab Controversy and French Republicanism: Critical Analysis and Normative Propositions', *French Politics* 7, no. 2 (2009): 167–193.
12 Laborde, Cécile, 'Républicanisme critique vs républicanisme conservateur: repenser les "accommodements raisonnables"', *Critique internationale* 3, no. 44 (2009): 19–33.
13 Ibid., 27.
14 Ibid., 32.
15 Stasi, *Rapport*, 50.
16 Ibid., 45.
17 Ibid., 18.
18 Ibid., 45.
19 Ibid., 40.
20 Ibid., 15.
21 Ibid., 56.
22 Ibid., 14.
23 Ibid., 41.
24 Ibid., 44.
25 Ibid., 57.
26 Ibid., 58.
27 Ibid.
28 Ibid.
29 Jennings, 'Citizenship, Republicanism and Multiculturalism in Contemporary France'.
30 Coq, Guy, *Laïcité et république. Le lien nécessaire* (Paris: Éditions du Félin, 1995); Jelen, Christian, 'La régression multiculturaliste', *Le Débat* 5, no. 97 (1997): 137–143; Landfried, Julien, *Contre le communautarisme* (Paris: Armand Colin, 2007); Taguieff, Pierre-André, 'Multiculturalisme et communautarisme devant les principes républicains', *Le Figaro*, 17 July 2003.
31 Jelen, 'La régression multiculturaliste'; Landfried, *Contre le communautarisme*, 16.
32 Jennings, 'Citizenship', 589.
33 Jelen, 'La régression multiculturaliste'.
34 Landfried, *Contre le communautarisme*, 95.
35 Jelen, 'La régression multiculturaliste'.
36 Ibid., 143.
37 Ibid.
38 Taguieff, 'Multiculturalisme et communautarisme'; Landfried, *Contre le communautarisme*, 14.
39 Taguieff, 'Multiculturalisme et communautarisme'.
40 Finkielkraut, Alain, 'La sainte alliance des clergés', *Le Monde*, 25 October 1989, 2; Badinter et al., 'Profs, ne capitulons pas!' *Le Nouvel Observateur*, 8 November 1989; Coq, Guy, 'Espace laïque', *Le Monde*, 24 October 1989, 2; Coq, *Laïcité et république*; Landfried, *Contre le communautarisme*.
41 Coq, *Laïcité et république*, 133.
42 Coq, Guy, 'Un principe universel', *Hommes et migrations*, November–December, no. 1258 (2005): 11.
43 Coq, 'Espace laïque', 2.
44 According to Coq, between 1989 and 1994, the veil issue took on 'the dimension of a perfectly conducted Islamist operation', Coq, *Laïcité et république*, 277.
45 Ibid., 328.
46 Ibid., 282.
47 Ibid., 267.
48 Badinter et al., 'Profs, ne capitulons pas!'; Finkielkraut, 'La sainte alliance'.
49 Badinter et al., 'Profs, ne capitulons pas!'
50 Coq, *Laïcité et république*, 267.
51 Ibid., 200.
52 Coq, 'Espace laïque'.

53 Coq described this bulletin as 'very moderate' and 'well-reasoned', saying that its tone was 'fitting'. As for the idea that the 'main risk of proselytism' in schools is tied to the Islamic headscarf, Coq feels that 'this viewpoint is extremely sensible'. Coq, *Laïcité et république*, 282.
54 Ibid., 200.
55 Ibid., 328.
56 Ibid.
57 Badinter et al., 'Profs, ne capitulons pas!'
58 Ibid.
59 Coq, 'Espace laïque'.
60 Coq, *Laïcité et république*, 200.
61 Finkielkraut, Alain, 'La France disparaît au profit des tribus. Entretien avec Thomas Ferenczi', *Le Monde*, 13 July 1989, 13.
62 Stasi, *Rapport de la Commission*, 15.
63 Ibid., 17.
64 Ibid., 35.
65 Ibid., 36.
66 Ibid., 18.
67 Ibid., 46.
68 Ibid., 36.
69 Ibid.
70 Ibid., 18.
71 Ibid., 59.
72 Wieviorka, Michel, 'Le multiculturalisme est-il la réponse?', *Cahiers internationaux de sociologie* 105 (1998): 233–260.
73 Khosrokhavar, Farhad, 'L'universel abstrait, le politique et la construction de l'islamisme comme forme d'altérité', in *Une société fragmentée?*, ed. Michel Wieviorka (Paris: La Découverte, 1997), 113–151.
74 Wieviorka, Michel, 'Le mythe français', *La Presse*, 14 September 2013.
75 Wieviorka, Michel, 'Quand la gauche va-t-elle défendre le multiculturalisme?' *L'obs avec Rue89*, 2011.
76 Roman, Joël, 'Un multiculturalisme à la française', *Esprit*, no. 212 (1995): 145–160; Roman, Joël, *La démocratie des individus* (Paris: Calmann-Lévy, 1998).
77 Roman, Joël, 'Pour un multiculturalisme tempéré', *Hommes et migrations*, no. 1197 (1996): 20.
78 Ibid., 13.
79 Ibid.
80 Stasi, *Rapport*, 38.
81 Ibid., 39.
82 Ibid.
83 Ibid., 65.
84 Ibid., 62.
85 Ibid., 64.
86 Ibid., 67.
87 Ibid., 40.
88 Ibid., 65.
89 Ibid., 64. It is true that the report explicitly gives an incorrect definition of reasonable accommodations (ibid., 16), but it uses the concept correctly. In addition, the report emphasizes that accommodations are perfectly compatible with French republican tradition: 'The requirements of absolute neutrality are therefore tempered by "reasonable accommodations" allowing everyone to exercise their religious freedom' (ibid., 123).
90 Roman, 'Pour un multiculturalisme tempéré', 20.
91 Stasi, *Rapport*, 54.
92 Ibid., 52.
93 Ibid., 67.

94 For an analysis comparing the CID to the Round Tables on Interculturalism, which addresses in particular the context, assigned objectives, content, reception, follow-up and implementation of the recommendations established, see Leyva, Karel J., and Léopold Vanbellingen, 'Debating Intercultural Integration: From the Commission for Intercultural Dialogue to the Round Tables on Interculturalism', in *Public Commissions on Cultural and Religious Diversity: Analysis, Reception and Challenges*, ed. Solange Lefebvre and Patrice Brodeur (London: Routledge, 2017), 104–124. For an examination of its media reception by comparing it to the reception of other national commissions such as the Stasi Commission, the Round Tables on Interculturalism and the Parekh Report, see Lefebvre, Solange et al., 'The Commissions: Caught between Media Simplifications and Political Interests', in *Public Commissions on Cultural and Religious Diversity: Analysis, Reception and Challenges*, ed. Solange Lefebvre and Patrice Brodeur (London: Routledge, 2017), 127–150.

95 Dumont, Hugues and Xavier Delgrange, 'Le principe de pluralisme face à la question du voile islamique en Belgique', *Droit et société* 1, no. 68 (2008): 76–87.

96 Dassetto, Felice, 'Exorcisme interculturel', *La Libre Belgique*, 6 June 2005.

97 Delruelle and Torfs, *Rapport de la Commission du dialogue interculturel*, 43.

98 Ibid., 57.

99 Ibid., 77.

100 Ibid., 7.

101 Dassetto, 'Exorcisme Interculturel'.

102 Delruelle and Torfs, *Rapport de la Commission du dialogue interculturel*, 73.

103 Ibid., 61.

104 Ibid., 7.

105 Honohan, Iseult, 'Civic Republicanism and the Multicultural City', in *Migration and Cultural Inclusion in the European City*, ed. William. J.V. Neill and Hanns-Uve Schwedler (Basingstoke: Palgrave Macmillan, 2007), 63–73.

106 Tully, James, 'La conception républicaine de la citoyenneté dans les sociétés multiculturelles et multinationales', *Politique et sociétés* 20, no. 1 (2001): 123–146.

107 Kymlicka, Will, 'Liberal Theories of Multiculturalism', in *Rights, Culture and the Law*, ed. Lukas H. Meyer, Stanley L. Paulson and Thomas W. Pogge (Oxford: Oxford University Press, 2003), 51–75.

108 Delruelle and Torfs, *Rapport de la Commission du dialogue interculturel*, 54.

109 Ibid., 93.

110 Ibid., 73.

111 Fraser, Nancy, 'Social Justice in the Age of Identity Politics: Redistribution, Recognition, and Participation', *The Tanner Lectures on Human Values* (Stanford University, 1996), 5.

112 Ibid., 24.

113 Delruelle and Torfs, *Rapport de la Commission du dialogue interculturel*, 77; Stasi, *Rapport*, 69.

114 Stasi, *Rapport*, 67. Delruelle and Torfs, *Rapport de la Commission du dialogue interculturel*, 90.

115 Stasi, *Rapport*, 55; Delruelle and Torfs, *Rapport de la Commission du dialogue interculturel*, 89.

116 Stasi, *Rapport*, 63; Delruelle and Torfs, *Rapport de la Commission du dialogue interculturel*, 92.

References

Akan, Murat, 'Laïcité and Multiculturalism: The Stasi Report in Context', *The British Journal of Sociology* 60, no. 2 (2009): 237–254.

Badinter, Élisabeth, and Régis Debray, Alain Finkielkraut, Élisabeth de Fontenay, and Catherine Kintzler, 'Profs, ne capitulons pas !', *Le Nouvel Observateur*, 8 November 1989.

Behr, Timo, *France, Germany and Europe's Middle East Dilemma: The Impact of National Foreign Policy Traditions on Europe's Middle East Policy* (Baltimore, MD: The Johns Hopkins University Press, 2009).

Coq, Guy, 'Espace laïque', *Le Monde*, 24 October 1989, 2.

Coq, Guy, *Laïcité et république. Le lien nécessaire* (Paris: Éditions du Félin, 1995).

Coq, Guy, 'Un principe universel', *Hommes et migrations*, November–December, no. 1258 (2005): 6–11.

Dassetto, Felice, 'Exorcisme interculturel', *La Libre Belgique*, 6 June 2005.

Delruelle, Edouard, and Rik Torfs, *Rapport de la Commission du dialogue interculturel. Rapport final et livre des auditions* (Brussels: Ministère de l'Égalité des Chances, de l'Intégration sociale et de l'Interculturalité, 2005).

Dumont, Hugues, and Xavier Delgrange, 'Le principe de pluralisme face à la question du voile islamique en Belgique', *Droit et société* 1, no. 68 (2008): 75–108.

Finkielkraut, Alain, 'La France disparaît au profit des tribus. Entretien avec Thomas Ferenczi', *Le Monde*, 13 July 1989, 13.

Finkielkraut, Alain, 'La sainte alliance des clergés', *Le Monde*, 25 October 1989, 2.

Foblets, Marie-Claire, and Christine Kulakowski, *Les Assises de l'Interculturalité* (Brussels: Ministère de l'Emploi et de l'Égalité des Chances, 2010).

Fraser, Nancy, 'Social Justice in the Age of Identity Politics: Redistribution, Recognition, and Participation', *The Tanner Lectures on Human Values* (Stanford University, 1996).

Freedman, Jane, 'Secularism as a Barrier to Integration? The French Dilemma', *International Migration* 42, no. 3 (2004): 5–27.

Gaspard, Françoise, and Farhad Khosrokhavar. *Le foulard et la république* (Paris: La Découverte, 1995).

Heine, Sophie, 'The Hijab Controversy and French Republicanism: Critical Analysis and Normative Propositions', *French Politics* 7, no. 2 (2009): 167–193.

Honohan, Iseult, 'Civic Republicanism and the Multicultural City', in *Migration and Cultural Inclusion in the European City*, ed. William. J.V. Neill and Hanns-Uve Schwedler (Basingstoke: Palgrave Macmillan, 2007), 63–73.

Jelen, Christian, 'La régression multiculturaliste', *Le Débat* 5, no. 97 (1997): 137–143.

Jennings, Jeremy, 'Citizenship, Republicanism and Multiculturalism in Contemporary France', *British Journal of Political Science* 30, no. 4 (2000): 575–597.

Khosrokhavar, Farhad, 'L'universel abstrait, le politique et la construction de l'islamisme comme forme d'altérité', in *Une société fragmentée?* ed. Michel Wieviorka (Paris: La Découverte, 1997), 113–151.

Kymlicka, Will, 'Liberal Theories of Multiculturalism', in *Rights, Culture and the Law*, ed. Lukas H. Meyer, Stanley L. Paulson and Thomas W. Pogge (Oxford: Oxford University Press, 2003), 51–75.

Laborde, Cécile, 'Républicanisme critique vs républicanisme conservateur: repenser les "accommodements raisonnables"', *Critique internationale* 3, no. 44 (2009): 19–33.

Landfried, Julien, *Contre le communautarisme* (Paris: Armand Colin, 2007).

Leane, Geoffrey W.G., 'Rights of Ethnic Minorities in Liberal Democracies: Has France Gone Too Far in Banning Muslim Women from Wearing the Burka?' *Human Rights Quarterly* 33, no. 4 (2011): 1032–1061.

Lefebvre, Solange, Karel J. Leyva, Giomny H. Ruiz and Mathilde Vanasse-Pelletier, 'The Commissions: Caught between Media Simplifications and Political Interests', in *Public Commissions on Cultural and Religious Diversity: Analysis, Reception and Challenges*, ed. Solange Lefebvre and Patrice Brodeur (London: Routledge, 2017), 127–150.

Leyva, Karel J., 'Multiculturalisme et laïcité en France: les trois républicanismes du rapport Stasi', *Dialogue: Canadian Philosophical Review / Revue canadienne de philosophie* 54, no. 4 (2015): 647–684.

Leyva, Karel J., and Léopold Vanbellingen, 'Debating Intercultural Integration: From the Commission for Intercultural Dialogue to the Round Tables on Interculturalism', in *Public Commissions on Cultural and Religious Diversity: Analysis, Reception and Challenges*, ed. Solange Lefebvre and Patrice Brodeur (London: Routledge, 2017), 104–124.

Roman, Joël, 'Un multiculturalisme à la française', *Esprit*, no. 212 (1995): 145–160.

Roman, Joël, 'Pour un multiculturalisme tempéré', *Hommes et migrations*, no. 1197 (1996): 18–22.

Roman, Joël, *La démocratie des individus* (Paris: Calmann-Lévy, 1998).

Stasi, Bernard, *Rapport de la Commission de réflexion sur l'application du principe de laïcité dans la République* (Paris: La documentation française, 2003).

Taguieff, Pierre-André, 'Multiculturalisme et communautarisme devant les principes républicains', *Le Figaro*, 17 July 2003.

Thomas, Elaine R., 'Keeping Identity at a Distance: Explaining France's New Legal Restrictions on the Islamic Headscarf', *Ethnic and Racial Studies* 29, no. 2 (2006): 237–259.

Tully, James, 'La conception républicaine de la citoyenneté dans les sociétés multiculturelles et multinationales', *Politique et sociétés* 20, no. 1 (2001): 123–146.

Wieviorka, Michel, 'Le multiculturalisme est-il la réponse?' *Cahiers internationaux de sociologie* 105 (1998): 233–260.

Wieviorka, Michel, 'Quand la gauche va-t-elle défendre le multiculturalisme?' *L'obs avec Rue89*, 12 February 2011.

Wieviorka, Michel, 'Le mythe français', *La Presse*, 14 September 2013.

Part III

Reasonable accommodations revisited

Quebec's Bouchard-Taylor Commission

9 The Bouchard-Taylor Commission and beyond

Cultural and religious diversity in Quebec*

Jocelyn Maclure and Sébastien Lacroix

Introduction

Canadian academics in the humanities and social sciences are arguably primarily known for their research on themes such as identity, diversity, citizenship, nationality, aboriginality, language, and federalism. Since the country is founded on the interaction between people of different cultural origins and that it contains several types and layers of diversity, it was to be expected that Canadian scholars would zero in on the challenges raised by what Charles Taylor called "deep diversity".[1] As political philosophers, we usually tackle a conceptual or normative question, and offer arguments pro and con a given position. That is not the purpose of this essay. Since this volume seeks to compare and contrast four national contexts, we will rather offer an interpretation of what happened in Quebec since 2006 with regard to the accommodation of religious diversity.

One thing that was not expected, even by those of us who were working on identity politics in Canada, was that secularism and the status of religion in the public sphere would dislodge issues such as the Quebec–Canada constitutional question, language politics, multiculturalism, and aboriginal politics as the key identity-related issue, at least in the province of Quebec.[2] This change of focus was initiated by the "reasonable accommodations" controversy in Quebec. To everyone's surprise, the accommodation of religious diversity was perhaps the most heatedly debated public issue in Quebec in 2006–2007. Most observers agree that the Supreme Court of Canada's decision in the 2006 *Multani* case[3]—involving a Sikh schoolboy who was allowed by the Supreme Court to wear a kirpan at school, under some very strict conditions—kick-started the reasonable accommodation debate in Quebec. Many years later, more than 90 percent of the Quebec population still believed that the Supreme Court was wrong.[4] The decision was widely interpreted as another symptom of judicial activism, of the Supreme Court's propensity to overrule legitimate laws passed by Quebec's National Assembly, and of the imposition of Canadian-style multiculturalism (that is seen as encouraging ghettoization and cultural relativism) in Quebec, which in turn would conflict with Quebec's own integration policy, i.e. interculturalism.

A factor that amplified the sense that the accommodation of diversity was going too far and that it was threatening fundamental values is that the notion itself of "reasonable accommodation" was not well understood by the general public. Reasonable accommodation is a well-defined and circumscribed legal norm that stipulates that there is a duty on the part of an employer or of a public institution to offer an accommodation measure (such as an exemption) to someone who is adversely affected by a rule that seems *prima facie* neutral, but that in effect discriminates against the members of a minority group.[5] For instance, a rule that prohibits wearing headcovers at school does not target any specific group, but it infringes on the religious freedom of Muslims, Jews or Sikhs who need to wear a hijab, a kippah or a turban in order to conform to what they see as a religious duty or commitment. The general rule can be legitimate, but since a religious obligation has more moral and legal weight than a personal preference, accommodation measures are seen as required.[6]

That said, the *quid pro quo* about reasonable accommodation is not the end of the story. Our personal interpretation of why apparently minor events—individual accommodation requests—ended up at the center of a stormy public debate is that the opposition to the accommodation of religious diversity became the rallying point for groups that are usually in opposite camps.[7]

To cut a long story short, people of more liberal and civic inclinations became "strange bedfellows" with some conservatives and cultural nationalists, as they all had their own reasons to be disquieted by religious accommodations. On the more liberal side, many felt that the religious values or practices that required accommodation were incompatible with Quebecers' "shared values". One of the values that was constantly brought to the fore was gender equality. Either religion in general,[8] or Islam or orthodox Judaism in particular, were presented as inherently opposed to equality between men and women. Religious accommodations were also seen by many as inherently incompatible with *laïcité*. Accommodations that allow people to practice their religion in the public sphere (at work or in public institutions such as schools and hospitals) would put into question the public–private distinction and privatization of religious practice that many saw as a condition of peaceful coexistence in morally diverse societies. Fairness was also invoked: religious accommodations are seen by some as favoring religious conceptions of the good over secular conceptions. This would be a breach of the neutrality with regard to conceptions of the good that we expect from the liberal state.

On the more conservative side, Quebec "culture" and "identity" were seen as threatened. This discourse involves a revival of French Canadian nationalism. Cultural nationalists' main fear is that the French Canadian identity that fought to survive and that mutated into a Quebec civic identity and, in doing so, opened itself to cultural diversity, is not recognized as it should be—i.e. as the core culture to which other cultures should assimilate. They believe that French Canadians lost themselves as they opened up to diversity. Hence, both sets of allies felt a form of "identity-related anxiety".[9] These strange bedfellows united in a more formal and explicit way in the 2013–2014 "Charter of Values" debate, where people associated with cultural nationalism and republican secularism bracketed their disagreements and supported the Parti Québécois' legislative initiative.[10] This coalition, which can be found in several countries, generally works because the conservatives agree to incorporate Enlightenment values such as equality and political secularism as part of the "traditions" that need to be protected against (for the most part) Muslim immigrants. As Olivier Roy put it in the *New York Times*:

> Notably, these measures are being advocated in the name of protecting not Christianity but liberal secularism. The hijab is said to offend women's rights; circumcision, children's rights; ritual slaughter, animal rights. Oriana Fallaci and Ayaan Hirsi Ali, two radical spokespersons for the feminist resistance to Islam, became darlings of the conservative right in Italy (Fallaci) and the Netherlands and America (Hirsi Ali).[11]

The Bouchard-Taylor Commission

Feeling that the debate was getting out of control and that it was giving an edge to the right-wing Action Démocratique du Québec (ADQ), the (Liberal) Government of Quebec decided to create a high-profile public commission in February 2007, one month before calling a general election. The commission was co-chaired by the philosopher Charles Taylor and the historian Gérard Bouchard, two of the most respected Québécois scholars. Its mandate was to:

a) take stock of accommodation practices in Québec;
b) analyze the attendant issues bearing in mind the experience of other societies;
c) conduct an extensive consultation on this topic; and
d) formulate recommendations to the government to ensure that accommodation practices conform to Québec's values as a pluralistic, democratic, egalitarian society.[12]

The co-chairmen opted for a wide interpretation of their mandate. Rather than focusing strictly on the legal obligation to accommodate as it was defined by the Supreme Court of Canada, they gathered that citizens wanted to express themselves about large questions related to *laïcité*, the place of religion in the public sphere, immigration and integration, and the fate of Quebec identity in a globalized world.

The Bouchard-Taylor Commission was a "consultation commission". It organized hearings in all the administrative regions of the province from September to December 2007. The hearings were made of (1) pre-selected presentations of briefs and testimonies from 9:00 am to 5:00 pm and of (2) open public forums from 7:00 to 9:00 pm. Nine hundred briefs were submitted. In addition, four "national forums", 31 focus groups, and more than 80 meetings with NGOs and experts were held. The commission handed in its Final Report *Building the Future: A Time for Reconciliation* in June 2008.

The commissioners defended a version of an inclusive society resting on a pluralist yet robust "common public culture".[13] As pillars of this common public culture, "*laïcité ouverte*" (open secularism) and "interculturalism" needed, according to Bouchard and Taylor, to be better understood and formalized. The adoption of a formal policy statement on interculturalism and of a white paper on political secularism (*laïcité*) are the two most well-known recommendations. Most of the other recommendations had to do with (1) measures for fighting against discrimination against immigrants and members of minorities and (2) training and best-practices with regard to the management of accommodation claims for managers and frontline decision-makers in various milieux.

The commission's Final Report was a crucial contribution to public deliberation not only for its formal policy recommendations, but also for the hermeneutics of Quebec's political culture that it proposed. It came to the conclusion that there was a wide gap between the cultural anxiety felt by many Quebecers and the robustness of Quebec's public institutions and values. To put it briefly, the pillars of Quebec's public culture—the Charter of Rights and Freedoms, the Charter of the French Language (Bill 101), the (selective) immigration policy, the moderate form of *laïcité* or religious neutrality of the state—already contained the safeguards to ensure that basic principles such as gender equality, fairness, or the religious neutrality of the state were legally protected, and allowed the democratic majority to pursue reasonable collective projects such as promoting French as the common public language. In addition, the normative notion of "reasonableness" in "reasonable accommodation" was for the most part sufficiently clear and precise to screen out unfounded or excessive accommodation claims.

This gap is one of the reasons why the co-chairmen wrote that the "Reasonable Accommodation Crisis" was a "crisis of perception"[14] rather than a crisis at the level of practice or day-to-day encounters. The other reason, probably the main one, is that the commission's staff found that in 15 accommodation cases among the 21 that were discussed during the controversy, significant factual errors were found in the ways the media presented them. The *mea culpa* made by some journalists and columnists after the Bouchard-Taylor Commission was one of the highlights of the post-commission discussion.[15]

Although the report was criticized by the official opposition and several intellectuals, the Bouchard-Taylor Commission had a calming effect for about a year. The liberal government, after being re-elected to a minority government in March 2007, paid lip service to the process and to the main recommendations, remaining for the most part inactive. It immediately rejected the commission's proposal to remove the Catholic cross at the National Assembly, and then wished the entire issue away.[16] The report has become an inescapable reference in our collective debate on how to live together under conditions of diversity.

One of the worries of the commission's staff was about the open nature of its consultation process. The public forums held at night gave all citizens, not only those working with organized groups, an opportunity to express their point of view. Many observers thought that this was too risky. The jury is still out regarding the costs and benefits of including easily accessible public nightly forums.

That said, it should be kept in mind that one of the complaints expressed by many citizens was that decisions about identity-related issues had until then been made "behind closed doors" by elites sold to multiculturalism—a discourse that has now become commonplace given the rise of populism in Western democracies.

One the one hand, it cannot be denied that the very open and public character of the Bouchard-Taylor consultation process and of the ensuing discussion provided a springboard for racist, xenophobic, anti-Semitic (mainly directed at Hassidic Jews) and Islamophobic opinions. A study sponsored by the commission established that the vast majority of the interventions were acceptable, but racist or xenophobic views were also aired.[17] In addition, it is also true that Muslim citizens and Muslim women in particular were frequently targeted. This was even worse during the Charter of Values debate,[18] where Islamophobic acts were witnessed. For example, mosques were vandalized in Sherbrooke and Quebec City, and a pig's head was left at night in front of a mosque in Saguenay.[19]

On the other hand, the Bouchard-Taylor process gave a platform for voices that were heard for the very first time in the wider public, such as women who presented themselves as both feminists and Muslim, fighting both for the rights of Muslims and for the rights of Muslim women. NGOs founded by Muslim women became more visible. New voices among Quebec Muslims were heard also during the Charter of Values debate, and an NGO called The Arab and Muslim Association for Secularism was formed. It has been very demanding for Quebec Muslims in the past decade, but they now have a firm foothold in our public culture. Their NGOs are active, and they are present in intercultural organizations such as Quebec Inclusif or the Women's Federation of Quebec. The harm of Islamophobic speech should not be downplayed,[20] but new solidarities took form.

The Charter of Values debate

In the wake of the longest student strike in the province's history, the Parti Québécois (PQ) formed a minority government at the 2012 general election. The lesson that the PQ learned from the Bouchard-Taylor episode was that it should never allow itself to be pushed aside as the champion and defender of Quebec identity ever again. Indeed, after the 2007 election, the PQ was relegated to third-party status when the right-wing ADQ became the official opposition.[21] The PQ then toyed with different, more radical policies on language and citizenship. Such proposals did not withstand public scrutiny. Hardening Quebec's model of secularism turned out to be a more promising proposition, electorally speaking.

Having promised to draft a "Charte de la laïcité" (Charter of Secularism) and to fill the alleged normative deficit with regard to the regulation of religious expression in the public sphere, the PQ submitted Draft Bill 60 in November 2013. The Bill, often referred to as the "Charte des valeurs québécoises" (Charter of Quebec Values), was officially called "Charter Affirming the Values of State Secularism and Religious Neutrality and of Equality between Women and Men, and Providing a Framework for Accommodation Requests". The intensity of the debate within civil society reached pre-Bouchard-Taylor Commission levels. Although the PQ claimed that Bill 60 would have put a new regulatory framework in place, its main elements of novelty would have been to:

- be the first piece of legislation officially affirming the secular character of the Quebec state and its neutrality with regard to religion. It would have enshrined these principles in the Quebec Charter of Human Rights;
- ban full-face veils for civil servants and citizens wanting to obtain public services;
- ban all so-called "conspicuous" religious signs for all the employees of public and para-public organizations, including civil servants, health care professionals, teachers, professors, public daycare workers, judges, and so on.

There was in fact widespread agreement on many aspects of the proposed Charter. Most of the clauses of the Bill would not have changed much legally speaking. The frenzy was about one article, i.e. the ban of so-called "conspicuous" or "ostentatious" religious signs for all public sector employees. Supporters of the ban agreed with the distinction made by Bouchard and Taylor between public institutions and individuals. State secularism implies that public institutions ought to be religiously neutral and that citizens have the right to freely exercise their religion in their private and associative life.[22] However, for the Charter of Values defenders, public employees represent the state or public institutions and must accordingly embody the religious neutrality of the state, which entails that they ought to refrain from expressing their religious belonging when they are at work. They have a "duty of circumspection" with regard to their faith, just as they have one with regard to their political opinions. Bouchard and Taylor acknowledged in the report that the status of public employees was a difficult one. They reached a compromise: there should be no general ban for all public employees, but a limited one for those who embody to the highest degree the authority and coercive power of the state: the President of the National Assembly, judges, crown prosecutors, police officers, and prison guards.[23] A striking fact about the Charter of Values debate was that the actual disagreement was very specific: should public organization employees be allowed to wear easily visible religious signs? This specific question most probably became a proxy for a more general debate on the management of cultural diversity and on the status of Islam in Quebec society.

At the level of political strategy, the PQ arguably decided to use secularism and reasonable accommodation as a wedge issue: it defended a position that was more radical and controversial than it needed to be order to polarize the electorate and force everyone, including the other parties, to take a stand. The goal was not to build a consensus and to defend a principled and evidence-based position, but to find the issue that would radicalize the discussion, and hope that there would be more people on its side than on the other side at election day. It almost worked. Although the PQ lost the following election in April 2014, one can hardly blame the defeat on the Charter of Values, which is even today probably supported by a slim majority of the population.[24]

The PQ channeled the cultural anxiety described above and, more specifically, the contemporary moral panic about Islam for gathering the support for a majority government. Building on both the former liberal government quietism and the undeniable Islamophobia afflicting a large segment of the Quebec public,[25] it gave the impression that a government was finally prepared to take action. Populist statements about "putting one's foot down" or "taking our responsibilities" freed the voice of those who used to think until then that it was better to keep their opinions to themselves. It lowered the social cost of expressing anti-minorities attitudes. During the Charter of Values debate, multiple verbal aggressions against members of religious or ethnic minority groups were registered. The PQ leaders tried to play a balancing act by surfing on these sentiments without explicitly condoning them.

The current liberal government

In April 2014, the PQ lost the general election. The party was not able to impose the Charter of Values as the ballot question. The Quebec Liberal Party, led by Philippe Couillard, formed a majority government with 41 percent of the votes, whereas the PQ received only 25 percent of the ballots cast. Soon after being elected, the liberal government unveiled two draft Bills:

- Bill 59—*An Act to amend various legislative provisions to better protect persons*;
- Bill 62—*An Act to foster adherence to state religious neutrality and, in particular, to provide a framework for religious accommodation requests in certain bodies*.

The original version of Bill 59 was named *An Act to enact the Act to prevent and combat hate speech and speech inciting violence and to amend various legislative provisions to better protect individuals*. It provided for the prohibition of

> hate speech and speech inciting violence that are engaged in or disseminated publicly and that target a group of people sharing a common characteristic identified as prohibited grounds for discrimination under section 10 of the Charter of human rights and freedoms.[26]

It has, however, been modified by the government under public pressure. In dropping the provisions regarding hate speech, the government altered the objective of the Bill, which is now only to amend "a number of rules set out in the Civil Code of Québec regarding the solemnization of marriages and civil unions".[27] The original version of the Bill provided more means to fight hate speech directed at any group or category of the population, including minorities, but there were specific dispositions seeking to prevent radicalization among the youth. This was indirectly targeting young Muslims and their educational settings. The context was that about 30 young Muslims had left, or were thought to have left, for Syria or Iraq to join ISIS. The Bill was introduced in the wake of two attacks on Canadian soil (St-Jean-sur-Richelieu, QC, October 20, 2014; Ottawa, ON, October 22, 2014).

Bill 62 is a rerun of Bill 94.[28] It formally asserts the existing legal framework for managing accommodation claims, and gives legislative expression to the religious neutrality of the state (which had already been deducted from legally protected human rights and affirmed by Canadian courts). Its main element of novelty is that it would ban full-face veils for public servants and require that citizens who want to obtain a public service do so with their face uncovered. Some proponents of France-inspired secularism have criticized the Bill for its timidity. Alternatively, one can argue that it is a step in the right direction, as it would be the first piece of legislation which would officially affirm the religious neutrality of the Quebec state.

The Department of Immigration also released a consultation document on a renewed policy statement on integration, in which a civic and pluralist understanding of interculturalism is fleshed out.[29] There are (at least) two competing notions of interculturalism at play. The first one can be found in the Bouchard-Taylor Report and has been expounded by Bouchard since;[30] the other one may be found in the policy consultation document released by the Department of Immigration.[31]

(i) To simplify, for the first understanding, interculturalism is about interaction and mutual recognition between the cultural majority (Québécois of French Canada ancestry) and cultural minorities, all understood as communities. Since identities are plural, fluid, and internally contested, we doubt that this picture is accurate. Quebecers of French-Canadian origins do not form a monolithic whole with a unified collective consciousness and political will, and neither do the minority cultures. For instance, during the Charter of Values debate, the hard-line secularist Djemila Benhabib, who immigrated from Algeria, fought hand-in-hand with French-Canadian nationalists. On the other hand, some of the founders of the Parti Québécois who adhere to a civic and inclusive form of nationalism coalesced with Muslim citizens against their former party. Our identities, including our moral identities, are too complex for the communitarian view of interculturalism.

(ii) Interculturalism can be civic in Tariq Modood's sense[32] if it seeks to maximize the opportunities for contact, dialogue and cooperation between citizens of all origins, without assigning them to discrete cultural communities. The current liberal government officially champions such a civic understanding of interculturalism, but has yet to put forward an action plan to better realize it in practice.

The end of the "Bouchard-Taylor compromise"

The Charter of Values debate was highly divisive. The political parties have since then rallied behind the "Bouchard-Taylor Compromise" or the "Bouchard-Taylor

Compromise +". The "Bouchard-Taylor Compromise +" adds teachers and sometimes daycare workers to the list of positions included in the ban on visible religious signs. As mentioned above, the Bouchard-Taylor Commission proposed a ban for "employees who occupy positions that embody at the highest level the necessary neutrality of the State[, i.e.] the president and vice-president of the National Assembly, judges and Crown prosecutors, police officers and prison guards".[33]

The Bouchard-Taylor Report states that there are two very different lines of reasoning that one might follow to conclude that the "narrow" ban is appropriate. On one hand, one might believe this proposal "has a permanent character that goes beyond the current context, inasmuch as it embodies the principle of the separation of the State and the churches".[34] The position is defended on the basis of a semantic thesis: the correct interpretation of the principles of state secularism leads to the limited ban. On the other hand, one might accept the limited ban on more pragmatic and contextual grounds. On this view, the compromise is just that: a compromise between conflicting principled positions. The co-chairmen concluded that they did not "have to decide this debate since the two lines of reasoning lead to the same conclusion".[35]

In a 2011 book co-written with Jocelyn Maclure,[36] Charles Taylor abandoned the compromise he had accepted previously for pragmatic reasons and made no reference to the limited ban. Maclure fleshed out in a more complete manner the argument against the narrow ban in a recent article.[37] Legal scholar Pierre Bosset recently fortified their analysis by meticulously demonstrating that the justifications offered for banning the wearing of visible religious signs for judges, police officers, teachers, and elected officials would probably not withstand judicial review.[38] Since the Supreme Court of Canada generally dislikes blanket interdictions that disregard the differences between the parties affected, a precise contextual analysis of each role might be necessary to establish whether a ban on religious signs would be legally permissible in a particular case.

The so-called "Bouchard-Taylor compromise" was shattered recently when Taylor published an op-ed stating that the context had evolved since 2008 and that he no longer supports the limited ban.[39] After the tragic shooting in a Quebec mosque, Taylor believes that it is now time to show solidarity toward Quebec Muslims and focus on giving opportunities to Muslims and non-Muslims to grow closer to each other.[40] Gérard Bouchard held fast and reasserted his belief that the limited ban is both a logical implication of state secularism and a way to move toward greater social harmony.[41] After some hesitation, the current liberal government repeated that it has no intention to ban religious clothing for any public organization employees. As things stand, it is almost certain that secularism and religious accommodations will be at the forefront of the next 2018 provincial elections.

Notes

* Jocelyn Maclure and Sébastien Lacroix, Faculty of Philosophy, Université Laval.

1 C. Taylor, *Reconciling the Solitudes*, Montreal and Kingston, McGill-Queen's University Press, 1993.

2 A few cases outside Quebec also drew some attention: the so-called "Sharia tribunals" in family law in Ontario, polygamy in British Columbia, the launch of a Law School at the Trinity Western University, an Evangelical Christian institution in British Columbia, etc. See L.G. Beaman, "Opposing Polygamy: A Matter of Equality or Patriarchy?" in M.-P. Robert, D. Koussens, and S. Bernatchez (eds), *Of Crime and Religion: Polygamy in Canadian Law*, Sherbrooke, Éditions Revue de droit de l'Université de Sherbrooke, 2014, 131–157.

3 Multani *v.* Commission scolaire Marguerite-Bourgeoys, [2006] 1 S.C.R. 256, 2006 SCC 6.

4 See S. Fine, "Sikh Student Who Won Kirpan Case Now Considers Leaving Quebec", *The Globe and Mail*, 22 October 2013, at URL: www.theglobeandmail.com/news/national/sikh-student-who-won-kirpan-case-now-considers-leaving-quebec/article15014254/ (Consulted on March 15, 2017). According to this article, "[n]inety-four per cent of French-speaking Quebeckers and 79 per cent of non-French speaking Quebeckers were opposed [to the Supreme Court's decision]".

5 See Ontario Human Rights Commission *v.* Simpsons-Sears, [1985] 2 S.C.R. 536.

6 For the normative case in favor of a legal obligation to accommodate, see J. Maclure and C. Taylor, *Secularism and Freedom of Conscience*, Cambridge (MA), Harvard University Press, 2011.

7 J. Maclure, "Quebec's Culture War: Two Conceptions of Quebec Identity" in S. Gervais, C. Kirkey, and J. Rudy (eds), *Quebec Questions: Quebec Studies for the Twenty-First Century*, Oxford and Toronto, Oxford University Press, 2010.

8 The image of a heavy-handed Catholic Church, particularly oppressive towards women, is strong in Quebec's social imaginary.

9 G. Bouchard and C. Taylor, *Building the Future: A Time for Reconciliation*, Final Report of Consultation Commission on the Practice of Accommodations Related to Cultural Differences, Quebec, 2008, 185–186.

10 See *infra*, "The Charter of Values debate".

11 O. Roy, "The Closing of the Right's Mind", *New York Times*, June 5, 2014, at URL: www.nytimes.com/2014/06/05/opinion/the-closing-of-the-rights-mind.html?_r=2 (Consulted on March 15, 2017).

12 Bouchard and Taylor, *Building the Future*, 17.

13 Bouchard and Taylor, *Building the Future*, 105–107.

14 Bouchard and Taylor, *Building the Future*, 18.

15 See M. Potvin, "The Reasonable Accommodations Crisis in Quebec: Racializing Rhetorical Devices in Media and Social Discourse", *International Journal of Canadian Studies/Revue internationale d'études canadiennes*, vol. 50, 2014, 137–164, especially 138–139.

16 The only legislative initiative of the liberal government was Draft Bill 94, which was broadly seen as too timid. The bill would have prohibited public officials from wearing the burqa and the niqab and would have required women who wear such kind of veils to lift it while transacting with a civil servant. Bill 94 died on the order paper.

17 R. Antonius, *Méthodologie des relevés liés aux interventions lors des forums de la Commission de consultation Bouchard-Taylor*, Document No. 18 produced by the Commission, Montreal, December 2007.

18 See *infra*, "The Charter of Values debate".

19 This paper was written prior to the terror attacks on a Quebec City mosque in January 2017. Six people died that Sunday night, and many more were injured. An impressive show of compassion followed in Quebec City and Montreal, as thousands of people gathered in the streets to commemorate the victims and reject Islamophobia. For those who died, however, it was too little, too late.

20 See, more generally, Rae Langton, "The Authority of Hate Speech", forthcoming; and Jeremy Waldron, *The Harm in Hate Speech*, Cambridge (MA), Harvard University Press, 2014.

21 In early 2012, the ADQ merged with another right-wing party, the Coalition pour l'avenir du Québec (CAQ, or Coalition for the Future of Quebec).

22 For an opposite understanding of state secularism, see T. Modood, "Moderate Secularism and Multiculturalism", *Politics*, vol. 29(1), 2009, 71–76.

23 The limited ban was dropped in Maclure and Taylor, *Secularism and Freedom of Conscience*. See also J. Maclure, "Freedom of Conscience and the Charter of Quebec Values", *Policy Options*, vol. 35(1), 2014, 59, at URL: http://policyoptions.irpp.org/issues/technology/maclure/ (Consulted on March 15, 2017).

24 See S. Boivin, "Un sondage CROP confirme que les Québécois comptent sur une charte pour les protéger", *Le Soleil*, February 21, 2015, at URL: www.lapresse.ca/le-soleil/actualites/politique/201502/20/01-4846120-un-sondage-crop-confirme-que-les-quebecois-comptent-sur-une-charte-pour-les-proteger.php (Consulted on March 15, 2017). According to this article, 51 percent of people surveyed are favorable to a Charter reaffirming Québécois shared values.

25 In a March 2014 poll, 57 percent of the sample declared having a "bad" or "very bad" opinion of Islam. See Louise Leduc, "Le malaise musulman", *La Presse* +, section ACTUALITÉS, screen 2, November 16, 2014, at URL: http://plus.lapresse.ca/screens/4d02-dc68-53237404-ad87-7128ac1c6068%7C_0.html (Consulted on March 15, 2017). These polling results are sadly not unique to Quebec, but are rather consistent throughout Canada. See Maxime Dagenais, "Québec Charter Values", *The Canadian Encyclopedia*, April 3, 2015, at URL: www.thecanadianencyclopedia.ca/en/article/the-charter-of-quebec-values/ (Consulted on March 15, 2017). As much can be said about Canadian federal elections, where religious symbols and immigration were also recently used as wedge issues. It was the case in October 2015, when the then-governing Conservative Party of Canada introduced into the campaign the question of immigrant Muslim women wearing the niqab when pledging allegiance to their new country—all the while knowing this concerns no more than a handful of women each year. The Conservative Party still lost the elections to Justin Trudeau's Liberal Party, but the niqab played an important role during the campaign. The New Democratic Party suffered from its clear and strong refusal to ban the niqab during the pledge of allegiance.

26 See the original version of Bill 59, Explanatory Notes, at URL: www.assnat.qc.ca/en/travaux-parlementaires/projets-loi/projet-loi-59-41-1.html (Consulted on March 15, 2017).

27 See all versions of Bill 59, Explanatory Notes, ibid.

28 See *supra*, note 16.

29 Department of Immigration, Diversity, and Inclusion, *Together, We Are Quebec: Quebec Policy on Immigration, Participation, and Inclusion*, Quebec, Quebec Government, 2015.

30 Gérard Bouchard, *L'interculturalisme. Un point de vue québécois*, Montreal, Boréal, 2012.

31 See *supra*, note 29.

32 See *supra*, note 22.

33 Bouchard and Taylor, *Building the Future*, 151.

34 Bouchard and Taylor, *Building the Future*, 151.

35 "Such is our conclusion. We admit that we can achieve this end by following different lines of reasoning. For example, we can deem this proposal to be the most appropriate in the context of contemporary Québec society, although it is understood that this context can change over time. Or, we can also maintain that the proposal has a more permanent character that goes beyond the current context inasmuch as it embodies the

principle of the separation of the State and the churches. We do not have to decide this debate since the two lines of reasoning lead to the same conclusion". Bouchard and Taylor, *Building the Future*, 151.

36 Maclure and Taylor, *Secularism and Freedom of Conscience*.

37 J. Maclure, "The Meaning and Entailment of the Religious Neutrality of the State: The Case of Public Employees" in B.L. Berger and D. Moon (eds), *Religion and the Exercise of Public Authority*, Oxford and Portland (OR), Hart Publishing, 2016.

38 P. Bosset, "Réflexions d'un juriste sur l'idée d'interdire le port de signes religieux aux agents de l'État",*Vivre ensemble*, vol. 20(70), 2013, 12–20.

39 C. Taylor, "Le temps de la réconciliation", *La Presse* +, section DÉBATS, screen 4, February 14, 2017, at URL: http://plus.lapresse.ca/screens/36c5c72e-28b9-49df-ba29-514fc56d647a%7CpUtyV30bPPsb.html (Consulted on March 15, 2017).

40 J. Maclure and C. Taylor, "Après l'attentat de Québec, désarmorçons la haine", *Le Monde*, February 8, 2017, 20.

41 G. Bouchard, "Laïcité: On y était presque, dit Bouchard", *Le Devoir*, February 17, 2017, A1.

10 Cohesive fear

A comment on Maclure and Lacroix

Jean-François Gaudreault-DesBiens[*]

In Chapter 9, Jocelyn Maclure and Sébastien Lacroix sum up very well the events that led to the creation of the Bouchard-Taylor Commission, as well as the ones that followed after the submission of the commission's report. They show that debates over secularism or religion in Quebec, although not fundamentally different from those found elsewhere in the so-called Western world, reflect the influence of several variables that make this Canadian province a most interesting case study.

In many countries, "new" minorities are seen as threatening the fabric or values of host societies. However, this perception risks being particularly acute in "small nations", a term that designates societies which represent themselves as nations, on the one hand, and which, irrespective of their geographical size, experience some kind of collective precariousness, on the other.[1] This precariousness may sometimes, but not necessarily, be attributed to obvious factors such as the hegemonic appetites of bigger neighboring countries, but the roots of what is essentially a *feeling* of precariousness can also be traced back to factors that are not so easily identifiable or, to put it otherwise, that are less tangible. It is probably fair to surmise that considering the prevailing dynamics in the relations among Western democratic countries, which have been relatively stable in the past decades, these more elusive factors play a much bigger role than obvious ones in creating this feeling of precariousness. Thus, the term "small nations" essentially designates nations that share a sentiment of cultural anxiety or that repeatedly find themselves confronted with the same existential problems. In other words, small nations are those that cannot avoid asking themselves the three famous questions posed by Paul Gauguin in his famous painting: "Where Do We Come From? Who Are We? Where Are We Going?" Perhaps ironically, one could observe that some large nations are increasingly behaving as small ones.

Yet, small nations need not be nation states. The notion refers as much to national minorities or sub-state cultural groups that can reasonably be characterized as "sociological nations", as it does to nation states strictly speaking. This remark is especially relevant to an examination of the Canadian situation.

As a result, a nation that perceives itself as a small one in the sense described above will experience a collective feeling of insecurity that will very likely affect its relations not only with neighboring nations but also with the "rest of the world", and, most importantly, *within* its own world. More specifically, this perception will lead to the creation of obstacles in the establishment and continuation of these relations.

Perceived otherness risks being grasped through a reductionist, if not entirely negative lens. In that context, the Other, whoever he or she is, will often be seen as a threat to the group's integrity, survival or flourishing. A group that feels threatened is, however, more likely to adopt rigid positions, to refuse compromise, and to ignore the Others' openings to dialogue. At the same time – and paradoxically – it will depend heavily on these Others' presence for its self-definition. Therefore, the obstacles to which I alluded before serve essentially a protective function. Consciously or not, they are "created" to shield from external pressures the idea that a given group has of its identity. This explains why these obstacles complicate any form of complex knowledge and understanding of the Other's "real" condition, at least one that this Other may relate to. Thus, in a way, fear fosters, or reinforces, social cohesion – hence the title of this comment: "cohesive fear" – a phenomenon that can easily be instrumentalized by populist movements of all stripes.

Quebec – often designated as a "French island in North America's English-speaking ocean" – is, and behaves like a small nation in that sense. The methodological and monological nationalism that permeates political discourse in the province only increases the impact of the pre-existing feeling of precariousness, probably more so than in nations that cannot be characterized as small ones, since this monological nationalism induces a form of Jacobinistic identity reductionism. In that context, it is sometimes difficult to subjectively claim a dual, and fluid, Quebec and Canadian identity. Imagine when one dares to claim a Quebec *and* Muslim identity at the same time.

To a large extent, the identity anxiety that the presence of some visible Others triggers, has led, in Quebec, to a return of a form of ethnic nationalism. This nationalism has the dual effect of narrowing down the "collective", while inventing fantasized legal rights – those of the majority, which are political or moral at best but certainly not legal strictly speaking. In that narrative, the "rights" of the majority are constantly trampled by Others depicted as Trojan horses undermining the nation's core values, while minority rights – equality and freedom of religion – are understood as protecting personal preferences, and characterized, and trivialized, as privileges rather than as rights proper. Debates over reasonable accommodation in Quebec provide a most telling illustration of this dynamic. Facts that were *prima facie* constitutive of valid rights claims, such as facts giving rise to a freedom of religion claim (understood as requiring a subjective belief in one's obligation to do or not to do something), were indeed reduced to "personal preferences", as Maclure and Lacroix observe in their essay. All this took place, it bears noting, even before security issues really came to the fore.

As was recently noted by Susanna Mancini,[2] we are witnessing in the political discourse of many societies a vigorous return of Carl Schmitt's dichotomy between friends and enemies as defining and determining politics. This phenomenon sheds light on what appears to be an erosion of mutual trust in such societies. What is particularly interesting, as Maclure and Lacroix highlight, is that an essentially right-wing reaction has appropriated traditional leftist ideological bedrocks such as gender equality and secularism. In Quebec, very stringent restrictions of religious expression have indeed been supported both by old-school ethno-nationalists romanticizing the province's Roman Catholic past and by progressive citizens valuing the separation of church and state.[3] This convergence of strange bedfellows is not unlike the implicit alliance that was forged between radical feminists and religious activists in the anti-pornography campaigns of the 1980s and 1990s. As was then the case, a "false consciousness" argument was invoked against some women, in our case feminist Muslim women, particularly those claiming both their feminism *and* the right to wear a headscarf. Be that as it may, gender equality is perhaps the most easily instrumentalizable rationale invoked in support of stringent restrictions on religious rights, and of the corollary establishment of a formal hierarchy of rights, as intellectually and legally problematic as this project may be.[4]

In this "Schmittian" process, interculturalism, long heralded as Quebec's response to Canadian multiculturalism, ends up being construed as merely encouraging a dialogue between essentialized communities, and as presupposing the systematic predominance of the interests of Quebec's "old-stock" majority, i.e. Quebecers of French ancestry ("Québécois de souche"). This is more or less what Maclure and Lacroix refer to as interculturalism's "first understanding". As an example, some critiques of the much-discussed Ethics and Religion course,[5] which seeks to develop a broad, sociocultural, religious literacy among Quebec primary and secondary school pupils, reject the course for the reason that it allegedly undermines Quebec's Roman Catholic and Western heritage.

An interesting paradox must be noted here. Quebec's Charter of Rights and Freedoms[6] ("Quebec Charter") uses a much more communitarian language than the Canadian Charter of Rights and Freedoms[7] ("Canadian Charter") when addressing questions pertaining to minority cultural identity. To wit, s. 43 of the Quebec Charter states that "persons belonging to ethnic minorities have a right to maintain and develop their own cultural interests with the other members of their group", while s. 27 of the Canadian Charter posits that it "shall be interpreted in a manner consistent with the preservation and enhancement of the multicultural heritage of Canadians". Yet, in spite of s. 43 of the Quebec Charter, the reasonable accommodation of minorities is now deemed unreasonable by many when it applies to religious minorities. Moreover, in what could be characterized as a "Stasi move",[8] reasonable accommodation's meaning is often distorted, being perceived as imposing upon minorities the duty to adapt, at all costs, to the majority's needs and interests. Yet, empirically, a generational divide seems to appear as a new, to some extent "post-nationalist", generation emerges. The Parti Québécois, the oldest and most important secessionist party, which, when in government, tabled the Charter of Values project that Jocelyn Maclure and Sébastien Lacroix analyze, increasingly suffers from the disaffection of this generation.[9]

That being said, letting a problem rot is rarely a sustainable option. The Bouchard-Taylor Commission did not provide a long-lasting catharsis, and since it tabled its report (and the government by and large shelved it), exclusionary discourses have not abated. Arguably, Pollyanna-ish types of positions (in French, "angélisme") are enemies of multiculturalism or "interculturalism", especially in small nations.[10] Thus, as much as shallow justifications of restrictions to freedom of religion should be rejected, so too should rights absolutism: not every restriction on a right or a freedom is unwarranted or illegitimate.

In this respect, abstract intellectual satisfaction can hardly trump the need for compromises in some cases, especially given that liberalism does not imply complete inaction on the part of the state. And this is perhaps where philosophy and law diverge to some extent. From a comparative legal standpoint (assuming that the sample only comprises democratic states), one must acknowledge that there is a range of defensible options that simply cannot be depicted as manifestly and disproportionately infringing on freedom of religion, even when one opposes them from a philosophical perspective. This could for example be the case of bans of full-face veils in carefully tailored circumstances. Such options can reasonably be contemplated provided that ethnographic evidence seeking to ensure a complex representation of the religious claimants' condition is relied upon. This observation further points to the need to examine debates stirred by religious claims in light of domestic legal cultures, and not systematically ignoring such cultures' normative thresholds, provided that these cultures make reasonable efforts to regularly revisit these thresholds. Although silencing "majority rights" arguments are unacceptable, social interests broadly defined cannot and arguably should not entirely disregard majority interests, provided that the notion of majority is understood in an inclusive and fluid manner.

In this respect, what could be characterized as a "heuristic" of reasonable accommodation seems in order, as both religious claimants and organizations receiving accommodation requests are in need of it. When correctly applied, the doctrine of reasonable accommodation appears to offer a rather good compromise between the rights and interests of all parties involved. On the one hand, rights are *really* taken seriously, since the doctrine acknowledges that in liberal societies there are, to quote Seana Shiffrin, "social practices in which we absorb some of the costs of others' free, morally relevant choices", this "even if this behaviour is voluntary and the cost-absorption is not necessary in order to achieve luck-insensitivity".[11] On the other hand, it seeks to avoid externalizing costs that are disproportionate or uncontrollable while reiterating the idea that even most fundamental rights are not absolute.

Although some criticize this doctrine for its casuistry (at the expense, they say, of bright-line, general rules), it rather highlights, in my view, the potential of legal strategies revolving around the micro-management of religious diversity. Indeed, normative tools that are too broad or too vague may prove inefficient, and create disproportionate negative externalities for all parties involved. However, there are also limits to micro-management. Indeed, failure to elaborate legal frameworks composed of norms the meaning and scope of which are sufficiently pre-determined to help both decision-makers "on the ground" and religious claimants to efficiently work through accommodation claims, may undermine any attempt at seriously grasping the complexities of religious diversity. In any event, an overarching objective of any such framework should perhaps be to avoid as much as possible double standards, and to ensure that the allocation of burdens is not only fair but perceived as being fair. But to face such a daunting challenge, and to question strategies resorting to what I called "cohesive fear", context-sensitive and principled political leadership is needed, but often missing in (in)action.

Notes

* Dean of Law, Université de Montréal.

1 Milan Kundera gives the following definition to the concept of "small nation": "A small nation is one whose existence may be thrown in question at any time, one that can vanish at any time and that knows it. . . . A Frenchman, a Russian, a Briton do not question themselves questions about the survival of their nation. Their anthems speak of greatness and eternity. However, the Polish anthem begins by the following words: Poland has not perished yet". Quoted in A. Finkielkraut, *L'ingratitude. Conversation sur notre temps* (with Antoine Robitaille) (Montréal: Québec-Amérique, 1999), at 26. As such, Finkielkraut adds, "[c]e n'est pas sa taille ou sa superficie qui caractérise la petite nation, c'est son destin". See: A. Finkielkraut, "Qu'est-ce qu'une petite nation?", *Nationalité, citoyenneté et solidarité* (M. Seymour, ed., Montréal, Liber, 1999), 435, at 438.

2 S. Mancini, "The Tempting of Europe, the Political Seduction of the Cross: A Schmittian Reading of Christianity and Islam in European Constitutionalism", *Constitutional Secularism in an Age of Religious Revival* (S. Mancini and M. Rosenfeld, eds, Oxford, Oxford University Press 2014).

3 On this paradoxical convergence of interests, see: J.-F. Gaudreault-DesBiens, "Religious Challenges to the Secularized Identity of an Insecure Polity: A Tentative Sociology of Québec's Reasonable Accommodation Debate", *Legal Practice and Cultural Diversity* (R. Grillo, P. Shah and A. Hoekema, eds, London, Ashgate, 2009).

4 J.-F. Gaudreault-DesBiens, "Les hiérarchies passagères, ou de la contingence dans l'équilibrage entre droits fondamentaux", [2012] 4 *Revue québécoise de droit constitutionnel* 4.

5 This course has already withstood, for the most part, two constitutional challenges before the Supreme Court: *S.L.* c. *Commission scolaire des Chênes*, [2012] 1 S.C.R. 237; *Loyola High School* v. *Quebec (A.G.)*, [2015] 1 S.C.R. 613.

6 *Charter of Human Rights and Freedoms*, R.S.Q., ch. C-12.

7 *Canadian Charter of Rights and Freedoms*, Part I of the *Constitution Act, 1982,* which constitutes Schedule B of the *Canada Act, 1982* (U.K.), 1982, c. 11.

8 The Stasi report had inverted the meaning of the reasonable accommodation doctrine. Instead of referring to this doctrine as imposing upon organizations the duty to accommodate sincere religious claimants unless the accommodation requested imposes upon such organizations an undue hardship, it depicted the doctrine as imposing upon

religious believers the duty "to adapt", in certain circumstances, their beliefs to majority practices or expectations: Commission de réflexion sur l'application du principe de laïcité dans la République (B. Stasi, chair), *Rapport au Président de la République* (Paris, La documentation française, 2004), 16.

9 For a report sponsored by the Parti Québécois, and acknowledging this problem, see: P. Saint-Pierre Plamondon, "Osez repenser le PQ", on line: PSP https://osezrepenserlepq.quebec/wp-content/uploads/2017/02/osez-repenser-le-pq-rapport-preliminaire-8fev2017.pdf.

10 I elaborate on this in: J.-F. Gaudreault-DesBiens, "La religion et la diversité au péril des angélismes?", *Les Assises de l'interculturalité: regards croisés / Der Rondetafels van de Interculturaliteit: een Terugblik / The Roundtables on Interculturalism* (M.-C. Foblets and J.-P. Shreiber, eds, Brussels, Larcier, 2013).

11 S.V. Shiffrin, "Egalitarianism, Choice-Sensitivity, and Accommodation", *Reason and Value: Themes from the Work of Joseph Raz* (P. Pettit et al., eds, Oxford, Oxford University Press, 2004), 270.

Part IV

Belgium's experiment with interculturalism

The Round Tables experience

Part IV

Heidegger's engagement with interculturality

11 The Round Tables on Interculturalism

A successful experiment in deliberative democracy?

Marie-Claire Foblets[*]

Introduction

Belgium is one of the four countries under consideration here. In Belgium, the federal government established a commission, on two occasions to date, consisting of experts tasked with making recommendations on how to advance the peaceful coexistence (the French and Dutch expressions translate literally as 'living together') of cultures and worldviews. The first such commission was established in 2004 under the name 'Commission for Intercultural Dialogue' and published its *Report of the Commission for Intercultural Dialogue* in 2005.[1] A second body was created under an agreement signed by the outgoing Leterme II government on 18 March 2008, which called for a series of 'Round Tables on Interculturalism' to be organized: 'To this end a number of experts are to be tasked with making recommendations with a view to building a successful society that is based on respect for cultural differences, diversity and non-discrimination and is committed to implementing and sharing common values'.

In a sense, the Round Tables can be seen as the successor to the work of the Commission for Intercultural Dialogue (hereafter: the Commission). Their working methods were not the same, however.[2] The reports of the two bodies are therefore two clearly different documents, yet in terms of their contents they can be considered complementary.[3]

I will now proceed to describe briefly the working method adopted by the Round Tables. Then I will discuss the content of the Final Report of the Round Tables on Interculturalism (hereafter: the Round Tables Report) and summarize a few recommendations that can be considered representative of the guiding principles underpinning the Round Tables Report and that constitute a plea for a more open and positive attitude to cultural and philosophical or religious diversity. I will then look at reactions to the publication of the Round Tables Report and group these into four types. Finally, I seek to provide an answer to the question why the work done by the Round Tables, and above all the Round Tables Report, was met with such scepticism and reluctance. What lessons can be learned from the Belgian experience of the Round Tables on Interculturalism?[4]

Working methods: challenges and pitfalls of an unprecedented way of proceeding

Thirty Commission members from various sections of Belgian society

As mentioned above, the working method adopted for the 2008 Round Tables on Interculturalism differed considerably from that of the Commission for Intercultural Dialogue. A group of 30 members was initially set up by the government (hereafter: the Steering Committee). They came from different layers of society and, significantly for Belgium, from the three official linguistic communities. French-speaking and Dutch-speaking members, as well as one German-speaking member, took part in the work from the very beginning. It was obvious from the first gathering that the linguistic borders formed a dividing line and that social networks in Belgium operate principally within the boundaries of each individual linguistic community. What all members had in common, however – and brought to the group – was their practical experience with issues of cultural and philosophical or religious diversity: some were active in the educational sector, others in trade unions, and still others in the cultural sector, and there was also a rabbi and a bishop among us. Various members were from civil society organizations, and in that capacity were closely involved with minority affairs. A number of scientists had also been invited and agreed to take part.

The composition of the Steering Committee (tasked with drawing up the Round Tables Report) was at the discretion of the Deputy Prime Minister / Minister for Employment and Equal Opportunity. All members had agreed voluntarily to be part of the Steering Committee – yet the discretionary manner in which the Steering Committee was constituted was not without its effects on the rest of the work. This can be seen from at least three perspectives: first, the democratic 'deficit' automatically meant that the Steering Committee could not claim to be representative, not in terms of its reflection process, and even much less so in terms of the final outcome of the Round Tables' work. That could be regarded as a limitation, but on the other hand, it also offered space for a degree of creativity, a subject I will return to later. A second consequence of this selection process was that the members were not there as representatives of their organizations or employers, but took part in their own name. This consideration led certain members to step down from the Steering Committee during the course of its activities, because they did not wish to put their name to any recommendations that might – potentially or in fact – go against the official line or position of their organization, party or employer. And, finally, the Steering Committee had to accept the fact that a non-democratic composition could mean that outsiders would consider the outcome of the process relevant only to the members who took part in it, and no one else.

Five thematic sub-committees

Once the Steering Committee was set up, it devoted more than 12 months to carrying out its mandate. Five thematic sub-committees were set up, focusing on: (1) education;

(2) work and employment; (3) governance; (4) housing and health care (goods and services); and (5) culture and media, including associations. These sub-committees sought to gather as many testimonies as possible from experts and others with experience in each sector who could testify to what they considered to be the main problems and where policy priorities should lie. In other words, the method to be adopted was inductive and empirically based, and did not start out from a pre-determined or jointly agreed position. The choice of the five topics was motivated by an interest in giving priority to policy areas that are crucial for newcomers and minorities and in which equal opportunity and non-discrimination is or should be of central importance. The choices also reflect the concern to ensure that there was sufficient expertise within the Steering Committee to be able to handle the content of the testimonies and data covering these five thematic areas. Each thematic sub-committee then prepared a synthesis of the data gathered in the area for which it had taken responsibility. It was only once all five reports were completed that the Steering Committee could begin its own work, namely to reflect upon what conclusions to draw from the data and the findings.

Internal divisions among Commission members: how 'vigorously' should freedom of religion and belief be protected?

In order to make it possible for the members of the Steering Committee – all volunteers – to take an active part in the collective reflection process, it was internally agreed that the meetings of the Steering Committee would be held on Sundays. The work was spread over several weeks, and the discussions were often far from easy, not least because the initial standpoints and experiences of the members were very diverse. The intention was to draw up a list of priorities on the basis of the empirical data collected by each of the five thematic sub-committees, so as to be able to build further on these and ultimately to put forward concrete policy recommendations. In the process, a number of members stepped down, as mentioned, as they found it problematic to have to work in their own name on a report that in the end would be supportive of an orientation that would not be sufficiently in line with the viewpoint of their organization or employer. For some of these members, the protection of religious and philosophical identity that was proposed in some of the recommendations went too far, while others considered the working method to be too open-ended and felt that, in the end, as I will show below, too much was being expected of the majority society. In contrast, some members thought that the reflection had not gone far enough and that the Steering Committee was too readily accepting 'mainstream thinking'. For instance, when it came to the issue of integration requirements for new minorities, some worried that there were insufficient guarantees that these would be rewarded with effective equal opportunities. Another reason why some members stepped down was a gradual loss of confidentiality in the work environment the Steering Committee offered: although there had been a clear agreement among the members of the Steering Committee that each of them would observe the strictest confidentiality until the Round Tables Report was submitted to the government, some appeared unable to keep silent. Advance information was leaked to the press on several occasions, and it was clear that at least one member sought to torpedo the work of the Steering Committee. Some members, therefore, decided to resign in protest.

Ultimately, 22 members remained active to the end. Their perseverance most certainly forged a close bond among them, and when on 8 November 2010 the Round Tables Report was submitted to the government, it can be said that all the remaining members of the Steering Committee were prepared to take personal responsibility for its contents.

The contents of the Round Tables Report

Two fundamental principles

The Round Tables Report built upon two fundamental principles. The *first principle* is that of non-discrimination, which entails affording maximum opportunity for participation to anyone who resides legally in the country. The *second principle* is that this participation in society must be secured while respecting each person's religious, philosophical, ethnic and/or cultural identity, including that of newcomers and minorities.

To this day in Belgium, a high percentage of men and women with ethnocultural minority backgrounds face considerable socio-economic disadvantage. This reality runs counter to both principles. Persistent structural discrimination against certain groups, worryingly high unemployment rates, insufficient mastery of one of the three official languages, dramatic underrepresentation at higher levels of decision-making, etc., are unacceptable situations in a democratic state that claims to abide by the rule of law.[5] Empirical studies show that the frailty that is characteristic of some minorities in particular is due, in part, to a deficit in language skills and formal education.[6] Yet that frailty is not inevitable, nor is it the fault of the minorities themselves.[7] Many among them in Flanders request Dutch language classes, but the demand exceeds the supply.[8] In the Round Tables Report, several recommendations were therefore made to expand the language teaching available and to extend the offer to parents as well.

A study carried out in 2008 on behalf of the King Baudouin Foundation had already shown that Belgians of foreign origin who have a higher education degree are *twice* as likely to be unemployed as their native Belgian counterparts, and those of Turkish or Moroccan origin are *up to six times* as likely.[9] The Round Tables Report therefore called for a consistent anti-discrimination policy and, where necessary, for additional measures such as a temporary quota or reasonable accommodation measures. In the past, quotas have proven – at least to a certain extent – to be beneficial for the strengthening of women's positions in various decision-making processes and especially in politics.[10] The principle of reasonable accommodation has also proven essential to advance the position of disabled persons.[11] 'Reasonable accommodation' policies, once they are put in place, entail an obligation not only on the part of public authorities, but also on the part of employers and service providers in the private sector to contribute to increasing the opportunities for participation by physically disabled persons wherever this can be achieved through such accommodation. These may include making buildings more easily accessible, or providing an ergonomically appropriate work environment. The concept of reasonable accommodation thus places responsibility partly on the majority society: where it is possible and reasonable, the bar must be lowered for persons who, despite a physical disability, wish to participate in social life. This approach comes from North America.[12] In the final Round Tables Report submitted to the Belgian federal government in November 2010, the emphasis was mainly on measures that could increase the effective participation of members of particular minority groups that today are disadvantaged chiefly due to their ethnic and/or cultural background.

The second *fundamental principle* on which the Round Tables Report is built is that the participation of newcomers and minorities in society must not only be encouraged, but also secured with full respect for each person's identity. In this regard, the Round Tables Report recommended taking as a starting point the freedom of religion and belief guaranteed in Article 19 of the Belgian Constitution. Ever since the sixteenth century, Europe has opted to make this freedom the foundation of peaceful coexistence of different worldviews – religious and other – and to entrust public authorities with the major task of creating the conditions necessary to make such coexistence possible within the structure of a single state. Since the Enlightenment, the separation of church and state and the neutrality of public authorities are among the foundations of the state legal system in several countries in Europe.[13] The principle of neutrality does not prevent anyone from expressing his or her religious belief – that is a fundamental freedom – as long as doing so does not threaten the neutrality of public services or, even more importantly, the rights of others. A more far-reaching ban on such expression would be contrary to the constitutional tradition of a neutral (Belgian) state. The principle of neutrality aimed at safeguarding religious and philosophical pluralism also means that, in Belgium, the public authorities themselves do not profess or choose any specific (religious) belief, but neither do they go any farther than that. It is the latter stance in particular that is relevant to a plural social context. Within the Belgian context, neutrality is to be understood in a functional manner.[14] Neutrality must allow everyone – even civil servants, where applicable – to live according to their beliefs and convictions in a public manner, albeit on the condition that this is not in any way to the detriment of other constitutionally based fundamental rules, such as non-discrimination, separation of church and state, neutrality in the actions and services of public authorities, respect for the beliefs of others, etc.

The Round Tables Report submitted to the Belgian federal government in November 2010 built, as mentioned, on these two fundamental principles: it took as its starting point the necessity of a consistent anti-discrimination policy to serve as a lever for minority groups within society. It is unacceptable that, in practice, people are denied the basic right to take part in society on account of their religious and/or ethnic minority background.

Controversial recommendations? Three illustrations

The recommendations that made it into the final version of the Round Tables Report provide for manifold concrete ways to transpose these principles into effective policy; to this end, not only state authorities but all segments of civil society would have to be committed to working hard to put them into practice. The underlying reasoning for this is that fundamental rights and freedoms – civil and political as well as social and economic rights – apply to everyone. Any restrictions placed on them, such as may be necessary for reasons of public security, must remain the exception. Exceptions, i.e., any limit imposed on the exercise of any of these rights and freedoms, must, moreover, pass the classic threefold test as developed in the case law: it must be necessary in a democratic society and must be legitimate and proportionate. That threefold 'test' applies not only to the relationship between authorities and citizens, but also, for instance, to private employment relations and to access to goods and services.[15]

In what follows, I briefly discuss three particular recommendations that were included in the Round Tables Report of 2010 and that met with harsh criticism from some commentators at the time the report was issued: first, the right of pupils in public schools to express their religious or philosophical beliefs; second, the right of civil servants to do likewise; and third, a recommendation that the calendar of legal holidays be thoroughly reorganized.

The *first* recommendation concerned the issue of whether outward symbols of (religious) belief were permissible in the public schools run by Belgium's official linguistic communities. The Steering Committee took as its point of departure that it is best to prohibit outward symbols of (religious) belief as long as the children are not sufficiently mature to decide for themselves how they wish to live out their belief, but that one must proceed with great caution regarding the maintenance of such a ban once adolescents reach the requisite maturity. In the Round Tables Report, the boundary was set at the fourth year of secondary school, corresponding to the age at which a minor is considered to have the legal right to a say over his or her personal life, in particular as regards very private matters. That is the case, for example, in custody arrangements in the event that the child's parents divorce: at this age, the child has a say before the judge.[16] The same is true when it comes to the question of whether a minor may engage in sexual intercourse: the law does not require that an adolescent have reached the age of majority (18) in order to be able to decide this for him- or herself.[17] Since the publication of the report, a Belgian law adopted in 2014 provides that children above the age of 15 have a say in end-of-life questions, should they suffer from a serious and incurable illness.[18]

As the Round Tables Report argues, the freedom and trust that older adolescents enjoy under the law in not unimportant areas of life also serve as principles guiding matters of religion or belief among young people of the same age, and more particularly the way in which they wish to give it expression. It is, after all, a fundamental right. Certainly, what is at stake is the enjoyment of a freedom, and therefore decisions about expressions of religion must be entirely voluntary and without compulsion, while at the same time in no way threatening the freedoms of others. In exceptional circumstances, however, a school's governing body must have the right to impose limitations on this freedom, as this is not an absolute freedom. Moreover, for younger students, as mentioned, a ban may be justified, given that they still need to develop to the full their capacity to decide for themselves whether, and if so, how, they want to express their freedom of religion and belief.

The *second* recommendation – alluded to above – that also elicited much controversy was to allow religious and philosophical symbols in the public service sector, in particular for civil servants. The argument given is that a functional application of the principle of neutrality by the state and its institutions does leave room for the personal beliefs of civil servants, and the same is true for the way in which the latter live out and express that belief. The service provision to citizens, however, must remain consistently neutral.[19] The provision of public service does not distinguish among citizens based on their religion or belief. Civil servants are there for all citizens. In the Round Tables Report, this principle was cited and translated into the recommendation to allow civil servants to wear religious or philosophical symbols as long as doing so fits with a functional interpretation of the neutrality principle.[20] In cases where a civil servant has a power of coercion, as is the case with judges, police officers or army officers, the Steering Committee agreed that the wearing of religious or philosophical symbols would not be appropriate.[21]

The recommendation that religious or philosophical symbols be allowed in the public sector was also an (indirect) criticism of the personnel policies of various municipalities in Belgium which subject permission to wear such outward symbols to the condition that they not be visible to the user of the public service (that is, the wearers of such symbols should not come into direct contact with the public). In practice, many city administrations in Belgium impose a ban on such symbols, albeit a selective ban so that only persons who have no direct contact with service user(s) are allowed to wear religious or philosophical symbols at work. The faith of civil servants must, in other words, remain invisible to citizens in the context of service provision. Is the principle of neutrality applied correctly in this regard? This policy gives rise to indirect discrimination: while the criterion is generally applicable, it affects certain groups in society more than others, such as Muslim women who wish to wear a headscarf. The prohibition excludes the latter from not insignificant segments of the labour market, namely, all positions in the public service that involve direct contact with the citizens as service users. Such a policy could, moreover, put certain forms of religious expression under pressure. Doing so does not eliminate but, on the contrary, heightens the risk that people will not feel recognized by society and will withdraw and enclose themselves within their own minority identity. The result can over the long term bring about a hardening of divisions.[22]

A *third* recommendation that met with harsh criticism particularly among practising Catholics and Christians in general was the suggested rearrangement of the calendar of (paid) public holidays. The reasoning underlying this recommendation in the Round Tables Report is the following: how can a calendar of paid public holidays that is based largely on Christian celebrations or days of observance be justified in today's religiously diverse society, leaving no room for any other choice of such days? Employees who practise other religions are not given the opportunity to take days off on another date unless they are prepared to utilize their allotted holidays or take unpaid leave. Not only does this practice violate the principle of neutrality, but it can also be regarded as unfair treatment.[23]

One of the 69 recommendations contained in the Round Tables Report sought to provide for an alternative regulation. The recommendation was that the current legislation – which lays down ten fixed holidays – be replaced by a more flexible system in which five of the ten existing holidays would be retained, while two days would be replaced by 'floating days' and the three remaining days would be designated as 'social days', determined in consultation among the social partners (representative bodies of the trade unions and employers in the different sectors). Under this recommendation, the following five holidays would be retained: 1 January (New Year), 1 May (Labour Day), 21 July (Belgian National Day), 11 November (Armistice Day) and 25 December (Christmas). They are already fixed holidays under the current legislation. Easter and Pentecost would in any case remain holidays, since they fall on Sundays. In addition, two floating days would be added. Floating days can in theory be chosen freely by employees, but they need to be agreed with the employer in advance as they are paid by the latter. The right of an employee to take (paid) holidays on days he or she chooses must be balanced with the right of the employer to ensure the viability of the company, or in the case of employment in the public sector, the need to ensure continuity of service. The choice of the floating days must therefore be made in consultation between employer and employee. Employees can take their floating days only under these conditions. The choice of the day(s) may be dictated by religious reasons, but that need not be the case. Floating days are an example of 'reasonable accommodation' that makes it possible for members of minority religions to get days off for their own holidays. Finally, the recommendation specified that it would be up to the social partners to set the three remaining legal holidays, since the choice of paid holidays in Belgium falls within their competence. They could, therefore, decide that the current calendar of official holidays remain unchanged. The Steering Committee's recommendation did, however, urge a profound and principled rethinking of the holiday arrangements before opting for the *status quo*.

In sum, the recommendation retained the total number of holidays, while aiming to reorganize them in such a way as to bring the legislation closer in line with the reality of contemporary intercultural society in Belgium. So far, however, everything has remained as it was, confirming the clear preference on the part of the social partners not to introduce changes that might be perceived as too drastic. In the short run, that is of course the easiest solution, but at the same time the decision represents a missed opportunity to take a new approach to the regulation of paid holidays. Public holidays may help enhance the social fabric of society, since on such days people recall or celebrate a shared event, and are given the opportunity to plan activities outside the work environment. The recommendation of the Steering Committee may have sounded far-reaching at the time, but fundamentally it was an invitation to the social partners to give the issue further thought and look for days when people today, in an increasingly diverse social context, can in symbolic fashion form and (re)build society. A day of rest is an opportunity to do so. But as noted, the recommendation has to date not been adopted.

The 2010 report made other recommendations (69 in total) that, similar to the reshuffling of the calendar of (paid) legal holidays, did not target any particular group but, on the contrary, should be seen as an invitation to be more conscious of the new characteristics of an ever more plural society and the needs that come with this (new) reality. Were the 22 signatories of the Round Tables Report dreamers, as some have said dismissively when commenting on their work? In any event, they signed a document that expresses the conviction that there is an urgent demand on the part of minorities for equal opportunity and participation by everyone in Belgian society, including members of more recently arrived groups and communities. The demographic trends in Belgium today are such that it is primarily the latter communities who account for population growth;[24] the country thus faces a ticking time bomb if the authorities do not quickly change course and make equal opportunity and effective participation by every legal resident in the country a top priority. Any such change must of course take place in line with the principles of a democratic society governed under the rule of law. Those principles not only allow for diversity, but require respect for each person's identity, including their ethnic belonging, religion or belief, to name but a few characteristics which people hold dear.

Typology of reactions to the Round Tables Report

As soon as the Round Tables Report was made public, it elicited reactions from across the political and ideological spectrum. Some reactions were extremely critical, others astonishingly supportive, and yet others were nuanced in their evaluation. The reactions may be divided into four types.

The *first* type of reaction came from intellectuals who also took a position in the press and had a reputation for writing regularly on these subjects. They considered that the integration of new minorities in Belgium must first and foremost follow the basic rules of any democratic society, for that is the only way to secure respect for diversity.[25]

The *second* reaction was expressed mainly by members and representatives of cultural and religious or philosophical minority groups. To the question of how to deal fairly with the new ethnic and cultural diversity that has irreversibly become a feature of the social fabric in Belgium, they responded primarily from the position that in their view there is a right to cultural identity[26] and that the search for a (new) balance between the vested cultural and religious traditions of the majority society and the longings of more recent minorities is therefore a collective responsibility that must take the law as the measuring rod; in other words, the majority should not enjoy any acquired privileges.[27]

Somewhere in between those two positions – and more or less close to the first standpoint – is the *third* reaction, by those who, although prepared for a certain degree of reform, do not wish to allow it to happen at the expense of what has developed historically in Belgium and therefore has its own grounds for legitimacy: hard-won compromises that were negotiated long ago could not simply be thrown out. Examples of such historic compromises in Belgium are the freedom of education, the official recognition of (some) religions and, more recently, regulations concerning public cemeteries.[28]

The *fourth* reaction was by far the most critical; it, too, found great resonance in the press and social media: it was an overall rejection, both of the government's initiative to set up a commission tasked with making recommendations that would promote greater openness to cultural and religious diversity in the context of Belgian society, and of the Round Tables Report itself, with its recommendations. The majority of the 69 recommendations were read as excessive concessions on the part of the majority, without an equivalent effort being required on the part of the new minorities when it comes to abiding by the expectations of the majority society.[29] 'Il n'y en a que pour les minorités' ('there's nothing in it for anyone but the minorities') was a reservation frequently expressed. Closely linked to that view was the position that the Round Tables Report reversed the gradual secularization of Belgian society, in its place claiming more room (again) for religion in the public sphere, which would necessarily lead to renewed tensions between groups within society. The argument was that the report opened a Pandora's box, as it were, by making recommendations that risked once again 'pillarizing' Belgian society[30] and encapsulating people's identity in a form of groupthink, a new 'communitarism' so to speak, at the expense of personal autonomy and individual self-determination. Members of the Steering Committee were accused of being unrealistic daydreamers and of putting forward largely undesirable recommendations. Some received hate mail and people asked them whether they were 'living on Mars' when they approved the report.

It is often said that we learn from criticism. The fourth point of criticism strikes me in particular as an interesting source of insight and lessons to be learned. It is especially relevant for the working method chosen by the signatories of the Round Tables Report.

In retrospect: a critical assessment

As mentioned, the Round Tables Report was received at the time with great uproar in Belgium. Now that a few years have passed, it may be easier to draw up a retrospective assessment: seen from today's perspective – seven years on – should we have proceeded differently, and if so, what would have been a preferable way forward? With a different methodology, would the Round Tables Report have encountered less resistance? Why were some of the reactions so piercingly critical? I see two main reasons for this, although surely there are others. I will focus here on what I consider to have impeded a more positive reception, generally speaking, of the recommendations made in the report. A *first* reason has to do with the timing and the context in which, in 2010, public opinion viewed new waves of migration. A *second* reason is linked to the way in which the Steering Committee conducted its work, partly from its own initiative and partly in response to the non-representative way in which it had been constituted, as mentioned above. In what follows, I will briefly address both of these reasons in an attempt to draw some lessons from the exercise.

Published at the 'wrong time'?

It is certainly the case that the timing of a document that favours greater and more systematic openness to cultural and religious diversity, a position clearly reflected throughout the Round Tables Report, was critical in 2010. A not-to-be underestimated reason lies in the conception many people had at the time (and still have) of the future of European societies and of the role of recent migration, mainly from outside the European Union, in preparing for this future. The perception is that so far both the EU and its Member States have failed to develop an appropriate legislative framework to respond effectively to what is seen as endangering, in the long run, the welfare as well as the identity of European societies. As a result, against the background of such opinions, the chances of a positive response to a document written in support of an intercultural society but that does not explicitly address the issue of humanitarian migration and its manifold consequences, not least for social services and for the sustainability of the social welfare state model, were heavily compromised.

Public opinion in Belgium in the years leading up to the report was wary after successive crises in border control policy and the anything but consistent way that increasing numbers of new immigrants, mainly from outside the EU, had been granted permission to stay in the country. In the months preceding the publication of the Round Tables Report, several judges in Belgium ordered the state to pay hefty penalties in a number of individual legal cases, with a view to compelling the public authorities to assume their responsibilities towards homeless refugee applicants.[31] In addition, the report appeared against the background of a large number of regularizations (with the issuance of stay permits) that were granted in previous years under the discretionary power of the competent minister: the regularization policy in Belgium was and still is lacking in transparency,[32] and was therefore perceived by some as a policy of impunity for undocumented newcomers. As a consequence, an ever more frequently voiced opinion considered that a much more consistent policy of removal from the territory needed to be pursued.[33] The situation gave rise to fear as regards the future. That fear in turn affected the willingness, generally speaking, on the part of an increasing number of people to believe in the possibility of a sustainable, peaceful society with (in particular new) immigrants. That reaction was, in a sense, understandable: the number of groups from around the world who today seek refuge in Europe despite increasingly tight regulations has been increasing steadily these past few years.[34] They call for solidarity on the part of the society where they settle, but without there being in place a well thought-out sustainable social plan for the future. Many people within the majority society therefore link openness to a growing cultural and religious diversity in society with unbridled migration.[35] An inefficient, inconsistent and non-transparent migration policy cannot, however, be an excuse for avoiding the debate about what measures need to be put into place to secure peaceful coexistence among groups and communities in society.

Another major lesson that can be drawn from the experience of the Round Tables is the following: the fate of a document such as the Final Report of the Round Tables on Interculturalism is almost entirely tied to the willingness on the part of policymakers to put the topic to further discussion. It may sound cynical, but partly in light of the fact that by far the majority of the 69 recommendations made in the report concern matters for which competence in Belgium has now been handed over to the three official language communities and the regions, it was ultimately relatively easy – and politically speaking not high-risk – on the part of the federal government, which commissioned the Round Tables, to give its formal support to the overall content of the report. And so it did, through Deputy Prime Minister Joëlle Milquet, with responsibility for Employment and Equal Opportunities, who on the occasion of the press conference spoke very highly of the work achieved by the Round Tables and forcefully applauded the position adopted by the drafters of the report, making it explicit that she, too, was in favour of greater openness to diversity. In any event, however, the federal government did not have to make any concrete commitments to implementing its recommendations. That would be by and large the task of the communities and regions, and so the government could draw satisfaction from a declaration of principle. It was much less certain, however, whether the communities and regions would *also* agree to take on board the concrete implementation of at least the most important recommendations made in the report, for there was nothing to compel them to feel bound by the results of an initiative of the federal government. Ultimately, the federal government was accused in a number of sharply formulated critiques of having exceeded its sphere of competency.

Other players have a (continuing) role when it comes to following up on the Final Report of the Round Tables initiative. A federal institution that has, at first reluctantly but over the years increasingly effectively, put its shoulders to the wheel in support of the report is the federal Centre for Equal Opportunities and Opposition to Racism (since renamed Unia).[36] At first, because of significant political disagreement between the directorship of the Centre and the competent minister, the Centre preferred in the months following the publication of the Round Tables Report not to associate itself too closely with its contents, i.e. the recommendations it advances. Yet, from a brief reaction to this contribution by Patrick Charlier,[37] presently co-director of Unia, one learns that in the meantime, even if discreetly, a considerable number of suggestions advanced in the Round Tables Report have been given concrete attention and some have also been implemented. Not only does Charlier identify around 25 recommendations that have received the full support of Belgium's equality body, but he also mentions that Unia has taken action with regard to issues addressed in no fewer than 30 recommendations advanced by the Round Tables Steering Committee, albeit the to the report itself is not necessarily made explicit. That is of course a most encouraging observation, for without any effective follow-up, which after all had been the original intention, a document such as the Round Tables Report would have remained a dead letter. One could of course regret that this follow-up – and the credit for the work done within the framework of the Round Tables – has not received any public attention. Taking account of the controversies that the Round Tables raised throughout the trajectory, i.e. from its launch until the publication of its final report and even thereafter, in the long run there is probably more to gain by appreciating the efforts as sketched out by Charlier in his comment below than by merely criticizing the lack of greater visibility given to these efforts. All in all, the efforts point to a considerable achievement.

Deliberative democracy: the right format for questions regarding a multicultural society?

A second factor that can help explain the initial resistance to the overall contents of the Round Tables Report, and to some recommendations in particular, is undoubtedly the fact that when the text of the report was made public, many people were taken by 'surprise'; the text was dismissed as too radical. Looking back, the (negative) surprise effect may well have been a consequence, directly or indirectly, of the *working method* adopted by the Steering Committee members who served as drafters (and later signatories) of the Round Tables Report. As explained above, the report built on empirical data without a clearly defined normative standpoint being agreed in advance among those who would later be the signatories. The only restriction was of course that the draft text should remain within the framework of the Belgian Constitution and the principles laid down in international human rights law.

It could be said that the Steering Committee developed a somewhat *ad hoc* working method: the 69 recommendations in the Round Tables Report were the result of collective reflection, carried out over many weeks by initially 30 and ultimately 22 members of the Steering Committee. They are a good indication of what can be achieved by proactive and at the same time creative thinking about the challenges of building a sustainable society in a plural context. A differently composed Steering Committee would doubtless have led to a different list of recommendations.[38] This may be seen as a weakness, but it can also be regarded as an inherent advantage: in this case 22 individuals, each with his or her own viewpoint, and who might otherwise not have met each other and/or not had the opportunity to discuss in depth a number of topics, were brought together with a relatively open mandate to develop ideas that could promote the peaceful coexistence of different groups and cultures within Belgian society. When outsiders were given access to the final document and read it in greater detail, the full text of the report, the entirety of the 69 recommendations, could come across as far-reaching. This is because the recommendations focused on the need for a renewed approach to cultural diversity and for addressing more forcefully some of the structural disadvantages faced by minorities, including indirect discrimination, and showed that there was still a great deal to be done in a wide range of areas and at many different levels of society.

Another key reason for the surprise effect was that the report had not sought to link up with the programmes of specific political parties, trade unions, associations, or religious or philosophical groups. Instead, the signatories had opted for the decision-making process known under the term 'deliberative' or 'discursive' democracy to designate a form of careful weighing and balancing (deliberation) that works from the bottom up with maximum openness to what can emerge from conversations and discussions conducted with the greatest respect for each other's standpoints. The result is a form of direct democracy: the outcome of such discussions are often unpredictable compromises, but can also be refreshingly new.[39]

Partly because questions to do with a multicultural society are politically speaking extremely sensitive, it was far from straightforward to appear out of the blue with 69 recommendations that, taken as a whole, did not fit with anyone's political and/or ideological agenda in particular. And in some cases, if a person could not agree with one or other of the recommendations, that was considered reason enough to dismiss the entire work of the Steering Committee. An understandable reaction, but it essentially comes down to throwing the baby out with the bathwater.

One might argue that it is unrealistic to expect to use the tools of deliberative democracy to develop solid public policy, much less to take action with a view to addressing the problems of contemporary multicultural societies in Europe. Some questions are simply too complex and the conflicts of interest too great for this to be possible. Yet it may be more realistic to accept that this working method lends itself better to finding solutions for certain problems than for others.[40]

Conclusion: peace returns?

What the experience of the Round Tables on Interculturalism has in any case made clear is that it is possible for individuals from various religious and philosophical perspectives to make a sustained effort to think together creatively about measures that could help advance harmony among cultures and beliefs within society and increase the possibilities of social advancement for minorities in particular. But what the Round Tables have also clearly shown is that such an effort is but a vain gesture if it does not enjoy sufficiently strong support from established institutions and the surrounding society. After all, the Round Tables, like the other expert commissions discussed in this volume, were an *ad hoc* commission with a limited life span and no direct means of implementation beyond the influence of the members. It is therefore comforting to know that the Belgian equality body, Unia, eventually took up the baton and has made concerted efforts to follow up on a number of recommendations. Yet it remains unfortunate that, despite this follow-up, the perception that the Round Tables of 2010 led to a radical, even unrealistic final result has remained, up to now, unchanged. The Round Tables, and the discussion that followed the publication of the Round Tables Report, significantly disrupted public opinion for a brief time. In the long run, however, it is the absence of greater preparedness among the wider society to work on more sustainable chances of success of a multicultural society that is profoundly disquieting and ought to be a very real ground for unease, for it bodes ill for the future.

The effect of the Round Tables Report: follow-up by Unia

Patrick Charlier, co-director of Unia

Marie-Claire Foblets, co-director of the Round Tables, concludes that in her view, perhaps the Round Tables on Interculturalism (Assises de l'interculturalité 2010) have been *un coup dans l' eau*, in other words, a missed opportunity. In this brief comment, I want to address the impact and (ongoing) influence of the Round Tables' recommendations in the daily work of Unia (previously the Interfederal Centre for Equal Opportunities and Opposition to Racism). To a certain extent, I adhere to Foblets' conclusions, and see two main reasons for that:

First, the Round Tables were a very personal political initiative. Even in the federal government, Minister Joëlle Milquet, then Deputy Prime Minister / Minister for Employment and Equal Opportunities, was unable to convince her colleagues that it was a necessary and useful project and not just a personal pet project. Our understanding has been that if the impetus for such a project is to come from the political world, there must be a principled distance between the (political) decision makers and the body tasked with drafting recommendations; that is a very preliminary condition for independent work. In this case, that condition was not met, or at least that is the way it has been perceived.

Second, such an initiative must be comprehensive and participative. In the federal Belgian state, working on issues related to interculturality requires that the communities and the regions be involved. To justifiably address such issues, especially within the framework of an initiative taken at the level of the federal government, the communities and/or regions need to be part of the proceedings, given that most of the issues clearly come under their purview (education, public housing, culture, media). At the very beginning of the process, even before the Round Tables were launched in November 2010, Minister Milquet had solicited Unia to organize and manage the Round Tables. One of the main reasons why Unia declined the request was that the regions and the communities were not involved. Unia made a counter-proposition which was quite different in nature (a different timetable[41] and involvement of the regions and the communities), but with the same objectives. The minister, however, maintained her position and launched the Round Tables, without first trying to federate all stakeholders and institutional bodies directly and indirectly concerned in order to work towards a single purpose.

Yet, despite these hiccups at the start, it is also important to note that the conclusions and recommendations contained in the Round Tables Report have not been without influence. While still today, several years later, explicit references to the Round Tables Report as such have remained pretty rare, it is fair to say that a considerable number of its conclusions and recommendations have *de facto* been on the political agenda, and several among them have also been implemented. At the same time, one notices that Belgian society has changed since 2010, and we can assume that the way some recommendations are formulated would be significantly different today. This being said, what matters is less the wording as such, but the spirit and the idea behind the concrete expression.

Speaking for Unia, the federal equality body in Belgium mandated to support and follow up on the recommendations contained in the Round Tables Report, I would summarize the work achieved thus far as follows:

As Foblets explicates, a total of 69 recommendations were included in the Round Tables Report.

With regard to ten of the recommendations, Unia has not adopted any official position (e.g., the reform of the calendar of fixed holidays, multilingualism at school, the proposed extension of the Belgian Holocaust denial law of 1995, etc.).

With regard to around 25 recommendations, while Unia can totally subscribe to them, the organization is not actively engaged in advocating them (e.g., the issue of fatherhood in the case of children born from mixed unions involving at least one ascendant from Congo, Rwanda or Burundi; more visibility to be given to diversity in the media, etc.)

Unia supports and plans to be active with regard to around 30 recommendations included in the report (even if the link with the Round Tables is not systematically made explicit). These include:

- inclusive diversity management in a school context: In particular, Unia will publish, in 2017, the barometer of diversity in education (in 2012, it focused on employment; in 2014 on housing);
- the recommendation regarding the protection of the freedom of religion and belief of pupils and, more particularly, regarding the wearing of religious symbols in schools;
- positive actions and policy measures;
- the negotiation of cooperation agreements with social inspection officials with a view to combating more efficiently discrimination in the vast area of employment;
- research on the NEETS in the Walloon region.

A positive observation, one that should not be overlooked, is that several recommendations have already been implemented, some partly, while others entirely:

- socio-economic monitoring (first edition in 2013, second edition in 2015 and third edition planned for December 2017);
- circular 13/2013 by the college of the general prosecutors on the investigation and prosecution policy regarding discrimination and hate crimes;[42]
- the negotiation of several cooperation agreements between Unia and actors in the field of employment;
- 'interfederalization' of Unia (now given competence to act at federal, regional and community levels).

Finally, we refer to our contribution in this volume, addressing Unia's 'highest common denominator' approach. It offers an alternative approach to the recourse to 'reasonable accommodation' so strongly supported in the Round Tables Report when it comes to finding the balance between the principle of non-discrimination and respect for diversity. Unia gives a clear preference to the 'highest common denominator' approach, and will continue to assess the pros and the cons of this approach, taking into account the potential pitfalls and scrupulously following the developments that will occur in the broader social and legal contexts. Unia will in particular follow the impact on the ground (e.g., for the labour market in Belgium) of the *Achbita* and *Bougnaoui* (CJEU) cases, also mentioned by Foblets.[43]

Notes

* Director of the Department of Law and Anthropology at the Max Planck Institute for Social Anthropology in Halle/Saale, Germany and Professor of Law at the Catholic University of Leuven, Belgium. Marie-Claire Foblets was, from September 2009 to November 2010, the co-chair of the Round Tables on Interculturalism in Belgium.

1 *Commissie voor Interculturele Dialoog. Eindverslag en getuigenissen* [Commission for Intercultural Dialogue. Report and Testimonies] (Brussels, 2005).

2 The work of the Commission consisted chiefly of reflection by academics, who subsequently also held a series of consultations. Prof. Edouard Delruelle served as the Commission's rapporteur. Soon after the Commission's work ended he became co-director of the federal Centre for Equal Opportunities and Opposition to Racism (now called Unia).

3 The reporter for the Commission for Intercultural Dialogue was also involved in the Round Tables, thus ensuring a link between the two.

4 For a number of reactions to the report, see: *Les Assises de L'interculturalité/De Rondetafels van de Interculturaliteit/The Round Tables on Interculturalism*, ed. M-C. Foblets and J.-P. Schreiber (Brussels: Larcier, 2013).

5 G. Turan, 'Discriminatie blijft structureel probleem in België' [Discrimination remains a structural problem in Belgium], *Knack*, 31 March 2014; see also *Equal Is Not Enough*, ed. D. Cuypers and J. Vrielink (Antwerp: Intersentia, 2016).

6 I. Adam, 'Immigrant integration policies of the Belgian regions: Sub-state nationalism and policy divergence after devolution', *Regional and Federal Studies* 23/5 (2013), 547–569.

7 Council of Europe, *Living Together as Equals in Dignity* (White Paper on Intercultural Dialogue), CM (2008)30 final, 2 May 2008; D. Cuypers and J. Vrielink, *Equal Is Not Enough*.

8 Vlaamse Parlement [Flemish Parliament], Verslag Commissievergadering, Commissie voor Bestuurszaken, Binnenlands Bestuur, Inburgering en Stedenbeleid [Report of the Meeting of the Committee for Governmental Affairs, Domestic Governance, Integration and Urban Policy] (2015–2016) C209, 3 May 2016.

9 L. Van Hamel and R. Darquenne, *Een andere kijk op hardnekkige jeugdwerkloosheid. Aanbevelingen en succesfactoren bij de inschakeling van laaggeschoolde jongeren /Un autre regard sur les jeunes enlisés dans le chômage: Recommandations et facteurs de réussite pour l'insertion professionnelle des jeunes peu qualifié* [Another look at persistent youth unemployment: Recommendations and success factors for the entry of low-skilled youth into the workplace] (Brussels: King Baudouin Foundation, 2008).

10 J.E. Goldschmidt, 'Gelijkheid als fundamenteel recht: de spanning tussen gelijke behandeling en sociale cohesie in een veranderende samenleving' [Equality as a fundamental right: The tension between equal treatment and social cohesion in a changing society], in *Rechten, plichten deugden* [Rights, obligations, virtues], ed. P.B. Cliteur and H.M.Th.D. ten Napel (Nijmegen: Ars Aequi, 2003), 95–111; J.G. Polavieja, 'Socially embedded investments: Explaining gender differences in job-specific skills', *American Journal of Sociology* 118/3 (2012), 592–634.

11 D. Ferri, et al., *Reasonable Accommodation for Disabled People in Employment: A Legal Analysis of the Situation in EU Member States, Iceland, Liechtenstein and Norway*, European Commission, Directorate-General for Justice and Consumers, 2016; see also, on reasonable accommodation measures in the workplace: K. Alidadi, *Religion, Equality and Employment in Europe: The Case for Reasonable Accommodation* (Oxford: Hart, 2017).

12 P. Bosset, 'Droits de la personne et accommodements raisonnables: le droit est-il mondialisé?' [Rights or the person and reasonable accommodation: Is law globalized?], *Revue interdisciplinaire d'études juridiques* 62/1 (2009), 1–32; see also: K. Alidadi, *Religion, Equality and Employment in Europe*, 2017.

13 See, among others: J. Ringelheim, 'State religious neutrality as a common European standard? Reappraising the European Court of Human Rights approach', *Oxford Journal of Law and Religion* 6/1 (2017), 24–47; J. Temperman, *State–Religion Relationships and Human Rights Law: Towards a Right to Religiously Neutral Governance* (Leiden/Boston, MA: Martinus Nijhoff, 2010); J. Baubérot, 'The two thresholds of laicization', in *Secularism and Its Critics*, ed. R. Bhargava (Oxford: Oxford University Press, 1998), 94–136; *Liberal Neutrality*, ed. R.E. Goodin and A. Reeve (London: Routledge, 1989); V. J.-M. Woehrling, 'La laïcité de l'etat' [The secularity of the state], in *Droit français des religions [French law on religions]*, ed. F. Messner, P.-H. Prélot and J.-M. Woerhrling, 2nd ed. (Paris: Litec, 2013), no. 999; *Scheiding van Kerk en Staat of Actief Pluralisme?* [Separation of church and state or active pluralism?], ed. P. De Hert and K. Meerschaut (Antwerp: Intersentia, 2007); *Church and State in Contemporary Europe: The Chimera of Neutrality*, ed. J.T.S. Madelay and Z. Enyedi (London: Frank Cass, 2003); W. Van der Burg, *Het ideaal van de neutrale staat. Inclusieve, exclusieve en compenserende visies op godsdienst en cultuur* [The ideal of the neutral state. Inclusive, exclusive and compensatory visions of religion and culture] (Erasmus Law Lectures, 18) (The Hague: Boom Juridische Uitgevers, 2009); for a critical analysis of the implementation of the principle of state neutrality in Belgium see, among others: J-P. Schreiber, *La Belgique, etat laïque. Du principe à la réalité* [The secular state: From theory to practice] (Brussels: Espaces de Liberté, 2014); L. Francken and P. Loobuyck, 'Is active state support for religions and worldviews compatible with the liberal idea of state neutrality? A critical analysis of the Belgian case', *Journal of Church and State* 55/3 (2012), 478–497.

14 See, among others: M. Roy, 'L'etat belge, etat laïc' [The Belgian state, a secular state], in *En hommage à Francis Delpérée. Itinéraires d'un constitutionnaliste* [Essays in honour of Francis Delpérée. Itineraries of a constitutional lawyer] (Brussels/Paris: Bruylant/LGDJ, 2007), 833–845; R. Torfs, 'Church and state in France, Belgium and the Netherlands: Unexpected similarities and hidden differences', *Brigham Young Law Review* 11/1 (1996): 945–946.

15 On the threefold test see, among others: F. Tulkens, 'L'évolution du principe de non-discrimination à la lumière de la jurisprudence de la Cour européenne des droits de l'homme' [The development of the principle of non-discrimination in the light of the case law of the European Court of Human Rights], in *L'étranger face au droit* [The foreigner in the face of the law], ed. J.Y. Carlier (XXes Journées d'études Juridiques Jean Dabin, Louvain-la-Neuve, 2008) (Brussels: Bruylant, 2010), 193–210; with a focus on Belgium: B. Renauld and S. Van Drooghenbroeck, 'Le principe d'égalité et de non-discrimination' [The principle of equality and non-discrimination], in *Les droits constitutionnels en Belgique* [Constitutional rights in Belgium] (Vol. 2), ed. M. Verdussen and N. Bonbled (Brussels: Bruylant, 2011), 553–605.

16 R. Vasseur, 'De verblijfsregeling van kinderen van gescheiden ouders: een verruimd wettelijk kader' [Custody arrangements for children of divorced parents: An extended legal framework], *Tijdschrift voor Jeugd- en Kinderrechten* 1 (2007), 11–20; S. Brouwers, 'Wijziging van de verblijfsregeling: criteria en (civielrechtelijke) sancties' [Amending the residency regulations: Criteria and (civil law) sanctions], in

C.B.R. (ed.), *Verblijfsregeling* [Regulations governing the right of residence] (Antwerp: Intersentia, 2008), 145–182; E. Van Der Mussele, 'Bijstand, spreekrecht en rechtsingang voor minderjarigen' [Assistance, right to speak and access to the courts for minors], in *Kinderrechten in België* [Children's rights in Belgium], ed. W. Vandenhole (Antwerp: Intersentia, 2008), 95–112.

17 Art. 372, 374 and 375 Belgian Penal Code.

18 Law of 24 February 2014 amending the Law of 28 May 2002 on euthanasia, to make euthanasia possible for minors, *Moniteur Belge/Belgisch Staatsblad* [Belgian official gazette], 12 March 2014.

19 For Flanders, see: Circular BZ 2011/6 Code of ethics for the staff of the Flemish government; point 4.4. Objectivity (www.vlaanderen.be/regering).

20 For a much stricter interpretation see, among others: C. Ruet, 'Interdiction du port des signes religieux par les agents du service public: La combinaison subtile de l'arrêt Ebrahimian' [Ban on the wearing of religious signs by public servants: The subtle combination found in the Ebrahimian judgment], *La revue des droits de l'homme*, Actualités Droits-Libertés, accessed 12 September 2016, http://revdh.revues.org/2516; DOI: 10.4000/revdh.2516; see also the open letter signed by around 30 scholars: 'Aan het loket noch God noch partij/Ni Dieu ni parti au guichet' [Neither God nor party at government offices], *De Standaard*, 2 May 2013.

21 More recently on the wearing of religious symbols by police officers, see: K. Lemmens, 'Hoofddoek bij politie: religieuze symbolen zijn ook aan anderen gericht' [Headscarves in the police force: Religious symbols are also messages to others], *Knack*, 7 June 2017.

22 More recently (on 14 March 2017), the Court of Justice of the European Union (ECJ) went as far as ruling that in the private sector as well, companies can prohibit religious symbols on the ground of neutrality (*Achbita and Anor v. G4S Secure Solutions NV*; case C-157/15; and *Bougnaoui & ADDH v Micropole SA*; Case C-188/15). A company's wish that employees dress 'neutrally' is legitimate and therefore allows internal rules banning political, philosophical or religious symbols, on condition that the rules treat all employees in the same way, notably requiring them all, generally and without any differentiation, to dress neutrally. In the first case, Samira Achbita had been a receptionist for the Belgian branch of G4S, the London-listed outsourcing and security company when, after three years at the firm she decided she wanted to start wearing the headscarf at work for religious reasons. She was fired in June 2006 for refusing to take off her scarf. In the opinion of her employer she had broken unwritten rules prohibiting religious symbols. In the second case, Asma Bougnaoui, a design engineer, was fired from an IT consultancy firm, Micropole, after a customer complained that his staff had been 'embarrassed' by her headscarf while she was on their premises to give advice. She had been told before taking the job that wearing a headscarf might pose problems for the company's customers. The neutral image of a business towards its customers or clients thus in a sense takes precedence over the freedom of religion and belief of the persons who may wish to enjoy the right to express, within the company, their belonging to a religious community, for example by wearing a headscarf. If for certain jobs one can easily understand that a dress code needs to be strictly regulated or subjected to discretionary conditions or even restrictions, the acceptance of the principle of the legitimacy of the ban on certain forms of dress in the name of neutrality risks having deleterious effects on the labour market over the medium term in the sense that it may well lead to the exclusion of a not insignificant group of jobseekers, in this case clearly that of Muslim women. By so doing, the ECJ 'give[s] greater leeway to employers to discriminate against women – and men – on the ground of religious belief', according to a critical comment by John Dalhuisen, Director of Amnesty International's Europe and Central Asia programme (Press release, 14 March 2017). Similarly, Jewish men who wear kippas, as well as Sikh men who wear turbans, and Christians who wear crosses in the workplace, risk being affected by the ruling. To Steve Peers, the ECJ ruling looks all the more awkward when set against the ECtHR case law that is more protective of

the employee's right to manifest freedom of religion (Steve Peers, 'Headscarf bans at work: explaining the ECJ rulings', EU Law Analysis blog, 14 March 2017). In its ruling, the ECJ has not referred to this case law or attempted to deal with the distinction between freedom of religion and non-discrimination. The impact of the 14 March 2017 rulings is likely to differ across the different European jurisdictions. To Steve Peers, the bottom line of the two judgments 'do not constitute a workplace headscarf ban, but merely permit employers to establish such a ban – subject to limits which might prove difficult to comply with in practice'. Peers, 'Headscarf bans at work'.

23 See, among others: M. van den Brink, 'Equals in faith – faith in equality: Equality as an additional standard for decisions on the accommodation of religious practices', in *Religious Pluralism and Human Rights in Europe: Where to Draw the Line?*, ed. T. Loenen and J.E. Goldschmidt (Antwerp: Intersentia, 2007), 211–217; G. Caceres, 'Reasonable accommodation as a tool to manage religious diversity in the workplace: What about the 'transposability' of an American concept into the French secular context?', in *A Test of Faith? Religious Diversity and Accommodation in the European Workplace*, ed. K. Alidadi et al. (Farnham, Surrey: Ashgate, 2012), 283–316.

24 For most recent data, see the information provided by the European Statistical System (ESS): Population (Demography, Migration, Projections) http://ec.europa.eu/eurostat/statistics-explained/index.php/Migration_and_migrant_population_statistics.

25 See, among others: M. Martiniello, *La démocratie multiculturelle. Citoyenneté, diversité, justice sociale* [Multicultural democracy: Citizenship, diversity, social justice] (Paris: Presses de Sciences Po, 2011). For a very supportive review, see: M.-F. Rigaux, 'Les assises de l'interculturalité. Quelques réflexions en marge d'un rapport' [The round tables on interculturalism: A few reflections on the margins of a report], *Journal des tribunaux*, 6414 (27 November 2010): 705–706.

26 On this question, and to what extent there is indeed such right, see for Belgium: *Le droit et la diversité culturelle* [Law and cultural diversity], ed. J. Ringelheim (Brussels: Bruylant, 2011).

27 See: N. Charkaoui, 'Minority position on the Round Tables', in *Les Assises de L'interculturalité/De Rondetafels van de Interculturaliteit/The Round Tables on Interculturalism*, ed. M.-C. Foblets and J.-P. Schreiber, 579–581.

28 A position in that sense, but that precedes the publication of the report: H. Dumont and X. Delgrange, 'Le principe de pluralisme face à la question du voile islamique en Belgique' [The principle of pluralism in the face of the question of the Islamic headscarf in Belgium], *Droit et société* 68/1 (2008), 75–108.

29 See for example, the reactions by Bart De Wever (www.n-va.be/nieuws/opinie/eenzijdige-onzin) and Jean-Marie Dedecker (www.jmdedecker.com/2010/11/de-ronde-tafels-van-de.html).

30 On the concept of the 'pillar' and its historical and sociological meaning in the Belgian context, see, among others: L. Huyse, *De verzuiling voorbij* [Beyond pillarization] (Leuven: Kritak, 1987).

31 L. Slingenberg, 'Recht op onderdak voor illegaal verblijvende kinderen? Analyse van een beslissing van het Europees Comité voor Sociale Rechten' [Right to housing for illegal resident children? An analysis of a decision of the European Committee for Social Rights], *Tijdschrift voor Vreemdelingenrecht* 4 (2010), 337–343.

32 E. Somers, 'De beoordeling van 'regularisatie'-aanvragen (art. 9bis Vw.) met criteria uit de instructie van 19 juli 2009 na de Raad van State-rechtspraak: rien ne va plus?' [The assessment of 'regularization' requests (Art. 9*bis* of the Law on Residence) with criteria from the instruction of 19 July 2009 after the Council of State decision: End of the road?], *Tijdschrift voor Vreemdelingenrecht* 3 (2013), 210–223.

33 See, among others: T. Francken, S. Smeyers and D. Duméry, *Migratie, een uitdaging. De totaalvisie van de NV-A* [Migration, a challenge: The total vision of the NV-A party], 2012.

34 http://ec.europa.eu/eurostat/statistics-explained/index.php/Migration_and_migrant_population_statistics.

35 For a critical analysis of this view: B. Benyaich, *Islam en radicalisme bij Marokkanen in Brussel* [Islam and Radicalism among Moroccans in Brussels] (Antwerp: Van Halewyck, 2013).

36 www.unia.be (see in particular: Unia 2015 annual report: 'Living together put to the test').

37 See below. This comment builds on Patrick Charlier's presentation at the international conference convened by the Department of Law and Anthropology of the Max Planck Institute for Social Anthropology: *Religious and Cultural Diversity in Four National Contexts*, 26–27 June 2015, Halle/Saale, Germany. Charlier kindly agreed to report on Unia's follow-up on the recommendations contained in the Round Tables Report.

38 An observation that is also made by Patrick Weil in this volume; Weil served as a member of the French Stasi Commission.

39 On deliberative democracy see, among others: B. Kingsbury, 'Global administrative law and deliberative democracy', in *The Oxford Handbook of Theory of International Law*, ed. A. Orford, Fl. Hofmann and M. Clark (Oxford: Oxford University Press, 2016), 526–544; U. Ehrling, *Deliberative Global Governance: legitimes Regieren durch Recht und Zivilgesellschaft* [Legitimate governance via law and civil society] (Wiesbaden: Springer, 2016); S.A. Ercan, 'Democratizing identity politics: A deliberative approach to the politics of recognition', in *Europe at the Edge of Pluralism*, ed. D. Gozdecka and M. Kmak (Antwerp: Intersentia, 2015), 13–26; C. Lafont, 'Religious pluralism in a deliberative democracy', in *Europa mit oder ohne Religion? Der Beitrag der Religion zum gegenwärtigen und künftigen Europa* [Europe with or without religion? The contribution of religion to Europe today and in the future] (Vol. 1), ed. K. Appel, I. Guanzini and A. Walser (Goettingen: V&R unipress, Vienna University Press, 2014), 39–55; B. Bashir, 'Accommodating historically oppressed social groups: Deliberative democracy and the politics of reconciliation', in *The Politics of Reconciliation in Multicultural Societies*, ed. W. Kymlicka and B. Bashir (Oxford, Oxford University Press, 2009), 48–69.

40 Some years ago, the French sociologist Bruno Latour (B. Latour, *Politics of Nature: How to Bring the Sciences into Democracy*, Harvard University Press, 2004) noticed that, when it comes to radically rethinking major societal issues – in his case environmental degradation and the possibility of maintaining a sustainable ecological equilibrium – despite all concerns of policymakers and other stakeholders such as non-governmental organizations and pressure groups, democratic decision-making is paralysed by established categories of thought. In his view, only a radical rethink will enable us to grasp the import of new ideas and launch a refreshing approach to the maintenance of a tolerable life. The observation probably also applies when it comes to rethinking our approach to issues that revolve around the coexistence of an increasing number of groups and communities within the political framework of contemporary democracies and more particularly, the search for a sustainable balance between, on the one hand, respect for their distinctive cultural, religious, ethnic, linguistic or other characteristics and, on the other, the need of unity.

41 In the initial timetable, the deadline set for submission of the final report was extremely tight, and the work was supposed to be done within a timeframe of less than six months. That was not realistic.

42 Joint Circular no. COL 13/2013 of the Minister of Justice, the Minister of the Interior and the College of prosecutors at the court of appeal on the investigation and prosecution policy regarding discrimination and hate crimes (including discrimination on gender grounds).

43 See note 22.

12 Religious and cultural diversity in Belgium

Finding the common denominator

*Patrick Charlier and Nathalie Denies**

Introduction

The accommodation of differences should be an exercise in social and political judgement that falls within the scope of common sense and reason. Unfortunately, though, it lies at the source of an escalation of strong rhetoric, emotions and contradictions, with the result that it is often implemented simply as a gesture of goodwill or through legislation.

Over the past few years, social workers, leaders of associations and trade unions, human resource managers in private companies in the commercial sector and in public services have been faced with professionals and/or users asking for their religious beliefs to be taken into account. It seems that there is an even greater emphasis on the religious component of identity than other components, such as professional or family identity, which have been weakened by the socio-economic context and changing family dynamics.[1]

The approach we advocate here is one that searches not for accommodation but for the "common denominator", relying on the work of Dounia Bouzar and Nathalie Denies. This diverges from the approach the Round Tables Report takes on a variety of issues. We are concerned with the following questions: how can diversity lead to unity? To what extent can we agree to make adaptations in accordance with the rights and the law (legal guidelines), while taking into account everyone's responsibilities (constraints and organisational mandates)? How can we manage religious diversity without yielding unconditionally to all the particularisms?

Professionals are often concerned with codes of conduct or belief systems that put the emphasis on religion. The lack of sociological, philosophical and legal reference points on this subject forces them to take a stance based on their own ideology, emotions or background.[2] Without reference points, many business leaders and decision makers allow their subjectivity, or even the balance of power, to win.

Religious diversity appears to be giving rise to a growing number of questions with regard to employment, professional training, health, youth, and so on. In 2015, 19 per cent of the cases opened by Unia (the Interfederal Centre for Equal Opportunities)[3] involved the criteria of religious or philosophical beliefs.

Specifically, Unia opened 330 cases about discrimination on grounds related to religious or philosophical beliefs, 11 per cent more than in 2014, and 91 per cent of these cases involved Islam. Out of all of these cases, 14 per cent had to do with employment and 11 per cent with education.

Leaders working in the field are turning to pragmatic solutions in response to the current situation. The search for pragmatic solutions is also the focus of much of Unia's work, as litigation is only pursued in a small minority of situations.

Unia has been active for many years on the issue of religious diversity: by offering pathways to solutions, but above all, offering a method based on dialogue, aimed at promoting free speech, and eradicating the legitimate concerns and fears expressed by many stakeholders in the social, economic and cultural life of the country.[4] We believe that there is an urgent need to create a space for dialogue, within which any question may be asked, and any debate conducted, with the respect it deserves. This goal represents the key challenge of our approach. But questioning another on his beliefs and habits on the one hand, and listening and understanding the confusion or resistance on the part of the other, on the other, calls for restraint, methodology, and respect.

Freedom of religion and the highest common denominator: towards the benefit of majority

Belgian law, like international law, recognises everyone's freedom of religion or belief.[5] Ensuring the freedom of religion means allowing individuals to believe – believing in what they want or not want to believe, as well as expressing these beliefs through a variety of observances and practices.[6] However, the right to freedom of religion is not limitless.[7] Where should we place the cursor?

Asked to intervene on several occasions concerning issues of religious diversity (forbidden foods, availability of a place of prayer, wearing religious symbols, mixing of men and women, etc.) raised by various institutions, Unia began research on these issues with the help of Dounia Bouzar, a French anthropologist specialising in religion. For a year and a half, a highly diverse group of participants, composed of representatives from the sectors of early childhood and youth, health and retirement homes, vocational training, trade unions, public services, education and school mediation, came together to hold discussions on situations they had encountered throughout their respective careers. Based on the legal framework, and on the culture of social dialogue that is so rich in Belgium, they discussed their viewpoints topic by topic in order to be able to expand the system of reference on the management of religious diversity.[8]

Their results are based on the concept of the "highest common denominator" (HCD).[9] This implies that the answer to an individual request (change of hours, specific food, etc.) must not only satisfy the person making the request but also be of benefit to everyone else. It has the advantage of avoiding two common stumbling blocks: imposing one world view as the superior and universal standard (which can lead to indirect discrimination and encourage certain ethnocentric approaches) or, on the contrary, introducing specific treatment for one demographic group (which can lead to division between workers or users on the basis of their belief system, the withdrawal of communities, or even the attribution of identities). It is based on the idea that we must ensure that the proposed solution can be applied, and be of benefit, to the greatest number of people without indirectly discriminating against the most recent arrivals. This can be done by neutralising a request based on religious motives in order to allow everyone to take advantage of it, by adopting general and neutral provisions, which have been the subject of prior consultation, and not distorted by a religious or cultural prism. The HCD can be ascertained by considering what brings people together and unites them, in order to work on the basis of similarities rather than having to reason by taking into account distinct communities or particularities.

In legal terms, non-discrimination implies treating individuals without any consideration for their origin, religion, gender, etc., i.e. without taking into account their ethnic, religious, or sexual identity.[10] The principle of non-discrimination is contingent upon processes of de-identification to prevent individuals from being unequivocally identified as Muslims, Jews, Christians, etc. in the work environment, as a way to invite their colleagues to regard them as an employee like any other.[11] Adopting this approach means choosing not to treat requests linked to religious beliefs as requests for recognition based purely on identity. Issues concerning discrimination at work relating to religious beliefs must be dealt with and resolved, first and foremost, like any other interpersonal social conflict: by taking into account the realities of an increasingly diverse society.

Religious diversity in the workplace appears to be growing increasingly problematic. Certain private companies, particularly in sectors employing large numbers of Muslim women, have adopted the principle of neutrality, which has been under discussion in relation to public service workers.[12] In fact, legally speaking, invoking neutrality for a private company is open to question.[13]

On the other hand, invoking neutrality is a fundamental aspect of public service in Belgium. Public service representatives must behave in an impartial, non-discriminatory and loyal manner.[14] These are statutory requirements.[15] Neutrality is a means of guaranteeing that identical service will be provided to all users. It is a matter of avoiding any risk of discrimination or partiality towards participants or users of any public service.[16] However, matters become much less clear when one broaches the subject of neutrality in appearance. To be sure, there can be no neutrality without the appearance of neutrality,[17] but there is no legislative text defining exactly what is to be understood by an appearance of neutrality.

Should neutrality in public service be understood as exclusive, in the sense that it requires relinquishing the use of any visible signs of religious affiliation, or as inclusive, in the sense that it is manifested by the visible diversity of religious convictions?

As this matter has never been explicitly addressed in Belgian law, the exercise of autonomy by administrations poses a challenge for the neutrality of the state.[18]

Case studies: requests for dietary accommodation, adjustment of working hours or prayer space under the HCD

The studies carried out under the aegis of the Interfederal Centre for Equal Opportunities[19] show that in a large majority of cases, the request made by a Muslim employee to be able to pray or have special meals *is not* a request to be recognised as a Muslim but, on the contrary, a request for "normalisation". Their requests amount to measures that would allow them to work in accordance with the tenets of their religion, but without being overtly recognised (either negatively or positively) as a Muslim, and without being distinguished from the other employees on the basis of religion or religious belief. Experience also shows that when companies deal with this issue from the point of view of identity (for instance, by seeking the assistance of an imam, a rabbi or a priest, and by promoting measures taken in the name of cultural diversity), they often stoke tensions that they thought they had appeased.

Among the concrete examples in the system of reference is that of the employer confronted with requests from various employees to have their working hours adjusted. Some people don't want to work on Friday afternoons, others would want to have their work hours adjusted so that they can be off on Friday afternoons or cut short their lunch break in the month of Ramadan. Still others want to be able to leave earlier on a more regular basis to deal with daily family commitments.

The system of reference proposes dealing with this type of request within a collective framework, based on the principle that discussion between the employer and all the employees is fundamental.[20] Only through discussion is it possible to assess the feasibility of the request according to context, the jobs of the employees concerned, and the organisation of the department. It is also important to assess whether this benefits the majority of the employees rather than leads to chaos within the company. This is how HCD is implemented.

By applying this principle, it is possible to establish a slot of fixed working hours and a slot of flexible working hours for all employees. Hence, there would be a common core for everyone (for instance, 9:00 to 12:00 am), and flexible hours according to individual needs, whatever they may be. This system allows all employees to benefit from flexible hours. Within this framework, someone who wants to have Friday afternoons off can organise their time so that they accumulate enough hours for the flexible period in order to be able to finish their work before the end of the week.[21] This solution has the advantage of giving everyone the same rights and freedoms. Employees need not justify how they organise their work[22] as long as they respect the fixed-hours slot. The manager only intervenes if the hours due have not been worked or if the fixed-hours slot was not respected. Instead of establishing the same working hours for everyone, the working hours could be divided into fixed and flexible periods across the board, and this would offer everyone the same margin of freedom.

It should be noted that if the company's sector of activity or its situation does not allow this type of organisation, it is better to organise group consultations within teams, in order to ask all the employees if they would like to take advantage of the concession granted over the period of a month, for instance, by way of exception, to shorten the lunch break in order to finish earlier.

Another example is that of the employer who has been approached by a small group of Muslim employees and asked to provide them with a room for their daily prayers. This group of workers considers that being able to pray during their break time "would improve their well-being at work", and would lead to the workers in the group "being more committed to their work".

Here again, the method of group consultation allows the workplace context (type of job, profession that requires relating to death, relations between workers within the company, etc.) to be assessed on a case-by-case basis, as well as the management of time, space, the organisation of teams, etc. On the basis of a collective request pursuing a particular interest, it also makes it possible to suggest a response that will be of benefit to all employees and meets the requirement of general interest.

The system of reference underlines the fact that this consultation requires a few markers and several stages.

On one hand, it allows everyone to formulate a request and express it. The role of the union delegation is of utmost importance in this respect to communicate these requests to the company's consultation bodies.[23] On the other hand, the principle of individual freedom must be taken into account (religious or other) since it is a fundamental freedom.

Nevertheless, the employer has the duty to assume his/her responsibilities regarding the proper functioning of the company and to ensure the wellbeing of all the employees, without creating a situation of discrimination and/or iniquity.

If feedback reveals that favouring one particular interest and allowing some employees to have a dedicated room for worship may impede the team's cohesion and sense of equity, it may be in everyone's interest, in order to create a balanced compromise, to set up a rest area that would serve as a shared space for all employees. Everyone would be able to use this room for a quiet activity and to do what they wanted (read, meditate, sleep, pray, etc.), insofar as there is no noise and they only stay there for a limited amount of time. Quietness is the common denominator of all the activities of the employees who use this space.

If an employee wants to pray, they can while another colleague chooses to rest or read the newspaper. The space belongs to everyone and is open to all. In some companies, this type of place is part of their institutional culture. The rules of co-existence are negotiated by consultation bodies (health and safety committee), decided on collectively, and communicated to everyone, so that every employee can act autonomously in compliance with these rules. Ascertaining whether the time spent in this space could be considered work time is also a decision made by these consultation bodies.

However, the system of reference emphasises that, in some companies, the human management of this experience may require providing greater care and support for existing practices. In any case, the best solution is to draw up and display any internal rules that concern the users of these types of spaces.

Based on negotiation and consultation, the approach developed in the above-mentioned examples helps to rethink the overall structure by picking out, within the context of certain situations, the elements that link individuals and unite everyone's interests. Without denying the interests of a minority group, it is a matter of reaching a practical consensus "by stepping outside the box" and exceeding the ideological divide that pitches the supporters of a republican universalism against those who defend a certain idea of multiculturalism.

Outside of the sphere of employment, the HCD method also allows for pragmatic solutions that can be supported by all. How is it possible to fully respect the religious beliefs of certain people, for instance, at a cafeteria meal organised within the context of continuing education activities, without hindering the freedom of conscience of others? How to guarantee access for all without stigmatising the religious practices of certain individuals? How to find a way to "eat together" without imposing a specific single standard on all of the participants? How to ensure that the respect for the individual standards does not lead to "divisions" within the group, when the goal is, in fact, to "bring everyone together"?

By choosing "all halal" or "all pork", or by identifying tables according to a religious standard, such as the "halal table" – "non-halal table", there is a risk of assigning individuals to a certain religion (or lack of religion). This effectively forces them to define themselves and accept being perceived as Muslim/Atheist/Christian/Jew in order to take part in the meal.

And furthermore, how to guarantee a "mixture of cultures" and to avoid that each "religious community" clusters around "their own table"?

Finally, by organising "tables by religion", are those responsible for education not implicitly supporting a worldview according to which there is such a thing as a notion of "purity" and "impurity" which separates the "pure" from the "impure" by a type of symbolic boundary?

In order to be able to offer the corresponding diversity of foods without establishing separate spaces, the educational team should try to encourage users to do so without religious labelling or stigmatisation *at a common table*. Both alcoholic and non-alcoholic beverages, halal and non-halal cold meats, fruit, chips, cheese, eggs, and grilled meats (halal merguez or pork sausages) could be placed on the same table. Also recommended are "shared dishes" based on fish or eggs, or vegetables, which can be eaten by everyone. In case of organisational difficulties, these "shared dishes" could represent the highest common denominator (HCD) on which everyone could agree.

It must be emphasised that taking an individual request into account and considering it legitimate does not necessarily mean that one is going to create individually tailored responses; in all cases, equity must be sought and equal treatment guaranteed. The standard should therefore be expanded to include all workers (so as to provide vegetarian meals, and not just certified halal meals; a silent room, and not just a prayer room; breaks for everyone, and not only prayer breaks, etc.).

It should be clear that the HCD approach lies at the opposite end of the spectrum to the formal recognition of a right to "reasonable accommodations", which, based on the application of individual fundamental rights, offers individual and particularistic answers to requests linked to religious beliefs that, we argue, are likely to lead to the exacerbation of differences and even the attribution of identities.[24]

However, although negotiation is involved, in the context of continuing education as well as in the workplace, one must not lose sight of certain rights, values, and principles that are non-negotiable. The individual's right to actively express his religious or philosophical convictions must therefore be a part of an overall context that encourages dialogue and the construction of the most harmonious possible social cohesion.

Also, the right to individual freedom cannot come at the expense of a collectively constructed dialogue between citizens and without fully reflecting the intangible standards that underpin our rule of law and democratic state: respect for human rights, the promotion of individual freedoms, the fight against discrimination, equality between men and women, and the rules establishing processes of managing diversity.

To conclude, we would like to emphasise that we regard that no specific policy for combating discrimination and promoting diversity can reach its goal if it is not combined with "general" policies regarding employment, certainly, but also housing and, of course, education and training.

Notes

* Patrick Charlier is co-director of the Belgian equality body Unia (Interfederal Centre for Equal Opportunities, which was the former Centre for Equal Opportunity and Opposition to Racism). Nathalie Denies is a jurist at Unia and specialist in the field of discrimination in employment. This essay draws from Dounia BOUZAR and Nathalie DENIES, *Diversité convictionnelle, comment l'appréhender, comment la gérer?* (Académia- L'Harmattan, 2014) and action research conducted by the Brussels Centre for Intercultural Action. [Unia was tasked with the follow-up of the recommendations in the Belgian Round Tables Report.]

1 Postulate of the survey published under the title: Dounia and Lylia BOUZAR, *La République ou la burqa, les services publics face à l'islam manipulé* (Albin Michel, 2010).

2 For the premise of the survey, see Dounia and Lylia BOUZAR, *La République ou la burqa*.

3 Unia, 2015 Annual Report, "Living together put to the test", www.unia.be.

4 To present these initiatives chronologically, we could cite:

- the work done in the framework of the Commission on Intercultural Dialogue in May 2005;
- the participation of the Centre in the Conferences on Interculturality in 2010, particularly through the research conducted at the initiative of the Centre by A. REA and I. ADAMS: *La diversité sur le lieu du travail* (2010) as well as the minority opinion submitted personally by Edouard Delruelle calling for practical adjustments to be made by the social actors working in concert;
- creation of the website "signes convictionnels" (religious symbols) in November 2011, which directly offers a method for dialogue based on principles of harmonisation according to which,

the fundamental principle is freedom of expression and manifestation of one's beliefs. No freedom, however fundamental, is absolute. But any limitations must be circumscribed, justified and proportional. To define these limits, the Centre recommends the application of the 4 principles of harmonisation:

1) the principle of **consultation** at each stage of the decision-making process
2) the principle of the **necessity** of the ban ("is the ban truly necessary?")
3) the principle of the **proportionality** of the ban ("is the measure proportional to the desired goal?", "Who is affected – directly or indirectly – by the measure?", "What are the situations in which it would be applied?")
4) the principle of **transparency** and the motivation of the ban.

5 Article 19 of the Belgian Constitution provides that: "freedom of worship and freedom to express one's opinions in all matters shall be guaranteed except for the punishment of offenses committed in connection with the use of these freedoms", while article 21 states "the State has the right to intervene neither in the appointment nor in the installation of ministers of any kind of worship".

6 *Kokkinakis v Grèce*, 25 mai 1993, § 31, série A no 260-A; and *Buscarini v Saint-Marin* [GC], no 24645/94, § 34, CEDH 1999-I; "No one shall be compelled to disclose his thoughts or adherence to any religion or belief", Human Rights Committee, General Comment nr 22, art 18 CCPR/C/21/Rev. 1/Add (1993).

7 *Sviato-Mykhaïlivska Parafiya v. Ukraine*, no 77703/01, § 146, 14 June 2007.

8 This system of reference will be published by Académia L'harmattan, and was presented to the public at a symposium organised by the Centre on 23 October 2015.

9 D. and L. BOUZAR, *Allah a-t-il sa place dans l'entreprise ?* and *La République ou la burqa, les services publics face à l'islam manipulé*, in two editions (Albin Michel, 2009 and 2010). HCD (highest common denominator) is used to define a management philosophy that will offer "everyone a benefit based on a special request". This concept can also be applied to manage users of a public service or employees in a company.

10 See the laws of 10 May 2007 on combating certain forms of discrimination and theirs comments in C. BAYART, S. SOTTIAUX and S. VAN DROOGHENBROECK, *Les nouvelles lois luttant contre la discrimination* (La Charte, 2008), 603.

11 See also Andrea REA and Ilke ADAM, "Diversité culturelle en milieu de travail. Pratiques d'accomodements raisonnables en Belgique", IES-ULB Germe September 2010, 68, available via the website of the Center for Equal Opportunities, www.diversite.be, and Stéphane JONLET, *Les pratiques islamiques et le monde du travail, Perspectives des musulmans de la région de Liège*, Centre Interdisciplinaire d'Études de l'Islam dans le Monde Contemporain (CISMOC) Essays and Research Collections (Catholic University of Louvain, 2010); see also Younous LAMGHARI, *L'Islam en entreprise: la diversité culturelle en question* (Editions Academia-L'Harmattan, 2012); A. YERNAUX, *Les convictions du travailleur dans l'entreprise. Etude pratique du droit social* (Kluwer, 2014), 63.

12 S. VAN DROOGHENBOECK, "Les transformations du concept de neutralité de l'État. Quelques réflexions provocatrices", in Collection Centre des Droits de l'Homme, J. RINGELHEIM (ed.) *Le droit et la diversité culturelle* (Bruylant, 2011), 75–121; J. JACQMAIN, "Les lois anti-discriminations, services publics compris", in BAYART, SOTTIAUX and VAN DROOGHENBROECK, *Les nouvelles lois luttant contre la discrimination*, 603.

13 The CJEU has now issued a decision on this issue in Case C-157/15, *Achbita and Centrum voor Gelijkheid van kansen en voor racismebestrijding v G4S Secure Solutions NV*, 14 March 2017.

14 The obligation to respect this neutrality derives from the protection of the rights and freedoms of others and more precisely from an imperative to guarantee to all citizens equal access to the public service. The Council of State also recognises that

> the neutrality of the public authorities is a constitutional principle which, although not included as such in the Constitution, is nevertheless intimately linked to the prohibition of discrimination in general and to the principle equality of public service users in particular. In a democratic State under the rule of law, authority must be neutral because it is the authority of all citizens and for all citizens and must, in principle, treat them equally without discrimination based on their religion, belief, or preference for a community or party. For this reason, it is therefore possible to expect officials of the public authorities to observe strictly the principles of neutrality and equality of users
>
> (Opinion No. 44.521 of 20 May 2008 of the Legislative Section of the Council of State on the Proposal for a Law to apply the Separation of the State and Religious and Philosophical Non-denominational Organizations and Communities, Doc. parl. Senate, sess.ord. 2007–2008, No. 4-351 / 2, p. 8)

15 The Federal Royal Decree of 14 June 2007 concerning the status of state agents, which states in article 8:

> A State official strictly respects the principles of neutrality, equal treatment and respect for laws, regulations and directives. When in the course of his duties, in contact with the public, the agent of the State avoids any speech, attitude or presentation which might be such as to shake public confidence in his total neutrality, in its competence or in the dignity of his office. It also avoids any situation where, even by intermediary, it could be associated with occupations contrary to the dignity of its functions.

16 Stéphane GUERARD, "La liberté religieuse dans les lieux publics", *CRDF*, 4 (2005), 49–70.

17 For an original approach to the question see Bruno MARTENS, *Les guerres puritaines. Signes religieux et vêtements pol(Ys)émiques* (Editions L'Harmattan, 2011).

18 Edouard DELRUELLE, "Neutralité dans les services publics: exigence éthique ou obligation juridique", CERAP, ULB, October 2010, www.cerap.be/spip.php?article321. An example is the situation prevailing at the CPAS in Mechelen. The officials of this administration have decided to allow all of their workers to wear a uniform in which the headscarf is intended. In this way, the wearing of the headscarf is not only limited to the use of Muslim women but could also be used for workers who have just undergone chemotherapy. Another example is the municipal administration of the city of Ghent in June 2013, and just recently in September 2017 of the city of Leuven, which abolished the employment regulations that prohibited the wearing of convictions for all workers. Today the agents of the communal administration are therefore free to carry a religious sign.

19 Cf. Andrea REA and Ilke ADAM, "La diversité culturelle sur le lieu du travail. Pratiques d'aménagements raisonnables en Belgique", IES-ULB Germe, September 2010, available on the website of the Centre for Equal Opportunities, www.diversite.be.

20 This collective approach implies questioning certain management practices within the company in order to prevent conflicts related to demands that are described as "convictions" and to preserve unity between workers within the company. It is mainly found in collective labour agreements dealing with the fight against inequality (TCCs Nos. 38 and 95) and in legislation on welfare at work. This collective approach involves not the demand for the right but the negotiation of applications within the social consultation bodies. It is based on the principle of equality (which also forms the basis of the second approach) but introduces the notion of solidarity between workers in the legal debate. This approach is part of a collective approach to the prevention and resolution of labour disputes

21 For more on this, also see LAMGHARI, *L'Islam en entreprise*. This work discusses the experience of STIB. In this company, a consensus was reached concerning employee holidays: a system of exchange for days off, insofar as it is possible, to satisfy both Muslim and non-Muslim workers. In other words, Muslim workers who want to work during Christmas/New Year can inform the company about that by putting their names down on a list, thus allowing more workers for whom these occasions are important, to take time off during this period. In exchange, the latter work on Muslim holidays.

22 While being obliged to inform the manager of their religious beliefs.

23 Its objective is to defend the interests of workers with the employer. In this context, it has competence in the field of labour relations, ensuring the application of the employment regulations and contracts, the right to be heard by the entrepreneur in the event of any dispute or dispute of a collective nature, that of assisting workers in the event of individual complaints, and of having prior information as to changes likely to affect working conditions or remuneration. More broadly, union delegation in the company plays an essential role in detecting and preventing latent tensions and conflicts between one or more workers and management or other workers.

24 About this notion, see B. DUARTE "Enjeux juridiques soulevés par les accommodements religieux", in B. DUARTE and Frédérique AST, *Les discriminations religieuses en Europe, droit et pratiques* (L'Harmattan, 2012) and Emmanuelle BRIBOSIA, Julie RINGELHEIM and Isabelle RORIVE, "Aménager la diversité: le droit à l'égalité face à la pluralité religieuse", *Quarterly Review of Human Rights*, 78 (2009).

13 "Making a difference"

Dialogue and compromise in the Belgian Federal Round Tables on Interculturalism

Nadia Fadil[1]

In Belgium, the Federal Round Tables on Interculturalism (hereafter: FRI), initiated by the federal Minister of Equal Opportunity Christian Dupont in 2008 and continued by his successor Joëlle Milquet, were presented as a final governmental attempt to settle a number of controversial issues that were derived from what many perceived as a multicultural "crisis" in the country.[2] The initiative succeeded the previous *Commission du Dialogue Interculturel*, organised by the Minister of Equal Opportunity Marie Arena (PS) in 2003 as a way to mitigate and circumvent a number of societal tensions linked with cultural diversity that had flared up.[3] Other initiatives had also seen the light in this period, either at the federal or regional level, which aimed to define a number of core principles and values for policy purposes.[4] Yet from the very outset, the initiators of the FRI sought to differentiate themselves from the previous initiatives by explicitly foregrounding a bottom-up and open-ended approach. A second line of demarcation was towards the *Commission du Dialogue Interculturel*. Although the work of the "Arena Commission" – as it had come to be known – was praised for its open view on cultural diversity, it was also criticised for its failure to directly impact political decision-making.[5] None of the 30 recommendations had been realised. Additionally, on a number of thorny issues, such as the presence of religious signs at schools, the report had remained inconclusive. Many commentators therefore perceived the FRI as a "second chance" and a new opportunity to make a difference on a myriad of key themes that divided cultural and political life on issues related to multiculturalism. The work of the commission started in the autumn of 2009 and lasted for a year. A steering committee of initially 30 members[6] was convened, and chaired by Marie-Claire Foblets and Christine Kulakowski. The dialogues were organised in two stages. A first stage consisted of hearings and socio-cultural activities organised at the initiative of the commissions. The respective commissions worked quite independently from each other on a number of selected themes: schooling, employment, governance, public goods, civil society, culture and media. The minutes and findings of the different commissions were, in a second stage, synthesised into a general document that served as a draft for the final report. This draft was then discussed in a retreat that took place in Spa in September 2010 and where the members of the steering committee and the two chairs gathered to reach an agreement on a common text that would figure as the final report.

This essay offers an account of the conversations that took place during this final retreat and that would result in the 69 measures proposed by the FRI. I had been invited by the chairs to participate and supervise the final discussion as a moderator, together with another (Francophone) civil actor. My role during the debates consisted mostly of making sure that people were given the space to participate, introduce the themes, offer a synthesis, and raise a few questions. I also avidly took notes on my laptop during the discussions, which resulted in a 29-page document that offers an account of the dynamics of the conversation during the meeting. My notes also served as a basis for the minutes of the meeting written by the administration.[7] In reviewing the conversations that took place during this final retreat, the aim of this essay is twofold. The first aim is to demonstrate how – in opposition to some of the existing initiatives in a European context – the FRI was characterised by a consistent awareness that its bottom-up approach and the views held by many members of the commission deviated from the "dominant view" that existed within Belgian society on cultural diversity. Many commission members held an open and positive view on "multiculturalism" and defended positions that were explicitly understood as marginal. It was also one of the outspoken commitments of the chairs of this initiative to allow for a fair representation of such views and to consider the commission as a space wherein such positions could be expressed. Yet this self-awareness of the commission also occasionally conflicted with its desire to "make a difference" by adopting views that were seen as more restrictive than those initially endorsed by several commission members. Some of the social hierarchies the commission sought to avoid would be reproduced as its work gradually came to anticipate a more "public" character (i.e. via the final report). Through a description of the debates on religious signs at schools and in public offices, which was one of the most contentious themes, this essay will show how the work of the commission continuously vacillated between a desire to give space for "minority" perspectives and inverting existing hierarchies – thus functioning as a liminal space[8] – and the desire to find a consensual position that was deemed readable for "the public". Rather than addressing the effectiveness of the measures proposed by the RFI, this essay seeks to understand how the necessity of making compromises and the anticipation of a particular public worked as a regulatory semantic of a seemingly open-ended conversation.

The intercultural dialogues in a European perspective

The FRI were part of an international trend of similar initiatives in Western Europe (France, Germany, UK) and Canada at the turn of the twenty-first century. Several dialogue-initiatives were set up around that period that revolved around the experienced "challenges" of the multicultural composition of these societies. The starting point for these initiatives was in all cases similar: the integration of migrants (mostly with a Muslim background) was seen to provoke a number of tensions. The perceived threat of "radical Islam", the poor socio-economic advances of migrant communities or the occasional outbreak of riots (like the ones in the French *banlieues* in 2005) triggered continuous debates on multiculturalism and Islam since at least 2001. A series of dialogues and consultations between distinguished members of the academic and political world, social and cultural life, and representatives from the concerned religious or cultural communities were organised. Through these, the commissions came to be seen as an extra-political space that were not subjected to the same kind of regulatory mechanisms as those within political life (e.g. the rhythm of the elections, being representative of a constituency etc.). The assumption was that the work of the commission would transcend the existing political cleavages. By establishing parallel and lengthy deliberation processes (usually between six months and one year) the initiators hoped that these commissions could formulate a series of guidelines that would inspire and influence political decision-making.[9]

In reviewing the Round Tables and dialogues that took place on cultural diversity, many academics have, however, underscored their depoliticising and disciplining nature, calling attention to the fact that some of these initiatives often served to reinforce – rather than challenge – particular cultural norms and views on cultural diversity.[10] The French Stasi (2003) and Gerin (2010) commissions are, for instance, good examples of initiatives that have facilitated the adoption of a number of nation-wide restrictive regulations on religious signs (e.g. the *hijab* and the *niqab*), although the intentions of the members were often more complex and aimed at a larger series of questions.[11] Some analysts have also noted how initiatives with a less direct disciplining character have also contributed to both reproducing the majority/minority hierarchies and to silencing certain voices. In her analysis of the Bouchard-Taylor Commission, Gada Mahrouse states, for instance, that the public hearings, which encouraged ordinary Canadians to share their views on cultural diversity, often resulted in public moments of interpellation of ethnic minorities and immigrants by white French-Canadians.[12] Political theorist Schirin Amir-Moazami also observes in her analysis of the *Deutsche Islam Konferenz* (DIK) – a series of conferences initiated by the German state in 2006 that aim to foster a dialogue between Muslim civil actors and state officials – that a set of regulatory scripts predetermined the ways in which such dialogue could ensue and which actors were seen as legitimate interlocutors.[13] Authors like Levent Tezcan or Frank Peter have also argued that initiatives like the DIK or other intercultural dialogues partially reproduce the imaginary of European Muslims as "*other*".[14] Rather than understanding these dialogue initiatives as open-ended, numerous analyses have therefore stressed how several of these initiatives actually reinforced existing power dynamics by relying upon and re-enacting majority and minority dichotomies.

The federal Round Tables were, from the very outset, understood as a "forum" and conceived as an open-ended and bottom-up initiative. The first phase consisted of a wide-ranging series of conferences, meetings and cultural gatherings that were organised with the support of the steering committee. Differing from the round tables discussed earlier, the members of the steering committee were ideologically diverse, with a blend of representatives from civil society and the academic world. Additionally, the steering committee membership was largely composed of people who either self-identified or were identified as a religious or ethnic minority (11 out of the 20) and who were selected on the basis of their reputation and representativeness (some of them were quite controversial). The work of the different commissions resulted in several intermittent policy briefs and reports, meanwhile some members had also commissioned university scholars to write reports on assigned topics.

This essay examines the negotiations and dynamics within the steering committee as it proceeded to adopt the final report. The aim of this essay is twofold. A first aim is to contribute to the existing scholarship on commissions, round tables and dialogue-initiatives by not only relying on a governmentality perspective, but also by apprehending the *potentiality* of (some of) these initiatives through the lens of ritual and performativity studies.[15] Building upon an anthropological literature that considers rituals as performance, I want to suggest that the work of the FRI can be understood as a *liminoid* that temporarily suspends and challenges the existing hierarchies, yet without completely subverting them. In *The Forest of Symbols*, Victor Turner describes the liminoid as an inter-structural moment that is typical for initiation rites or "rites de passage".[16] This seminal concept, which builds on the work of Arnold Van Gennep, describes a stage in the ritual where the individual is no longer firmly embedded in their initial social position, nor has it reached its "new" social position. Liminoid refers to a state which temporarily eludes existing social norms, and which is also marked by a structural invisibility. Turner's initial interest in this concept was to understand how small-scale societies deal with transitions. He also believed that this period of in-betweenness contained a creative potential, something he will explore further in his work on communitas or anti-structure.[17] But what is often overlooked is that Turner's theorisation of liminality is equally a reflection on how power works, and that the periodic suspension of reigning power relationships by no means signal their exemption. Liminality figures more as an attempt to reflect on the possibility of temporarily escaping the existing hierarchies without entirely overcoming them. It is a performance of a subversion that is enabled by virtue of the reenactment of the existing structures.

A second aim of this essay is to therefore understand how, despite its open character, a certain regulatory mechanism nevertheless structured these interactions – especially as the commission moved towards drafting the final report. The early work of the Round Table consisted of diverse activities, and the chairs played an important role in maintaining this openness. Ethnic minorities were also quite vocal in setting the agenda and the dominant concern with "integration" was strongly critiqued throughout the discussions.[18] While the highly decentralised structure of the FRI and the continuous disruption of existing majority/minority dynamics turned it into an attractive space for many civil representatives, ostensibly demarcating it from the initiatives organised elsewhere in Europe, some of its initial open-ended nature eroded as the commission gradually moved to drafting a position text. A lot of this had to do with the desire shared by several commission members, I want to suggest, to "make a difference" by adopting a unanimous position on a number of controversial issues. The question of veiling in schools and public offices, in particular, was one of the issues with not only a strong public character but which also divided the commission members. Various attempts were therefore set at reaching a "consensus". In what follows, I will try to show how these attempts at reaching out to the public likewise enabled a re-enactment of certain majority/minority dynamics, which had been temporarily suspended. Yet these attempts at reaching closure remained inconclusive. Disagreements continued to haunt the on-going discussions, resulting in the ultimate failure to adopt a consensual view on certain controversial issues. I want to argue that it is precisely the unsettled character of the FRI that gives the Belgian initiative a distinctive and unique character. As such, the work yielded by the commission lays not so much in its ability of reaching compromises, but rather precisely in its *inability* to do so. Understanding the work of the FRI as a *non-performative*, as coined by Sara Ahmed, is therefore a more fruitful way to apprehend its political and ethical work.[19]

The FRI as anti-/structure

The concluding meeting of the steering committee of the FRI took place in the hilly town of Spa in September 2010. The participants gathered on a Saturday morning in a meeting room of the hotel where all commission members resided for the weekend. The members sat behind tables that were organised in a circle, and faced each other throughout the discussion. About 11 members of the steering committee were present. A first draft of the final report with an initial series of recommendations had been written by a journalist, and served as the basis for the discussion. The members of the steering committee had been asked to read the draft carefully and to come prepared with their comments for the final discussion.

The meeting opened with some introductory remarks by one of the chairs. She explained the main purpose of this retreat and reiterated the general aims of this initiative. "The starting point in our conversation is that we live in an intercultural society", she insisted. That, she explained, had to be taken for granted. The next question was therefore how one was to properly organise such a reality: "Are the regulations we have sufficient, do they draw on valid premises, or should we maybe even reconsider the premises that inform our legal system?" The chair also insisted that while it was important to strive towards a set of coherent standpoints, the possibility of having dissenting views should be upheld. After the introduction, the different members were invited to share their general impressions of the text. From the very outset, however, it became clear that some members strongly disagreed with the final text, either because it was "too" multicultural or not multicultural enough. Some individuals from a minority background felt that the report was not inclusive enough and also felt that the bottom-up approach and rich set of insights that had informed the hearings had somehow been lost in the final result. Some also felt there was an absence of a few important themes. For example Amin,[20] an African member of the commission, deplored the absence of an explicit reference to the colonial history in the national storytelling. Khalid, who represented an anti-racist organisation, also shared his sentiment that the report seemed to cater primarily to the sensibilities of the majority, and that the effort of "integration" was mostly expected from the minorities: "but interculturalism is not only about minorities. It should go both ways. It should also concern the dominant society". All attendees did not, however, agree with this conclusion. Others alternately viewed the report as "unbalanced" and too "radical" in its defence of the ethnic minorities. One participant felt that "society and political life" were too easily blamed, and not enough attention was given to the responsibilities of the ethnic minorities. Another member, Antoine, even expressed a general sentiment of "malaise" with the report as a whole:

> The introduction presents the report as intercultural. Yet what we have here is 100 per cent multiculturalism. Some positions in the report are described as "majoritarian", but I don't recall a discussion or a vote in the commission where we agreed on that. We need to bring in more nuance, the different positions have to be clarified.

Once the first positions were settled, the commission proceeded by going through the different themes addressed within the report and inviting everybody to list their remarks. As the discussion advanced and the different members were sharing their comments on education and labour, Khalid suddenly intervened. He expressed his frustration regarding the adopted method and felt that too much attention was wasted in discussing small and futile details that stood in the way of a more profound discussion of the central topics that still divided the commission members such as the headscarf, the principle of reasonable accommodation or the adoption of the principle of quota.[21] He suggested that the primary focus should be on these "central issues", because answers concerning these questions were to be expected from the Round Tables. Khalid's remark did not fall upon deaf ears and most members agreed to proceed in this manner, although some – like Antoine – insisted that one should also have the time to offer more precise formulations on all the themes included in the report.[22]

A central theme that divided the commission members was the question of visible religious signs or the "*headscarf*" – as it was called throughout the discussions – in public institutions. This question had led the previous *Commission du Dialogue Interculturel* (CDI, 2004) to abstain from deliberating on the matter. The CDI members decided instead to outline the different views that existed within the commission on this matter. If the members of the FRI, however, agreed on one thing, it was that the CDI's reservation and failure to decide on the matter ("*trancher*") had been a missed opportunity. The FRI members thus saw this commission as a new chance to settle the matter once and for all.

One of the moderators started by drawing the commission members' attention to what he saw as a methodological mistake in the initial draft: the text put students and teachers on an equal footing. For him, the right to wear a headscarf was a very different matter depending on whether it concerned students or teachers. Teachers were confronted with issues of neutrality, a principle that didn't apply to students. He therefore suggested that the commission members treat veiled students separately from the teachers, beginning first with the students' right to wear

Table 13.1 Students' right to veil

Students' right to veil	*Votes*
Total ban	2
Partial ban	1
No ban	5
Abstention	1

a headscarf at school, and then narrowing the question further to focus on high school students. The commission members agreed, and the moderator proposed to start the discussion with a first round of votes. All committee members were aware that the group was divided on this issue. A vote would allow the members to have a clear idea of how the opinions varied on this question. The commission members agreed and the moderators (myself included) invited the commission members to cast their vote on three positions that corresponded with the ones identified in the CDI: (a) in favour of a total ban on the veil, (b) opposed to any kind of ban or (c) in favour or a partial ban. The first round of votes revealed the position shown in Table 13.1.

The chair observed that there was a pronounced disagreement among the commission members and wondered how to move forward. As shown in Table 13.1, five were opposed to any kind of ban, while a minority was in favour of a partial or total ban. The question was then raised on how to proceed: should the commission members strive for a consensus, or should they just list the existing positions and disagreement? The commission members were divided. At that moment, Khalid intervened. He expressed his disagreement with the whole procedure:

> I don't like the way the debate started, through a vote. The stronger we can unite behind a standpoint, the easier it will be to defend it [to the outside world]. I want to reach a consensus. I can make compromises, if that means sending a strong signal to the outside world. There is a *de facto* norm, since most schools already ban the headscarf.[23] And it is towards that norm that we need to position ourselves. I am ready to make a compromise that will work for us all.

Antoine, who was in favour of a total ban on religious signs in schools, seconded this opinion. He argued that a "structural solution" was needed, because certain schools use the ban on the veil to select students on social grounds: "I think there should be one rule for everyone, even if it is one I cannot agree with. The political world needs to take its responsibilities on this matter". Not everybody was convinced, however, by the need for a compromise. One of the chairs questioned the necessity of one position on this issue and suggested that simply listing the different positions – and their effects – could be equally beneficial. Amin, from the African platform, refused the idea of banning the veil on more principled grounds: "a prohibition can limit a girl in her growth. You cannot defend the freedom of religion and similarly limit the girls in their self-development. It's not enough to allow for three days of religious holidays, you also need to allow people to flourish as individuals". He also insisted that one should not "overestimate" the potential of this commission: "we are not going to change the world. What people expect from us is an opinion. Dissenting opinions could serve as a solution. I don't want to compromise on this question". Miriam, the coordinator of a platform for ethnic minorities, also shared this view. She argued that the prohibition on religious freedom in no way guaranteed the disappearance of phenomena such as peer pressure:

> If you adopt that line of reasoning, you can also extend it to other practices. Can you guarantee that there won't be any peer pressure on youngsters to fast during Ramadan for instance? What do you do then, ban Ramadan altogether? Such prohibitive measures will disproportionately impact the chances and talents of women in schools or on the labour market. If you want to emancipate women, you need to let them make their choices by themselves.

But despite these reservations, several commission members were convinced that the commission ought to endorse one viewpoint if it were to have an impact. Several members, like Khalid, reminded the commission members that previous initiatives, such as the *Commission du Dialogue Interculturel*, had failed exactly because they didn't manage to solve their internal disagreements. He also argued that making such a compromise was not easy and would go against his most fundamental convictions, which was that veiling was an elementary right. But he also felt that certain sacrifices ought to be made in order to reach a consensus. The position he proposed was therefore of a partial ban, where visible signs were prohibited until a particular age. This position would, however, also presuppose a total freedom to veil after that age:

> When we look at age, it is useful to look at the moment when a child can testify in front of the court. That's at the age of 12. Sexual maturity is reached at the age of 14. It seems illogical to me, then, that a veil should only be allowed at 16 – as has been suggested by some.

He thus proposed to take the age of 14 as the limit. However, not all members agreed, and the discussion concerning what age might be considered the threshold continued.

The work of the commission resumed the next morning. Diverging from the day before, some of the more principled voices seemed to have changed their minds and were more willing to consider a position of compromise. Khalid reminded attendees that making a compromise is never easy: "but we have to figure out something that can be understood as a third way for us all. I therefore propose to adopt a prohibition until the third year of high school and to allow it from the fourth grade onwards". Amin responded that he felt we shouldn't compromise, but instead come up with a new standpoint. But he was also inclined to accept the position proposed by Khalid, under the condition that this solution was understood as being temporary. The other commission members agreed, and Isabelle, a Congolese member saw this move towards compromise as acceptable since she often interacted with autochthonous people in her professional life and she felt that it was important to also give their views some space: "when it concerns interculturalism, it is important to give all opinions some space, also those who oppose the headscarf". After a very brief deliberation, the attendees decided to adopt a motion where the idea of a partial ban was defended. The moderators explained that a proposition would be drafted and sent to the different commission members, who could then formulate their remarks if necessary.

***

The negotiations described above reveal a complex set of concerns that go beyond ideological differences or an opposition between ethnic minorities and an ethnic majority. A first important observation is the clear disagreement among commission members on the "multicultural" nature of the report. While some minority members considered the report as not inclusive enough, others saw it as too multicultural or, as Antoine had argued, "100 per cent multicultural". Several of the commission members and chairs, furthermore, repeatedly stressed their desire to be "*avant-garde*" on these matters by being as inclusive as possible. They considered this commission as a space that would produce a new vocabulary and perspectives on diversity, often to the dissatisfaction of some (white) members. This was also visible through the limited support that the idea of a total headscarf ban initially had among the commission members: a clear majority opposed the idea altogether. This position reflects a minority view in the Belgian context, for types of regulations and restrictions that exist within the Belgian context and are upheld by schools, institutions and the different political parties.

Yet this acute awareness of being "*avant-garde*" on a number of key themes also reinforced the impression that compromises would be needed if the commission were to reach out to the larger society. This ratio was not only deployed by the members of the commission who were initially in favour of a ban, but also promoted and asked for by some individuals with a minority background (and most notably Khalid) and who opposed the idea of a ban. Several commission members shared their concern that if their position didn't at least partially respond to "public opinion", they would not be taken seriously by policy makers and other stakeholders. They also insisted that the commission should avoid falling into the same trap as the CDI by not deciding on the issue ("il faut trancher"). Achieving a compromise on some of the thorny issues that divided the commission members – such as the question of the headscarf – was seen as crucial for the success and impact of the commission. This need for a compromise was justified in two ways. First, by calling for a "general solution" to the headscarf problem that would be implemented in all schools. The arbitrariness on this question, as also shown above, was considered a problem as it allowed schools to decide for themselves whether to adopt a ban. This, they argued, worked in favour of elite schools who used the ban on the headscarf to reinforce discriminatory measures towards poorer students and to strengthen social barriers. By calling for an opinion that could be endorsed by the whole commission, several members (like Antoine) also stressed the need for a general regulation that could go beyond what many perceived as a policy of "*laissez-faire*". Regulating on the headscarf thus figured here as a matter of public governance. But this need for a compromise was also, and repeatedly, justified with reference to the ability to reach out to the different members of the group and to remain legible to the "outside world".

By invoking this idea of a compromise, the attendees reflected a cultural principle that is at the heart of Belgian political life. Political decision making in the country has often been labelled "la politique du compromis". This principle is particularly applicable to the three fault lines political scientists have come to identify as being central for the country: the linguistic barrier (Dutch/French), confessional cleavage (Catholic/free thinking) and economic differences (wealthy Flanders/impoverished Wallonia).[24] The idea of compromise has therefore been considered as an important technique that enables political decision-making in a highly decentralised and complex country. But besides figuring as a powerful idiom that captures a distinct political tradition, the idea of compromise also seems to tacitly function as a regulatory technique in this specific context. In an essay entitled "The Morality in/of Compromise" (2004), the Tunisian-Belgian anthropologist and sociologist Mohamed Nachi calls for a social scientific position that takes the pragmatics of compromise seriously.[25] He starts off by describing how this idea of compromise has often been distrusted and neglected because of its general association with positions where one's principles or values are left aside (*compromission*).[26] Yet by re-invoking the ethical value of a compromise, relying partially on the work of Golding, Boltanski, Thévenot and Ricoeur, he also seeks to demonstrate how this "pragmatic of agreement" can be understood as a "bulwark against violence in so far as it puts to the test a 'logic of reconciliation' and a 'space of abstraction'".[27] In his article, the social scientist delineates a number of guidelines that can help us determine the morality of compromises. One of them is the principle of reciprocity and exchange, as well as a "moral acknowledgement of the other party".[28] Nachi's perspective on compromise is informed by a pragmatic philosophical view and a desire to understand how such a perspective can offer a model for conflict resolution that relies on an older tradition and departs from the idea of a "consensus".[29] Compromise thus presupposes here the acceptance and acknowledgement of a number of pre-existing conflicting views. My interest in this idea of compromise is partially informed by Nachi's suggestion that compromise needs to be taken seriously as a pragmatic, though I do not necessarily approach it as a method of conflict resolution, signalling a departure from Nachi's assertion. Rather, I am interested in how the idea of "compromise" seems to act here as a regulatory discourse. This operated in a twofold manner. The first, which has been extensively illustrated in this section, was in considering a gap between the work and positions held by the members of the FRI and the anticipated "public opinion", which figured here as their addressee. Building upon Michael Warner (2002), one can argue that the idea of the public operates here as a conditioning fiction which partially regulates the kind of opinions that are considered legible and valid to the outside world.[30] The "public", as anticipated and imagined, was considered far removed from the life-world of the commission members. A speech act that would fail to account for this reality was thus treated here as a "failed" speech act, something the commission members wanted to avoid at all costs. Yet in addition to this pragmatic concern, another aspect was entailed in this ideal of compromise, which also touched upon an ethical dimension raised earlier by Nachi's conceptualisation. Compromise, so the argument goes, is not only good to reach a solution, but also says something about one's moral subjectivity. The next section, which addresses the negotiations around the headscarf for public officers, explores this point more deeply.

The moral virtue of compromise

Like with the students, different views existed on a teacher's – or public officer's – right to veil. Yet this topic was considered to be even more controversial than the previous one. The idea that civil servants should remain "neutral", also in their appearance, was a dominant norm in Belgium, which was challenged by only a few minority, anti-racist and feminist organisations, especially in Flanders.[31] The discussion started, like the previous one, with a first round of votes among the commission members. Four positions (two principled and two intermediary) were identified for which the members were invited to cast their votes: (a) totally in favour of visible religious signs for civil servants, (b) totally opposed to it, (c) restricting a ban to civil servants who were in contact with the public and (c) restricting the ban to civil servants who were in a position of authority (like police officers or judges).

Although opinions were more divided on this matter, as shown in Table 13.2, a majority of the attendees rallied behind a position that either partially or totally opposed a ban on civil servants, which clearly demarcated them from the public opinion. One of the chairs explained that while she understood the argument of the opponents, who stated that being neutral should be understood as a matter of conduct rather than appearance, she also noted that such a position could produce confusion in the legal domain: "People fail to make those distinctions. That's why I'm in favour of a ban for those who are in a position of authority. I don't think that the mentalities are ready to see a judge with a

Table 13.2 Civil servants' right to veil

Civil servants' right to veil	*Votes*
Allowed	5
Not allowed	2
Banned for functions visible to the public	2
Banned for authority figures	3

headscarf on". However, the idea that society wouldn't be ready was challenged by other members. Amin rejected this argument, comparing it to other modes of reasoning that were used to justify the discrimination of black people:

> During apartheid, people also said that black people were disturbing! We are using exactly the same mechanisms today. A judge can be a Muslim and issue a valid, and neutral, judgment. It is important to see someone as a citizen first. Religion should not limit my rights to fulfil a job. When we talk about interculturalism, it is also about the possibility of respecting oneself.

Several other members, however, insisted that while these principled arguments were correct in substance, the members should strive for a compromise. Especially Khalid, who insisted on compromise and reiterated his sympathy for the more principled positions: "We are all touched by it. But we have to make a compromise to acknowledge the more secular trends within our society . . . we still need to learn to trust each other". He therefore proposed an alternative solution that consisted of listing the professions that could be equated to holding a position of authority (such as judges, police officers etc.). Other members opposed to this idea also questioned the necessity of finding a compromise. Miriam stated:

> I respect all opinions that are present in the room, because they express a sincere concern. But I don't want to be seen as someone who doesn't want to make a compromise. We are talking here about discriminatory measures. You cannot make a compromise on everything!

She also reminded the commission members that they had previously unanimously endorsed the principle that no derogation to the official school programme would be tolerated, which was a reaction to certain tendencies among (Muslim) students to challenge evolutionary theory. "We didn't feel the need to come up with a compromise solution then. Well, we shouldn't for this one either". She also added that neutrality had to be understood as a learning process: "It's the same as for women and for black people. People just simply have to get adjusted to seeing that [veiled women]. It's a learning process we have to go through as a society".

These principled arguments, however, did not convince all commission members. Several felt that a compromise position was necessary to avoid alienating the rest of the Belgian population. The intermediary solution proposed by Khalid, to adopt a ban only for authority figures, gained the approval of several members. Philippe, a philosopher and academic, agreed with this solution and wanted to explore it further:

> Maybe we should come up with a list. We certainly have to add judges and teachers. And we also need to specify what we mean by distinctive markers. This compromise proposition is motivated by a desire to render an intercultural society workable for all. This also means that we cannot avoid creating a situation that will occasionally be experienced as uncomfortable.

Table 13.3 Civil servants' right to veil (final round)

Civil servants' right to veil	*Votes*
Allowed	3
Not allowed	2
Banned for authority figures	7

Miriam remarked that such a compromise would *de facto* encourage the introduction of a ban where it doesn't exist right now. Most members, however, rallied behind the proposition suggested by Khalid and a motion was adopted which advised for a partial ban among civil servants who were in a position of authority. A final round of votes was held, and a majority of the members supported this proposition (Table 13.3).

Some members, however, felt the need to express their reservations regarding their vote in favour of the partial ban. Mehmet, a scholar who had previously refused any ban on principled grounds, yet had agreed to support the compromise position, insisted that the authority figures should be very explicitly listed. He warned that if this didn't happen, he would withdraw his support. Others agreed with him, and Khalid added that while there seems a clear consensus for public functions like the police or a judge, this was less the case for teachers. Yet this didn't seem to satisfy Mehmet, who felt that the whole motion was premised on a fundamental distrust towards difference: "I find that problematic. We have to strive towards a society of trust. And this motion doesn't facilitate the construction of trust. I therefore withdraw my initial support". Some members were perplexed by his sudden change of mind. But other members, like Miriam, tried to appease these emotions by noting that it was already a good thing that a majority of the commission was opposed to the idea of a general ban. She stressed that this element should be highlighted in the final report.

***

The descriptions above show that a large part of the discussions were not only centred on the substantial arguments, as several commission members seemed to initially oppose the idea of a ban, but also revolved around the viability of a position that would oppose a ban altogether for civil servants. All members were able to express their views and, differing from the dominant views that circulated in the outside world, the discussion was premised on the idea that imposing a ban on civil servants was intrinsically problematic. Yet a large part of the discussion was also conditioned by the individual members' ability to make a compromise. This capacity of adopting a compromise position seemed to be far more persuasive in the final outcome of the discussions than the initial opposition several members had towards the idea of a ban.[32] Several members, as noted above, stressed their desire to avoid both alienating "the rest of society" and being considered as too excessive. They also stated that society was not ready for the measures proposed by the commission and stressed the need to find a position that could be considered as acceptable for the "secular segments" (*composantes laïques*) of the population – as Khalid called them. A second set of arguments did not so much pertain to the ability to remain in touch with "society" but rather referred to the members' capacity to sideline their own convictions for the betterment of all. Khalid, for instance, stressed that he felt conflicted about making such a compromise but eventually resorted to doing so for a greater good (i.e. having a report that would make a difference). Others, like Mehmet, were more ambivalent. While he initially accepted the conciliatory motion, he ended up withdrawing his support at a later stage. Some members, like Miriam, however, refused the dominant view that a compromise position was essentially a good thing in itself by reminding the attendees of other "principled questions" where commission members would never tolerate a compromise view (such as challenging evolutionary theory). Her intervention thus sought to problematise the necessary virtue of compromise by highlighting the power dynamics that undergirded such positions, in this case Belgian society's hostility to headscarves.

A tacit opposition was thus created between those who were open for a compromise and those who were not. Being able to find a way out of what many experienced as an impasse (i.e. the disagreement around religious signs) operated here, furthermore, not only as an important rhetorical devise, but also as a moral claim. Compromise was framed as a difficult position, and some of its defenders invoked an emotional rhetoric of self-sacrifice ("we don't take this lightly"). The affective register here highlights the ethical labour performed in the name of greater values and similarly acts as a subtext to the formation of a new community – i.e. a reasonable community, the community of the "*vivre-ensemble*" that is fermented by this ability to sidestep one's most intimate conviction. Cultural theorist Sara Ahmed has already suggested that emotions "do things, and work to align individuals with collectives – or bodily spaces with social space – through the very intensity of their attachments".[33] By being explicit about his difficulties and conflicted sentiments on a partial ban, Khalid was also inviting his fellow members to perform the same kind of ethical work as himself. Making a compromise thus not only acts here as a rational or deliberative effort, but becomes understood as a technology of the self (Foucault) or a virtue.[34] Being able to make compromises is linked with the performance of a particular moral self whose ethical subjectivity is conditioned by its ability to reach out towards the other. At the other side of this ethical spectrum, lies – in this suggested reasoning – the individual who refuses to make a compromise and remains locked up in his/her own truth.[35] Yet several commission members resisted the proposed framing. Both Mehmet and Miriam, for instance, refused to understand compromise as an essential moral good by stressing its discriminatory components. The ethos of compromise became thus contrasted with another ethos: anti-racism. This logic was acknowledged by Khalid, but became incorporated into a self-sacrifice logic that was deemed necessary for a greater good (i.e. publishing a report that would have a societal impact). These discursive manoeuvres were not, however, successful. A number of dissenting notes were drafted, which allowed members like Miriam to render their principled opposition to a partial ban on religious signs explicit. The proposed compromise, thus, did not completely succeed, and the minority notes acted as a reminder of the essentially unsettled nature of the suggested position.

Conclusion

In her essay on the FRI, co-chair Marie-Claire Foblets offers an overall negative assessment of the outcome of this initiative. Her assessment is based on two observations. The first one is the overall relatively limited appropriation and impact of the report on public policy in Belgium and its different regional governments, at least in the short term. The second reason informing her scepticism refers to the harsh critique the Round Table committee members received, especially regarding the measures which were considered highly controversial: the generalised authorisation of the headscarf for young girls after the age of 16, the selective – yet large – authorisation of religious signs for civil servants and the proposition to tend towards more flexible holidays that would also account for non-Christian celebrations. These propositions were very negatively received and some accused the FRI of playing the card of "communalism" (*communautarisme*), which would eventually also lead Marie-Claire Foblets to raise her concerns on the deliberative nature of the Round Tables. While she praises the chosen methodology and "*open ended*" character of the FRI, she also questions the effectiveness of such a decision-making process:

> One might argue that it is unrealistic to expect to use the tools of deliberative democracy to develop solid public policy, much less to take action with a view to addressing the problems of contemporary multicultural societies in Europe. Some questions are simply too complex and the conflicts of interest too great for this to be possible. Yet it may be more realistic to accept that this working method lends itself better to finding solutions for certain problems than for others.[36]

In proposing to adopt a performance perspective that does not account for the effective political results and outcome of the FRI but rather considers the type of social positions performed *during* these interactions, this essay has sought to elucidate the kind of space the FRI enabled. I have suggested that the FRI could be understood as a *liminoid* or an anti-structural moment as minority positions gained a large degree of agency and voices that are otherwise illegible in the public domain were considered as one of the default positions. The descriptions of the interactions displayed a stabilisation, and in some cases inversion, of existing hierarchies and the FRI could thus be viewed as a utopian performance of a "multicultural drama" wherein social actors were given a similar degree of legitimacy and power. Yet the descriptions above also showed that a second concern cut across these interactions, i.e. a desire to remain "legible" to the outside world and to make a difference. Although initiated as an open-ended process, the final phase of the report witnessed an especially renewed concern with the dominant hierarchies. This became more explicitly articulated through the desire to find a compromise position on certain thorny issue that could differentiate the FRI from its predecessor. Compromise figured here not only as an ethical and moral good, but also as a method that could create a foundation for a larger audience. I have tried to show how this rationality regulated the course of the discussions in two ways. The first was through the anticipation of a gap between the multicultural orientation of the commission and the outside world. The members of the commission largely saw themselves as deviating from the *doxa*, thus reiterating the abovementioned view equally expressed by Foblets that the commission held an *avant-garde* position. The second was through self-performance and self-fashioning as a reasonable subject. We saw how some of the conversation partners ended up adopting a conciliatory note for moral reasons. Yet these negotiated positions remained unstable and were also consistently and ardently challenged. While some members refused such a position, either because they deemed it too multicultural (like Antoine) or because it ran against their principles of non-discrimination (like Miriam), others withdrew their earlier support for a compromise position (like Mehmet). The achieved "compromise" was, consequently, permanently unsettled and the consensus never reached a position of closure. This was also rendered explicit in the minority reports published as an annex with the final report.[37]

As a final note, I want to argue that instead of viewing the unstable nature of the compromise and the inability of making a "difference" on the ground as evidence of a failure, this idea of a "failed speech act" could also be understood more productively. In making this point, I rely on cultural theorist Sarah Ahmed's vocabulary to suggest that the FRI could be understood as a "non-performative". In *On Being Included*, Ahmed develops the concept of non-performatives in order to account for a particular type of speech act that doesn't produce what it states. Within a non-performative, she contends, the failure is essential to how this speech act works:

> In my model of the non-performative, the failure of the speech act to do what it says is not a failure of intent or even circumstances, but is actually what the speech act is doing. Such speech acts are taken up as if they are performatives (as if they have brought about the effects they name), such that the names come to stand in for the effects. As a result, naming can be a way of not bringing something into effect.[38]

This inability of "bringing something into effect" is, however, not the end of it – for Ahmed also discerns a particular ethical value in these non-performatives. They are the enactment of a moral utopia that is not (yet) realised, but continues to be nurtured and exists by virtue of its performance. Understanding the FRI as a non-performative, whose force consists in temporarily unsettling existing norms rather than re-enacting them, is what characterises the value of this initiative. The inability of achieving a particular compromise, and the consistent openness that was given to minority positions to challenge attempts at responding to a "public" were the elements of subversion that distinguish – I want to suggest – the ethical value of the FRI from other initiatives that were deemed more "successful". It is perhaps the unsettled nature and open-endedness of these dialogues, rather than the failed attempts at reaching closure, that should be understood as the force and value of such Round Tables. For such initiatives indicate that in a highly multicultural and complex society the outcome might not reside in the type of compromises that are being made, or the consensus that is being created, but rather in the ability to offer different perspectives and to keep them open.

Notes

1 The author would like to express her gratitude to Mathias Delori, Marie-Claire Foblets and Katayoun Alidadi for their comments on earlier drafts of this paper.

2 Alana Lentin and Gavan Titley, *The Crises of Multiculturalism: Racism in a Neoliberal Age* (London: Zeno Books, 2011).

3 The "Arena Commission" – as it was called by many – was established as a result of a series of round tables that were organised in the fall of 2002 by the then Minister of Equal Opportunities Laurette Onckelinckx. Riots broke out in Antwerp after an Islamic teacher, Mohamed Achrak, had been killed by his neighbour who was known for having far-right sympathies. The riots also took place in a highly charged political climate, right after 9/11, in the context of enduring success for the far-right party Vlaams Blok and a growing assertiveness of minority organisation. It was at that time that the Arab European League, one of the most controversial political organisations led by minorities, gained public attention. The movement was also accused of being implicated in the riots that followed the murder of Mohamed Achrak. For a further account of this period see Mohamed Ludo Dewitte, *Wie is er bang voor Moslims. Aantekeningen over Abou Jahjah, etnocentrisme en islamofobie* (Leuven: Halewyck, 2004) and Mohamed Benzakour, *Abou Jahjah. Nieuwlichter of oplichter* (Amsterdam/Antwerp: L.J. Veen, 2004).

4 In Flanders, the "commissie voor normen en waarden" was established in 2004, whose main purpose was to establish a series of guidelines that would be used in the integration courses (*inburgering*) that were organised from that period onwards.

5 Edward Delruelle, one of the co-authors of the final report of the *Commission du Dialogue Interculturel*, categorically describes it as a "failure" (*un échec*), which he links with the absence of a clear Belgian national identity and the fragmented nature of Belgian political life. Edouard Delruelle, "Du Commissariat royal aux Immigrés aux Assises de l'Interculturalité: 20 ans de débats publics sur l'intégration" in *Les assises de l'interculturalité*, ed. Marie-Claire Foblets and Jean-Philippe Schreiber (Brussels: Larcier, 2013), 36.

6 A number of members left the steering committee before the work was concluded. See Chapter 11 in this volume, by Marie-Claire Foblets.

7 For reasons of confidentiality, all participants have been rendered anonymous. I am grateful to Marie-Claire Foblets for allowing me to use the notes.

8 Victor Turner, *The Forest of Symbols: Aspects of Ndembu Ritual* (Ithaca, NY: Cornell University Press, 1970).

9 Some commissions have also been criticised on this ground for corrupting the legal decision-making process. See for instance the work of Richard Ashby Wilson, "Anthropological Studies of National Reconciliation Processes", *Anthropological Theory*, 3 (2003): 367–387.

10 For the case of the French Stasi and Gerin commissions see Talal Asad "Trying to Understand French Secularism" in *Political Theologies: Public Religions in a Post-Secular World*, ed. Hent De Vries and Elisabeth L. Sullivan (New York: Fordham University Press, 2006), 494–526 and Jennifer Selby, "Islam in France Reconfigured: Republican Islam in the 2010 Gerin Report", *Journal of Muslim Minority Affairs*, 31

(2011): 387. For the case of Germany see Schirin Amir-Moazami, "Dialogue as a Governmental Technique: Managing Gendered Islam in Germany", *Feminist Review*, 98 (2011): 9–27.

11 See Jean Baubérot, "La Commission Stasi vue par l'un de ses membres", *French Politics, Culture and Society*, 22 (2004): 135–141. The members of the Gerin Commission also expressed a very pronounced and clear opposition to the face veil (*niqab*), which they judged to be incompatible with republican values. Jennifer Selby also notes that this hostility points towards a tacit desire by these commissions to promote an "acceptable Republican Islam", see Selby, "Islam in France Reconfigured", 387.

12 Gada Mahrouse, "'Reasonable Accommodation' in Québec: The Limits of Participation and Dialogue", *Race & Class*, 52 (2010): 89.

13 This occurred, she suggests, through the pre-determination of who counted as a valid interlocutor as well as through the themes that were selected. She shows, for instance, how secular Muslim feminists figured in the first round as preferred interlocutors, for they embodied the liberal, autonomous subjectivity that was promoted by the state and contributed to "the further legitimisation of the state's interventions into Muslim gender norms and sexuality" (Amir-Moazami, "Dialogue as a Governmental Technique", 17). This asymmetrical structure was also expressed through the agenda setting, which was relatively unilateral, and through the consistent interpellation of Muslims as subjects who needed to demonstrate their capacity to endorse fundamental German values. See Schirin Amir-Moazami, "Pitfalls of Consensus-Orientated Dialogue: The German Islam Conference (Deutsche Islam Konferenz)", *Approaching Religion*, 1 (2011): 8.

14 Quoted in Riem Spielhaus, "Germany" in *Handbook of Islam in Europe*, ed. Jocelyne Cesari (New York: Oxford University Press, 2015), 131.

15 For a further account of such an approach see Catherine Bell, *Ritual Theory, Ritual Practice* (Oxford/New York: Oxford University Press, 1992).

16 Turner, *The Forest of Symbols*, 93.

17 Turner defines the latter in a later text as: "a relation quality of full, unmediated communication, even communion, which arises spontaneously in all kinds of groups, situations, and circumstances", Victor Turner, "Variations on a Theme of Liminality" in *Secular Rituals*, ed. Sally Falk Moore and Barbara G. Myerhoff (Assen/Amsterdam: Van Gorcum, 1977), 46.

18 This position also resonates with a number of critiques on integration that have been circulating in the Belgian and Dutch context and which were formulated by academics. For the case of the Netherlands see Willem Schinkel, *Denken in een tijd van sociale hypochondrie. Aanzet tot een theorie voorbij de maatschappij* (Kampen: Klement, 2007). The case of Belgium and Flanders has been explored through the work of Jan Blommaert and Jef Verschueren, *Het Belgische Migrantendebat. De pragmatiek van de abnormalisering* (Antwerp: International Pragmatics Association, 1992) and Karel Arnaut et al., *Leeuw in een Kooi. De Grenzen van het Multicultureel Vlaanderen* (Antwerp: Meulenhoff-Manteau, 2009).

19 Sara Ahmed, *On Being Included: Racism and Diversity in Institutional Life* (Durham, NC: Duke University Press, 2012).

20 All names used in the text are pseudonyms as to guarantee the anonymity of the respondents.

21 "Aide-Mémoire débat, 19–20 Septembre 2010", Private Communication.

22 The topics that were discussed were: the terminology used in addressing ethnic cultural minorities, the question of visible religious signs in schools and administration, the reorganisation of the school calendars in order to incorporate the different religious holidays and the issue of reasonable accommodation. A profound discussion of these different topics would lead me too far for my purpose. I will therefore focus instead on the veil, because it was understood by all commission members present as the topic that divided them and which had highly charged symbolic weight.

23 Belgian schools are legally entitled to adopt regulations that restrict the presence of religious signs, impose a uniform or regulate in any other way the sartorial practices of

students. Since 1989, schools have thus been introducing bans in their internal regulation on a case-by-case basis. Yet in 2009, Flemish public schools (*Gemeenschapsonderwijs*) departed from this position by adopting a general ban for all its schools and thus introducing the first structural ban in the Flemish context. The prohibition has been legally contested but continues to hold up in the organisation of the schools that fall under the auspices of this institution. See also Gilly Coene and Chia Longman, "Gendering the Diversification of Diversity: The Belgian Hijab (in) Question", *Ethnicities*, 8 (2008): 302–321 and Paul de Hert and Caroline de Geest, "Raad van state hervindt grondrechtenlijn in arresten GO! hoofddoekenverbod. Naar een kader voor een 'school per school'-beleid", *Tijdschrift voor onderwijsrecht en onderwijsbeleid*, 5 (2015): 8–11.

24 For an early account of how this logic of compromise eventually lead to the development of a liberal Catholicism that stood at the base of several negotiations between the Catholics and the liberals in the founding moment of the Belgian state, see Olgierd Kuty, "Aux sources du compromise belge: l'invention du consensualisme et du pragmatisme (1828–1835)" in *Eloge du compromis. Pour une nouvelle pratique démocratique*, ed. Mohamed Nachi and Matthieu de Nanteuil (Brussels: Academia Bruylant, 2006), 175–204. For a more general account on the ideological fault lines in Belgium see Luc Huyse, *De Gewapende Vrede. Politiek in België na 1945* (Kritak: Leuven, 1980).

25 Mohamed Nachi, "The Morality in/of Compromise: Some Theoretical Reflections", *Social Science Information*, 43 (2004): 291–305.

26 See also Alain Fumurescu, *Compromise: A Political and Philosophical History* (Cambridge: Cambridge University Press, 2013).

27 Nachi, "The Morality in/of Compromise", 295, 297.

28 Goding quoted in Nachi, "The Morality in/of Compromise", 299.

29 In this pragmatics of compromise, Nachi and de Nanteuil insist on the necessity to abandon *a priori* deliberation procedures, and to highlight the context-driven and practical nature of the compromise. See Nachi and de Nanteuil, *Eloge du Compromis*, 18.

30 Michael Warner, "Public and Counterpublics", *Public Culture*, 14/1 (2002): 49–90.

31 Organisations such as Minderhedenforum, BOEH, VOK have successfully campaigned against and lifted the municipal bans on civil servants in Ghent in 2013.

32 In a reflective note on his role in the Stasi Commission, Jean Baubérot equally highlights how a mix of ideological arguments as well as dynamics within the commission eventually led to the quasi-unanimous adoption of a motion on religious signs in public schools. He argued that an impression was created, also by the chairs, that in refusing this motion, the commission would be seen as an organ that was not entirely endorsing the value of equality between men and women: "Ainsi, sans reel débat, il fut suggéré que l'on ne pouvait pas être véritablement pour l'égalité homme-femme et tolérer le foulard à l'école publique", Baubérot, "La Commission Stasi", 139.

33 Sarah Ahmed, *Cultural Politics of Emotion* (London: Routledge, 2004), 26.

34 Alisdair MacIntyre, *After Virtue: A Study in Moral Theory* (Notre Dame, IN: University of Notre Dame Press, 1984 [1981]).

35 Several books have recently tried to account for the emergence of this "moderate" or "reasonable" figure that is seen to be at the heart of liberal democracy. See in this respect Alberto Toscano, *Fanaticism: On the Use of an Idea* (London: Verso Books, 2010).

36 Chapter 11 in this volume.

37 See "Note Minoritaire de Edouard Delruelle" and "Note Minoritaire de Naima Charkaoui", 122–125.

38 Ahmed, *On Being Included*, 117.

Part V

Engaging retrospective insights, forging ways forward

14 Religion in the retreat of multiculturalism

Christian Joppke

Introduction

Already in the mid-1990s, a reluctant co-organizer of a major academic meeting on "multiculturalism, minorities, and citizenship" found this a trite and outspent topic.[1] In the resultant volume, which assembled mostly liberal supporters of multiculturalism, Will Kymlicka (1999:113) declared that "the multiculturalists have won the day", in having "successfully redefined the terms of public debate" towards favoring difference-conscious rules and institutions. Around that time, even skeptical observers found that immigrant integration across a number of Western societies occurred in terms of "multicultural integration", if only in the minimal sense that integration "shuns the assimilation of immigrants" (Joppke 1999:146). This was just before a wave of political pronouncements that multiculturalism was "dead", which gained momentum in the post-2001 Islamist terror attacks and in the context of apparently failing Muslim integration, particularly in Europe.

The death-sayings reached fever-pitch in late 2010 and early 2011, when the political leaders of Germany, France, and Britain spoke out "against multiculturalism", almost simultaneously and in near-identical terms (see Bowen 2011). But already before this striking coincidence, the writing was on the wall. Consider only two of the three previous flagships of multiculturalism in Europe, the Netherlands and Britain. The Netherlands, under the influence of the populist right gaining strength after the political murders of Pim Fortuyn and Theo van Gogh in 2002 and 2004 (see Buruma 2006), had long set out toward an increasingly hard-lining "civic integration" policy for immigrants. And Britain, under a Labour government that had previously been friendly of multiculturalism, had moved toward a policy of "social cohesion" and strengthening "Britishness" after domestic race unrests in northern England in the summer of 2001. These riots were attributed to a penchant for self-segregation and "parallel lives" among Britain's Muslim communities, against which multiculturalism, obviously, had offered no recipe (see Cantle Report 2001). If multiculturalism is "dead" or in "retreat" (as I argued in Joppke 2004), these have been multiple and protracted deaths or retreats, and also ones that have been immediately denied (e.g. Modood 2013). So what should we make of it?

John Bowen (2011) acutely dismissed the latest headline-grabbing denunciations of multiculturalism, by German Chancellor Merkel, British Prime Minister Cameron, and French President Sarkozy, as "effective, albeit irresponsible, populist politics" (p.8). When Chancellor Merkel found that multiculturalism has "utterly failed" there was no multiculturalism policy to blame, because such thing had never existed in Germany, at least not at the national level.[2] Amnon Rubinstein astutely observed that "the new concept of multiculturalism has manifested itself in Europe more by the absence of specific demands for integration than by granting specific collective rights" (2007:772). Through this lens, the most plausible way to make sense of the German Chancellor's abdication of multiculturalism is to associate "multiculturalism" with a previous *laissez-faire* approach to integration that was now found wanting. This could not be said about France, where "integration" had been a matter of statecraft at least since the late nineteenth century, when "peasants" were turned into "Frenchmen", in the world's classic nation-building exercise (Weber 1976). In this light, one is astounded by Sarkozy's complaint that too much fuss has been made about "the identity of those who arrive and not enough about the identity of the country that accepts immigrants".[3] After innumerous *foulard* affairs, each framed by solemn reaffirmations of *laïcité* as the source of Republican unity, this was plainly a nonsensical statement. And, as Bowen (2011:6) turns the tables against the accusation, the abhorred "communalism" was less to be found in France's Muslim community than on the "interlocking boards of major companies, its exclusive school system, and marriage practices designed to preserve the elite". Of Europe's Big Three, only the British Premier's declaratory move from "state multiculturalism" to a tougher "muscular liberalism" made some sense,[4] were it not to exaggerate the degree of "state multiculturalism" in the past and the alleged toughness of his alternative, not to mention that this move had started already under his Labour predecessor.

In reality, as John Bowen points out with respect to the prickly issue of Muslim and Islam integration, all three countries were committed to "long-standing, nation-specific ways of recognizing and managing diversity" (2011:2). Just when multiculturalism was found wanting by its leader, Germany was holding corporatist Integration Summits and Islam Conferences with designated representatives of the country's Muslim community, even establishing federally funded Islam faculties at several state universities, one of whose purposes is to educate imams. Religion, after all, enjoys public status under Germany's church–state regime of "open neutrality", from which Islam cannot in principle be excluded. France had assisted Muslims in federating nationally and in innumerous other ways, from mosque building to the production of halal meat and subsidizing prison chaplains, which continues a long French tradition of state management of religion, in defiance of the official rhetoric of strict state–religion separation. And Britain accommodated Islam within its public schools more than most other countries in Europe, and would continue to do so, from religious diet to relaxed rules on school uniforms, lately even having to support Muslim faith schools, simply because Christian or Jewish schools have long received state subsidies and are considered "legitimate sources of citizen education" (Bowen 2011:8). All this may be called "multiculturalism" of sorts, and one that continued against the rhetoric. However, it would be equally accurate to see it as a prolongation of long-established ways of dealing with religious diversity.

The next section takes a step back to ask what, in principle, may distinguish religion from other multicultural claims-makers, most notably language; from there I return to the question how and why Islam is involved in multiculturalism's current retreat in Europe; third, I further trace this retreat at the political level, with a specific look at Britain; finally, I point to a persistent form of legal multiculturalism beneath a changing political reality, which is the British sharia councils.

Religion as multicultural claims-maker

Religion and language are often considered the primary markers of multicultural claims-making (first Zolberg and Long 1999; more recently Brubaker 2013a). While both are "basic principles of vision and division of the social world" (Bourdieu quoted in Brubaker 2013a:3), religion differs from language as potential group-maker in interesting ways. In contrast to language, religion constitutes a "structure of authority" that often competes with the authority of the state (p.14). Compared to language, religion is simultaneously easier and more difficult to handle for the state. Religion is easier to handle because the state can distance itself from religion, and through its liberal neutrality mandate is even required to do so; by contrast, a public language has to be fixed as a medium of communication that is necessary for political rule. But religion is also more difficult to handle because of its ethical claims that may compete with or contradict the claims of the state. On the part of the individual, the situation is aggravated by the fact that monotheist religions, at least as practiced in the West, are exclusive, in that one can partake only in one religion at a time (and normally over a lifetime). From this follows the high level of protections for religious freedom in a constitutional state, which has no equivalent on the side of language rights. While being assimilationist on language, states are thus inherently pushed towards an "attenuated pluralism"[5] on religion (Zolberg and Long 1999:14).

Brubaker (2013a:7–15) summarized the distinctiveness of religious pluralism as being "more robust" and "deeper" than language pluralism. Religious pluralism is "more robust", because, among other factors, religious reproduction mostly happens informally, in the family, while linguistic reproduction requires "exo-socialization", i.e., a costly state apparatus that is not available in this respect to immigrants. There is thus no equivalent on the side of religion to second- or third-generation language assimilation; on the contrary, the descendants of immigrants often self-identify more ardently in religious terms than their parents. Religious pluralism also cuts "deeper" and is more divisive than language pluralism because of the ethical and moral content of religion. In language conflicts, both parties still "speak the same language" (p.15), as it were, while the different moral grammars and worldviews of religiously divided people may drive them much further apart, to the point of disrupting the bind of shared humanity. Brubaker darkly concludes that "religion has tended to displace language as the cutting edge of contestation over the political accommodation of cultural difference in Western liberal democracies" (p.16).

A peculiarity of religion is that it tends to be considered a voluntary affiliation. As Amy Gutmann put it, "most religious groups . . . are voluntary" (2003: 24), and she distinguishes them in this respect from "cultural" and "ascriptive" groups, which are involuntary. In Olivier Roy's account of increasingly deculturalized religions (as "new religious movements") in the contemporary world, religion becomes increasingly "choice", with "conversion" being "at the heart of the disconnection between the religious and the cultural" (2008:36). While one converts to religion, "one doesn't convert to a culture, one adopts it gradually, or one learns it" (p.52). But as pure choice, religion would be off the radar of multiculturalism, if we understand the latter as involving an element of ascription, if not discriminatory external labelling, which can trigger a politics of identity. There are already intrinsic reasons why a depiction of religion in terms of choice does not seem to get it quite right. Note that the true believer is more driven by her conscience than by choice, conscience being more akin to obligation or dictate as the exact absence of choice. This is the gist of Martin Luther's legendary *Hier stehe ich und kann nicht anders*, when he had to defend his heretical views before Emperor and Church in Worms, and it is the gist of contemporary fundamentalisms (see Stolzenberg 1993). The freedom of religion is thus better understood as protecting the freedom of conscience than merely a freedom of choice (see Sandel 1989). Still, in secular settings religious freedom tends to be understood in terms of choice. Some have argued that Islam, in particular, is irritating in a secular society that has unlearned the nature of religious obligation, because it emphasizes "self-realization as the realization of the transcendent will, and as dependent on . . . social and religious authority" (Fernando 2010: 25–26). Commenting on the notorious French headscarf controversy, Fernando thus finds that there is "no discursive space" for "pious" Muslims (p.30), and that the "consternation about ostensible Muslim unfreedom . . . helps to sustain a secular fantasy of personal autonomy" (p.30). It is not clear why for someone not subscribing to religion "personal autonomy" should be mere "fantasy". But a conception of religion as purely voluntary obviously misses something essential.

However, religion moves onto the radar of multiculturalism mostly not for internal "religious" reasons but because of its close association with an ethnic immigrant minority. This is the story of Islam in Europe, which for Zolberg and Long (1999) and many in their footsteps (e.g. Joppke 2015:ch.4) is the main cultural integration issue this side of the Atlantic. In this context, "Islam" is simultaneously the result of external labelling (often incorrectly and overgeneralizing) and of reactive self-identification (on the dynamics of labelling and self-identification of Islam and Muslims, see Brubaker 2013b). Specifically, for Zolberg and Long Islam has become the focus of the European immigration debate because "European identity . . . remains deeply embedded in Christian tradition, in relation to which Muslim immigrants constitute a visible 'other'" (1999:7).

This seems to exaggerate the vestigial nature of Christianity in Europe, and it misses too many other variables to be plausible—such as the high degree of secularism in Europe and the socioeconomic profile of Muslim immigrants, who were originally low-skilled and rural and even in the second and third generations have not advanced much on the social ladder (see Hansen 2012).

As shall be explored further, Europe's recent retreat from multiculturalism is intrinsically tied up with perceived or real problems of Muslim and Islam integration. However, the important takeaway message is that with respect to religion states lean toward the "pluralist end" of the assimilation–pluralism continuum (Zolberg and Long 1999:31). It would be wrong to mistake this for explicit "multiculturalism", at least on the part of the state. Even the corporate integration prong,[6] by means of which Islam as a collective enterprise becomes accommodated, is more a prolonging of historically established church–state regimes than anything "multicultural". If in Germany the long-sought recognition of Islam as "corporation under public law" may eventually occur, it is simply because failing to do so would put in question Germany's established church–state regime and its privileging of society's major religions through granting them public status (see Joppke and Torpey 2013:ch.3). The individual rights prong of Islam integration is even further removed from anything identifiably "multicultural", because in a constitutional state religious liberty rights cannot be limited to the followers of the majority religion but are equally available for all.

This is not to deny that on the side of religious claims-makers one is often dealing with declared "multiculturalism", that is, a "politics of recognition of difference or respect for identities" (Modood 2007:37). In fact, the whole point of Modood's religion-friendly variant of multiculturalism is to attack a "secularist bias" in previous multiculturalism theories, especially that of Kymlicka, which "puts religious groups and especially Muslims outside multiculturalism as a civic or policy idea" (p.30). While this may capture the activist side of the equation, it distracts from the distinctly non-exceptional, routine and mainstream ways in which minority religions and their people, *qua* liberal states' generic disposition to "attenuated pluralism" in matters of religion (Zolberg and Long 1999:14), have become incorporated.

Islam and the retreat of multiculturalism

At least in Europe, the questioning of multiculturalism is intricately tied to problems of Islam and Muslim integration. Will Kymlicka argued in this context that "liberal multiculturalism" in Canada is premised on two conditions: a de-securitization of state–minority relations and a liberal-democratic consensus shared by the minorities. Turning his eyes on Europe, whose Muslim immigrants are "often seen as both disloyal and illiberal", the perceived absence of both conditions "help(s) explain the partial retreat from multiculturalism" there (Kymlicka 2010:108). One account of the European "backlash against multiculturalism" listed 13 backlash-inducing moments or events between 2000 and 2006, 11 of which involved Muslims directly, while the other two involved them indirectly

(Vertovec and Wessendorf 2009:11–12). Other groups may have been central when doubts were raised about multiculturalism in other parts of the world, such as Asians in the perhaps earliest attack on multiculturalism, in 1980s Australia (Joppke 2005:80–90), or Hispanics in a more recent attack on American multiculturalism (Huntington 2004). But neither was reproached with illiberalism or disloyalty, which exposes the "principal issue" of the multiculturalism debates: "how to tolerate intolerant communities" or fifth columns (Rubinstein 2007:763).

It should be said that Muslims in Europe differ from other Europeans not so much with respect to trust in democratic institutions or even identifications with host societies, which are fairly similar to—if not sometimes stronger than—those of non-Muslims (Joppke 2015:156–157). Instead, Muslims differ with respect to moral and sexual issues, above all the status of women and homosexuality, where Muslims hold markedly more conservative and illiberal views than other Europeans (Joppke 2015:158–159). Accordingly, 55 percent of British Muslims polled in 2006 find that "wives should always obey husbands", and just 3 percent of British Muslims polled by Gallup in 2009 find pre-marital sex "morally acceptable" (against 82 percent of non-Muslim British) (Orgad 2015:36). And they differ with respect to the role of religion in public life, where Muslim views often clash with the principles of secularism. In Britain, almost 70 percent of Muslims polled in 2006 found that "people who insult Islam should be arrested and persecuted" (Joppke 2015:160). In Germany, whose Muslims mostly stem from laicist Turkey, and whom one might expect to hold more moderate views than Britain's overwhelmingly conservative Muslims from Pakistan and Bangladesh, nearly half of Muslims polled in 2005 ranked scriptural commandments as "more important" than abiding by secular state laws, and over 65 percent advocated censorship to "protect morality and order" (Joppke 2015:160). No wonder that 87 percent of German Muslim students described themselves as "religious" around 2006, while just 19 percent of non-Muslim Germans considered themselves thus—which shows the rather intense religiosity of the second generation in one of the world's most secular regions (Orgad 2015:38). And pulling the security trigger that is never far from debates surrounding Islam and Muslims in Europe, 20 percent of British Muslims asked in 2006 felt "some" sympathy with the "feelings and motives" of the bombers who in July 2005 attacked the London transport system (killing or wounding hundreds of people) (McGhee 2008:68). It is findings like these that moved the British government to distance itself from "multiculturalism". As the Secretary of State of the freshly created Communities and Local Government department put it in 2006, in somewhat stilted terms: "We have moved away from a period of uniform consensus on the value of multiculturalism, to one where we can encourage that debate by questioning whether it is encouraging separateness" (Ruth Kelly, in McGhee 2008:100).

Multiculturalism's political retreat: a closer look at Britain

Nevertheless, despite a shift to "community cohesion" and a stress on "Britishness" in recent British policy, there has been no real "end of multiculturalism" in Britain, even in a hardened context of counter-terrorism and worries about ethnic segregation. Derek McGhee (2008:145) finds a "reflexive multiculturalism" continuing at the local level in Britain, and a most recent survey of the policy elite found that "a more pragmatic multiculturalism" retains "considerable strength" (Taylor-Gooby and Waite 2014:285).

But not all is well. A good window into current sensibilities is an unintendedly acrimonious exchange between the British Communities Secretary, Eric Pickles, and British Muslim leaders over the implications of the terrorist attacks in Paris in January 2015. In the immediate aftermath of the slaughtering of the Charlie Hebdo cartoonists and the killing of Jewish hostages in the Hyper Cacher store by three Muslim terrorists, Pickles had written to British imams that "there is more work to do", and that there was need to "show our young people, who may be targeted, that extremists have nothing to offer them".[7] The letter, strategically co-written by a Muslim MP, was utterly polite in tone, and wrapped up in multicultural etiquette. "British values are Muslim values", one can read there, "Britain would be diminished without its strong Muslim communities", and detailed support is offered against the "thugs" of the English Defence League and Britain First, two far-right organizations. But British Muslim leaders rejected the letter as an exemplar of a far-right mindset itself and branded it as "ridiculous and unhelpful".[8] "Is Mr Pickles seriously suggesting, as do members of the far right, that Muslims and Islam are inherently apart from British society?" asked a spokesman for the Muslim Council of Britain, the country's major Muslim federation (for years the government's main interlocutor though known for advocating a theocratic state and refusing to take part in the annual Holocaust Memorial Day).[9]

It would be naïve *not* to assume that large portions of British Muslims and Islam *do* stand "inherently apart". Consider that nearly half of British mosques are run, and even more British imams schooled, in the South Asian Deobandi tradition, which is known for its extreme conservatism. These "market leaders" of British Islam (Bowen 2014), who are said to exert "overwhelming dominance in Islamic education in Britain" (p.34), advise that friendships with non-Muslims should "not touch the heart", and that Muslims should not adopt the "dress of the unbelievers". Its imams also often advocate the face veil (*niqab*) for women (who at the same time are not allowed to enter mosques). Certainly, Britain's mainline Deobandi Muslims are "apolitical" and "pietistic", and thus may pose a problem more for "social cohesion" than for "violent extremism" (p.34). This is the one fine line that the Pickles letter to British imams fails to make.

It is appropriate to look at the British case more closely, because it shows the twisted, if not contradictory reality of European multiculturalism, Muslims always being the central concern. To repeat, the conservative leanings of many British Muslims are beyond doubt. The other Islamic currents in Britain, like the Barelwis, Salafis, or Tablighi Jamaat, are not any more "liberal" or integration-minded than the mentioned Deobandi (see Bowen 2014; also Klausen 2005). But it is quite another question whether multiculturalism policies are to be held responsible for deficient minority integration. For this there is scant evidence. One study found that there has been "no effect" of multiculturalism policies on integration, because minority group members were found to "have positive orientations towards *both* minority and majority cultures" (Heath and Demireva 2014). This is on the assumption that British multiculturalism policies have been "thin" to begin with, not going much beyond exhortations of curricular diversity, religious exemptions from certain laws, the provision of religious diet in public canteens, and—perhaps most extreme—a belated support for minority (read: Islamic) faith schools.

The devolution of British multiculturalism was marked by two critical events, the first being the 2001 race unrests that brought the turn to "community cohesion", and the second the July 2005 domestic terror that brought British Muslims under the closer lens of counter-terrorism policy. Each of the two events tilted British policy in contradictory directions, away from *and* towards multiculturalism. Let me explain. The Cantle Report, investigating the causes of the Northern England race riots of the summer of 2001 (in which South Asians had clashed with white youth), famously found that "many communities operate on the basis of a series of parallel lives. These lives often do not seem to touch at any point, let alone overlap and promote any meaningful interchanges" (2001:9). This was a questionable diagnosis, as it ignored the socioeconomic underpinnings of the grievance. As a geographer argued about Bradford, one of the decrepit and now heavily Muslim-inhabited mill towns afflicted by the unrest, "spatial segregation in poorer neighborhoods largely reflects bounded choices constrained by structural disadvantage, inequalities in the housing market . . ., worries about racism, and . . . racial harassment" (Phillips 2006:34). The unrests were clustered in Oldham, Burnley, and Bradford city wards that were among the 20 percent most deprived in the UK—parts of Oldham and Burnley even ranked among the most deprived 1 percent (McGhee 2008:91). However flawed the diagnosis, out of the Cantle Report's cultural "contact hypothesis" the new emphasis on "community cohesion", "active citizenship", and "Britishness" was born, first pronounced under the feisty Labour Home Minister David Blunkett.

The Cantle Report's emphasis on "community cohesion" differed sharply from the tenor of the Parekh Report, *The Future of Multi-Ethnic Britain*, which had been published just the year before (in 2000) and whose depiction of Britain as a "community of communities" must be considered, in retrospect, as the apex of British multiculturalism, from which point on, the reverse direction was taken. However, the Cantle Report's stress on strengthening "common elements of nationhood" was qualified by rejecting a "dominant or monoculturalist view of nationality" (quoted in Uberoi and Modood 2012:3), and Blunkett, though often portrayed as an opponent of multiculturalism, also espoused a view of Britain as "a mongrel nation" (quoted in Uberoi and Modood 2012:9). In an influential Home Office report, *Strength in Diversity* (2004), which hammered down the government's "community cohesion" policy, Britain is referred to as a "modern multicultural democracy" (p.5), and "integration" is defined as "not about assimilation into a single homogenous culture", but making "space within the concept of 'British' for people to express their religious and cultural beliefs" (p.7). Meer and

Modood (2011) thus rightly called the exercise a "civic rebalancing" rather than abdication of multiculturalism. It is also an instance of what Nikolas Rose called "government through community", which requires identifying and recognizing ethnic interlocutors: "for to govern communities, it seems one must first of all link oneself up with those who have, or claim, moral authority in the 'black community' or 'the local community'" (Rose, quoted in McGhee 2008:61). In engaging "communities", even the post-multicultural "cohesion" strategy was inevitably couched in multicultural colors.

The same twisted logic undergirds the British government's response to the domestic terror of July 2005, committed by home-grown second-generation Muslims. This was, in Prime Minister Blair's terms, a "game-changing" event that notionally toughened up the government approach to minority integration. Now there was, for the first time, an explicit distancing from "multiculturalism". Interestingly, as in its "culturalist" response to the 2001 race unrest, the government rejected linking the terrorism to socioeconomic exclusion and similar "simplistic, mono-causal explanations" (of which a culturalist line, of course, is yet another example). Instead, it pointed to the fact that "active terrorists . . . come from prosperous families or (are) personally well-educated and successful" (quoted in McGhee 2008:62). The focus was all on the "faith" component of this terrorism, the government seeking to engage "faith leaders" to "build bridges to other communities" and "condemn, without equivocation, those of their co-religionists who advocate violence" (quoted in McGhee 2008:56–57).

This happened in the context of the so-called "Prevent" counter-terrorism strategy. The mixing of the counter-terrorism and integration agendas, under the umbrella of the same new Department for Communities and Local Government that had been created to push the "cohesion" agenda, turned out a dismal failure. Two critics described this approach as asking Muslims "to spy on each other in return for grants" (Fieschi and Johnson 2013). An in-depth study of five local authorities report "strong resistance" to the Prevent program in all of them, and the authors quote a senior local councilor: "When the document came out from the Home Office I was aghast. I threw it out. I said we not having this. It was racist. Quite clearly racist" (Yunis and Husband 2012:148–149). The authors conclude that "the Prevent agenda experienced strong resistance from Muslim communities and from local authority staff who saw it as Islamophobic and likely to undermine their existing relationship with Muslim communities" (p.147). Even a House of Commons report criticized the effort:

> The single focus on Muslims in *Prevent* has been unhelpful. We conclude that any programme which focuses solely on one section of a community is stigmatising, potentially alienating, and fails to address the fact that no section of a population exists in isolation from others.

Moreover, the House of Commons committee criticized the government's "pre-occupation with the theological basis of radicalisation, when the evidence seems to indicate that politics, policy and socio-economics may be more important factors in the process" (House of Commons 2010:3). The thin-skinned Muslim response to the 2015 mosque letter by the Communities Secretary exudes the "Prevent" suspicion of being "spied upon, wrongly accused and treated unfairly" (House of Commons 2010:8).

Yet the post-2005 government engagement with Muslims *also* yielded what Grillo (2010) called "faith-based multiculturalism", as *quid pro quo* for Muslims' help in the counter-terrorism effort. Two of the bullet points on the "Preventing Extremism Together" agenda, for instance, were "meeting Muslims' educational needs" and "tackling Islamophobia", which materialized in further support for minority faith schools and (controversial) religious hate speech laws.

At the same time, there is no gainsaying that at this moment the term "multiculturalism" fell into disarray, and that the "community cohesion" campaign acquired a rougher edge. The 2006 Commission on Integration and Community Cohesion, under Ruth Kelly, explicitly "intend(ed) to avoid using the term 'multiculturalism'" in its first interim report (published in 2007), and the word also does not appear in its more programmatic *Our Shared Future* statement (CICC 2007). For CICC chair Darra Singh, the term "multiculturalism" "belongs in 1967 not in 2007" (quoted in McGhee 2008:101). And there *is* a new policy direction laid out in *Our Shared Future*, which opposes "single group funding" for its furthering of "insularity and a sense of separation", and which rejects the old idea that minorities should have preferential access to certain government services, such as public housing—instead there is "greater understanding of how settled communities may feel that positive action for minorities has unbalanced the way services are being provided" (CICC 2007:98). Tory Premier David Cameron's noisy departure from "state multiculturalism" thus only continued a line that had been more silently laid by his Labour predecessor.[10]

Persistent legal multiculturalism: sharia councils in Britain

Beneath the ebb and flow of multiculturalism rhetoric in the political arena, persistently institutionalized legal multiculturalism continues, in terms of "sharia councils" that have operated with tacit government approval in Britain for several decades now. This perhaps most contested and vilified form of Islam-induced multiculturalism needs to be demystified. It amounts to only a limited form of legal pluralism that is not to be mistaken for the existence of full-blown "parallel legal systems" (Malik 2012:6). The authoritative paper on the subject, by Maleiha Malik (2012:6), confirms that "Minority Legal Orders in the UK mainly accept the supremacy of the state system".

Werner Menski coined the notion of *angrezi shariat* for "a new hybrid form of Shari'a" (2001:140), which operates in a hazy sphere of the British state not officially recognizing it but also not wanting to prohibit it. What "recognition" of sharia might mean is an unclear matter to begin with. John Bowen (2010:413) has argued for the English case that recognition could equally refer to Islamic jurisprudence (*fiqh*), the laws and legal practices of Muslim countries, or the procedures and decisions of sharia councils operating in England. A legal scholar depicted the British situation in unflattering words: "(O)ften fearful of accusations of racism, and lately of Islamophobia if cultural practices are questioned, (government actors) have tended to allow communities free rein to 'police' themselves in cultural matters" (Sardar Ali 2013:114).

As most sharia councils operate privately, it is not clear how many of them there are in Britain. The estimates range from 12 (Bowen 2016:62, quoting a religious scholar) to 37 (Bono 2012:85) and 85 (MacEoin 2009:69), whereby the middle figure appears the most realistic. There is agreement, however, that the great majority of cases before these councils are brought by women whose husbands deny them a *talaq* divorce. Their only recourse left is the *khul* divorce by an imam, which, however, forfeits the divorced wife's right to her dowry. As the sharia councils thus fulfill a positive function for British Muslim women, who would otherwise be ineligible to remarry within their community, one is inclined to a positive view of them (as in Joppke 2015:162–166). Maleiha Malik (2012:29) argues reasonably that prohibiting the sharia councils would "only alienate minorities", and that it was preferable to make them "more 'women friendly'", within a "progressive multiculturalism".

But the sharia councils' shadowy side should not be overlooked. Sardar Ali (2013) reports that "imams hailing from the villages and towns of rural Pakistan" apply extremely "conservative and literalist interpretations of Islamic family law" (p.131), which have even been abandoned in Pakistan.[11] Pakistani family law actually seems to be more progressive than the Islamic family law practiced by the diaspora in Britain. An example is the *talaq-i-tafwid* divorce, which was introduced in Pakistan family law as early as 1961. It may be exercised by the wife if delegated to her in the marriage contract. But, as Sardar Ali (2013:128) claims, it "has not been adopted within the Diasporic community in Britain" and is not "actively canvassed by Shariah Councils" or "encouraged" by Muslim leaders.

A peculiar legal uncertainty surrounds the few Muslim arbitration councils that claim to operate under the 1996 Arbitration Act[12]—which would make their decisions binding under British state law. In September 2008, the British government was reported to have "quietly sanctioned" the powers for these councils to rule on financial disputes and family matters, from divorce to domestic violence.[13] However, the interior minister, Jack Straw, denied that a delegation of legal power in family law had ever occurred: "Arbitration is not a system of dispute resolution that may be used in family cases" (quoted in Zee 2014:8–9). Indeed, divorce is a matter of personal status, different from a dispute between individuals that may be privately resolved—it requires state involvement, as marriage in the Western tradition is not just a private contract (as it is under Islamic law) but a publicly instituted status (even under English law, where the contractual element is particularly strong). Accordingly, Muslim arbitration councils cannot issue divorce certificates that are valid under civil law.[14] The dual world of *angrezi shariat* remains in place: "(V)irtually all ethnic minorities in Britain marry twice, divorce twice, and do many other things several times in order to satisfy the demands of concurrent legal systems" (Menski 2001:152). The very fact of having to do things twice confirms the superiority of the civil law system, which is the only one to have enforcement power.

By the same token, even the Muslim arbitration councils under the Arbitration Act have not given rise to "plural legal systems": "The legal system is still one and the same", argues Lorenzo Zucca (2012:ch.6) about the English case. This is because an act of the legal system, such as an arbitration law, is needed to invest (or not) authority in "a variety of adjudications". John Bowen's ethnography of British sharia councils concedes that "confusion reigns" (2016:156) in this domain, but he underlines that "these councils carry out no actions that have the force of state law" (p.176). "English justice abides by its own rules and principles", concludes Bowen (p.183).

The British state seems to prefer retaining the uncertainty surrounding the current arrangement, which gives the state maximum flexibility to "pick and choose" (Malik 2012:7). According to a legal scholar, "no formal interaction takes place between the state and a sharia council" (Sardar Ali 2013:125). Proof for the state's tacit approval of *angrezi shariat* is the fact that a private member "equality bill", proposed by Baroness Cox, to amend the Arbitration Act, never received any political support—this bill would have outlawed sharia councils' unequal testimony, inheritance, and property rules for women, and severely punish any claim that Muslim arbitration council decisions are "legally binding" (see Maret 2013:276).

Most importantly for our purposes, the case of sharia councils proves wrong Prime Minister David Cameron's notion that "state multiculturalism" is dead.[15] Beneath the official rhetoric, the state's "liberal multiculturalist policies" seem to persist (Sardar Ali 2013:214), if more in a passively endured than actively promoted mode.

Conclusion

Religion differs from merely "ethnic" forms of claims-making and resulting conflict, including that based on language, through its strongly normative thrust. Ethnic or nationalist mobilization turns only on the "form of the state", while religious mobilization is about its content, the "substantive regulation of public life" (Brubaker 2015:12). There are other elements that distinguish religious from ordinary ethnic claims: an intrinsic "hyper-commitment" and production of "extreme otherhood", which stem from its ethical and not just non-normative foundations; and its abhorring of "profanation", which stems from its grounding in the sacred–profane distinction as two spheres that must never touch (Brubaker 2015:17–24).

Because of its ethical reach religion is more strongly protected in the liberal state than merely ethnic markers; but its ethical infrastructure may also come to a head with the thin but indispensable ethical infrastructure of the liberal state itself, most notably its constitutive dependence on the values of freedom and equality. *All* religion may be in conflict with these values. But in Europe at present it is mostly *one*, which is Islam. This is also why Islam is so closely connected with multiculturalism's current crisis.

Notes

1 The conference under this name was held in April 1996 at the European University Institute in Florence. The "reluctant co-organizer" was Tony Judt of New York University. The results were published as *Multicultural Questions* (Joppke and Lukes 1999).
2 "Merkel: 'Multikulti ist absolut gescheitert'", *Süddeutsche Zeitung*, October 16, 2010.
3 "Le multiculturalisme est 'un échec', affirme Nicolas Sarkozy", *Le Point.fr*, February 10, 2011.
4 David Cameron, Speech to the Munich Security Conference, February 5, 2011 (downloaded from www.number10.gov.uk).
5 The religious pluralism in liberal nation-states is "attenuated" in comparison with the unqualified pluralism of empire, which consisted of separate (and unequal) legal regimes for minority religions (Zolberg and Long 1999:14).
6 For the distinction between corporate and individual-level Islam integration, see Joppke (2014).
7 The letter is reprinted in *The Independent*, January 19, 2015 ("Eric Pickles' Letter to Muslim Leaders: The Text in Full", www.independent.co.uk/news/uk/eric-pickles-letter-to-muslim-leaders-the-text-i . . .).
8 Ibid.
9 "Multicultural and Aggrieved", *The Economist*, January 24, 2015, p.28.
10 David Cameron, Speech to the Munich Security Conference (see note 4).
11 But see Bowen (2016) for a more sympathetic and differentiated view of British sharia councils.
12 It seems to be mainly the Muslim Arbitration Tribunal (MAT), run by a charismatic Sufi barrister from Nuneaton, a small town 40 km east of Birmingham. John Bowen called MAT a "legal-sounding framework for what is really an age-old process of seeking guidance from a spiritual guide" (2016:166).
13 Abul Taher, "Revealed: UK's First Official Sharia Courts", *The Sunday Times*, September 14, 2008.
14 See Joshua Rozenberg, "What Can Sharia Courts Do in Britain?" *Telegraph*, September 14, 2008.
15 David Cameron, Speech to the Munich Security Conference (see note 4).

References

Bono, Samia. 2012. *Muslim Women and Shari'ah Councils*. Basingstoke: Macmillan Palgrave.

Bowen, Innes. 2014. *Medina in Birmingham, Najaf in Brent*. London: Hurst and Company.

Bowen, John. 2010. "How Could English Courts Recognize Shariah?" *University of St. Thomas Law Journal* 7(3), 411–435.

Bowen, John. 2011. "Europeans against Multiculturalism", *Boston Review*, July/August (bostonreview.net/BR36.4/john_r_bowen_european_multiculturalism_islam).

Bowen, John. 2016. *On British Islam: Religion, Law, and Everyday Practice in Shari'a Councils*. Princeton, NJ: Princeton University Press.

Brubaker, Rogers. 2013a. "Language, Religion, and the Politics of Difference", *Nations and Nationalism* 19(1), 1–20.

Brubaker, Rogers. 2013b. "Categories of Analysis and Categories of Practice: A Note on the Study of Muslims in European Countries of Immigration", *Ethnic and Racial Studies* 36(1), 1–8.

Brubaker, Rogers. 2015. "Religious Dimensions of Political Conflict and Violence", *Sociological Theory* 33(1), 1–19 (the quotes are from the typescript, in author's possession).

Buruma, Ian. 2006. *Murder in Amsterdam*. New York: Penguin.

Cantle Report. 2001. *Community Cohesion*. London: Government Printing Office.

Commission on Integration and Community Cohesion (CISS) (UK). 2007. *Our Shared Future*. London: CICC.

Department for Communities and Local Government (UK). 2012. *Creating the Conditions for Integration*. London.

Fernando, M.L. 2010. "Reconfiguring Freedom: Muslim Piety and the Limits of Secular Law and Public Discourse in France", *American Ethnologist* 37(1), 19–35.

Fieschi, Catherine and Nick Johnson. 2013. "Islam and the Muslim Communities in the UK: Multiculturalism, Faith and Security", *The International Spectator* 48(1), 86–101.

Grillo, Ralph. 2010. "British and Others: From 'Race' to 'Faith'", in Steven Vertovec and Susanne Wessendorf, eds. *The Multiculturalism Backlash: European Discourses, Policies and Practices*. London: Routledge.

Gutmann, Amy. 2003. *Identity in Democracy*. Princeton, NJ: Princeton University Press.

Hansen, Randall. 2012. *The Centrality of Employment in Immigrant Integration in Europe*. Migration Policy Institute, Washington, DC, working paper.

Heath, Anthony and Neli Demireva. 2014. "Has Multiculturalism Failed in Britain?" *Ethnic and Racial Studies* 37(1), 161–180.

Home Office (UK). 2004. *Strength in Diversity: Towards a Community Cohesion and Race Equality Strategy*. London: Government Printing Office.

House of Commons (UK). 2010. *Preventing Violent Extremism: Sixth Report of Session 2009–10*. HC65. London: House of Commons, Communities and Local Government Committee.

Huntington, Samuel. 2004. *How Are We?* New York: Simon and Schuster.

Joppke, Christian. 1999. *Immigration and the Nation-State: The United States, Germany, and Great Britain*. Oxford: Oxford University Press.

Joppke, Christian. 2004. "The Retreat of Multiculturalism in the Liberal State: Theory and Policy", *British Journal of Sociology* 55(2), 237–257.

Joppke, Christian. 2005. *Selecting by Origin: Ethnic Migration in the Liberal State*. Cambridge, MA: Harvard University Press.

Joppke, Christian. 2014. "Europe and Islam: Alarmists, Victimists, and Integration by Law", *West European Politics* 37(6), 1314–1335.

Joppke, Christian. 2015. *The Secular State under Siege: Religion and Politics in Europe and America*. Cambridge: Polity.

Joppke, Christian and Steven Lukes, eds. 1999. *Multicultural Questions*. Oxford: Oxford University Press.

Joppke, Christian and John Torpey. 2013. *Legal Integration of Islam: A Transatlantic Comparison*. Cambridge, MA: Harvard University Press.

Klausen, Jytte. 2005. *The Islamic Challenge: Politics and Religion in Western Europe*. New York: Oxford University Press.

Kymlicka, Will. 1999. "Comments on Shachar and Spinner-Halev: An Update from the Multiculturalism Wars", in Christian Joppke and Steven Lukes, eds. *Multicultural Questions*. Oxford: Oxford University Press.

Kymlicka, Will. 2010. "The Rise and Fall of Multiculturalism? New Debates on Inclusion and Accommodation in Diverse Societies", *International Social Science Journal* 61(199), 97–112.

MacEoin, Denis. 2009. *Sharia Law or "One Law for All"?* London: Civitas (Institute for the Study of Civil Society).

McGhee, Derek. 2008. *The End of Multiculturalism? Terrorism, Integration and Human Rights*. Maidenhead: Open University Press.

Malik, Maleiha. 2012. *Minority Legal Orders in the United Kingdom*. London: The British Academy.

Maret, Rebecca E. 2013. "Mind the Gap: The Equality Bill and Sharia Arbitration in the United Kingdom", *Boston College International and Comparative Law Review* 30(1), 255–283.

Meer, Nasar and Tariq Modood. 2011. "How Does Interculturalism Contrast with Multiculturalism?" *Journal of Intercultural Studies* 33(2), 175–196.

Menski, Werner F. 2001. "Muslim Law in Britain", *Journal of Asian and African Studies* 62, 127–163.

Modood, Tariq. 2007. *Multiculturalism: A Civic Idea*. Cambridge: Polity.

Modood, Tariq. 2013. "The Strange Non-Death of Multiculturalism", *Max-Weber Lectures Series*, European University Institute, MWP 2013-03, February 20.

Orgad, Liav. 2015. *The Cultural Defence of Nations: A Liberal Theory of Majority Rights*. Oxford: Oxford University Press.

Parekh Report. 2000. *The Future of Multi-Ethnic Britain*. London: Runnymede Trust.

Phillips, Deborah. 2006. "Parallel Lives? Challenging Discourses of British Muslim Self-Segregation", *Environment and Planning D: Society and Space* 24, 25–40.

Roy, Olivier. 2008. *La sainte ignorance*. Paris: Seuil.

Rubinstein, Amnon. 2007. "The Decline, but not Demise of Multiculturalism", *Israel Law Review* 40(3), 763–810.

Sandel, Michael J. 1989. "Moral Argument and Liberal Toleration: Abortion and Homosexuality", *California Law Review* 77(3), 521–538.

Sardar Ali, Shaheen. 2013. "Authority and Authenticity: Sharia Councils, Muslim Women's Rights, and the English Courts", *Child and Family Law Quarterly* 25(2), 113–137.

Stolzenberg, Naomi. 1993. "'He Drew a Circle That Shut Me Out': Assimilation, Indoctrination, and the Paradox of a Liberal Education", *Harvard Law Review* 106(3), 581–667.

Taylor-Gooby, Peter and Edmund Waite. 2014. "Toward a More Pragmatic Multiculturalism? How the U.K. Policy Elite Sees the Future of Ethnic Diversity Policies", *Governance* 27(2), 267–289.

Uberoi, Varun and Tariq Modood. 2012. "Inclusive Britishness: A Multiculturalist Advance", *Political Studies* (advance publication), 1–18.

Vertovec, Steven and Susanne Wessendorf. 2009. *Assessing the Backlash against Multiculturalism in Europe*. MMG Working Paper 09-04, Max Planck Institute for the Study of Religious and Ethnic Diversity, Göttingen.

Vertovec, Steven and Susanne Wessendorf, eds. 2010. *The Multiculturalism Backlash: European Discourses, Policies and Practices*. London: Routledge.

Weber, Eugene. 1976. *Peasants into Frenchmen*. Stanford, CA: Stanford University Press.
Yunis, Alam and Charles Husband. 2012. "Parallel Policies and Contradictory Practices: The Case of Social Cohesion and Counter-Terrorism in the United Kingdom", *Collegium: Studies across Disciplines in the Humanities and Sciences* Vol. 11 (Helsinki: Helsinki Collegium for Advanced Studies), 136–56.
Zee, Machteld. 2014. "Five Options for the Relationship between the State and Sharia Councils: Untangling the Debate on Sharia Councils and Women's Rights in the United Kingdom", *Journal of Religion and Society* 16, 1–18.
Zolberg, Aristide and Long Litt Woon. 1999. "Why Islam Is Like Spanish", *Politics and Society* 27(1), 5–38.
Zucca, Lorenzo. 2012. *A Secular Europe*. Oxford: Oxford University Press.

15 Religion, culture and liberal democracy

The issue of majority cultural precedence

Geoffrey Brahm Levey[*]

Introduction

The established institutions and customs of a liberal democracy are a key theme in debates over the accommodation of cultural diversity. Some of these institutions and customs are political in the sense of being informed by liberal democratic values and norms. Others are more generically tied to the historical or dominant culture and are typically associated with a majority cultural group. The reports of the expert commissions on cultural diversity canvassed in this book and their public reactions reflect the force of these twin concerns.

The Bouchard-Taylor report's quest for 'reasonable accommodation', for example, was predicated on the non-negotiability of the priority of francophone language and culture in Quebec.[1] In contrast, the Parekh Report – the product of a non-government commission – reflected the approaches of its chairman (Bhikhu Parekh) and lead advisor (Tariq Modood) in seeking to challenge traditional notions of Britishness for being obstacles to an inclusive society.[2] The recommendations of the Commission on the Future of Multi-Ethnic Britain sparked extraordinary public controversy. While some of this reaction was the result of misunderstanding and polemical mischief, it was also due to a real sense that conventional understandings of British identity were being set aside. France provides a further contrast as it defines its national identity largely in terms of its foundational political principles. The Stasi Commission was tasked with exploring but also reaffirming *laïcité* as a core French institution.[3] And in a constitutively culturally divided society, Belgium's Round Tables on Interculturalism recommended reforms for accommodating diversity that would preserve state neutrality. Yet these were met with much public criticism for mollifying minorities at the expense of the majority and the historical *status quo* arrangements.[4]

Three broad schools of thought compete on the question of the appropriate status of a dominant or historical culture in relation to cultural minorities. One insists that the dominant cultural majority *rightfully* enjoys its dominance and privileging. Sometimes this position is framed in terms of ethnic or racial membership where minorities are simply excluded; sometimes it is cast in terms of cultural nationalism where minorities are expected to assimilate; and sometimes the argument is that minorities and their distinctive ways should be tolerated as long as they make no demands on the host society to adjust to their presence. Today, these approaches are generally discredited, which is not to say they are absent from real world politics.

A second school of thought begins from the opposite assumption, namely, disparities between a dominant majority and minorities breach liberal justice or democratic equality and should be eliminated. Here the positions divide over how best to avoid or eliminate these disparities. A traditional liberal-cum-libertarian approach holds that the state should pursue 'hands-off' neutrality, treating individuals only in terms of their common citizenship and remaining aloof from matters of cultural identity, whether of the majority or of minorities.[5] In contrast, a more recent, multiculturalist approach argues for a 'hands-on' form of neutrality in which the state accommodates and recognizes minorities in some rough parity or evenhandedness to that enjoyed by the dominant culture.[6] To be sure, even 'hands-off' neutralists accept that liberal states cannot be completely culturally neutral, the *lingua franca* and public holidays being the standard examples. However, their position is distinctive in that they tend to view these instances as merely pragmatic arrangements or unavoidable compromises. Their principled position is that hands-off neutrality is to be preferred wherever possible.

A third camp pursues an intermediate position. It accepts that a dominant culture may be foundational to a society *and* that some majority cultural precedence is legitimate even beyond the *lingua franca* and public holidays. However, it also holds that there should be decided limits on this privileging in the interests of honoring liberty, equality, and/or an inclusive society. There is wide disagreement though on where these limits should be drawn, with some being more accommodating of minorities than others. Advocates of this position often describe themselves as 'liberal nationalists';[7] however, Gérard Bouchard's and Charles Taylor's accounts of Quebecan interculturalism also fit this bill,[8] as do, arguably, the 'multicultural nationalisms' of Bhikhu Parekh and Tariq Modood.[9]

In this essay, I pursue this third approach. My working assumption is that if we wish to take seriously the cultural interests and attachments of minorities we must also take seriously the cultural interests and attachments of established majorities. As the third approach still credits majority precedence only up to a point, the question I want to address is, where and on what basis is a majority's cultural precedence legitimate? Such a large and involved question necessitates some delimitation of cases.

Many cultural controversies concern minorities seeking institutional accommodation for their particular cultural traditions and practices. However, another set concerns *majority* traditions and their institutional or symbolic privileging. In some cases, these are simply two sides of the same coin: the institutional privileging of the dominant culture disadvantages, directly or indirectly, a minority in pursuing its cultural practice. An example addressed by the Stasi Commission is how Muslim and Jewish school children have to miss classes in order to observe their key religious holidays, which occasioned the recommendation that schools do not hold classes on these days. Similarly, the Belgian commission noted how the established calendar of public holidays puts minorities at a disadvantage insofar as their pay is docked if they take time off to attend their own religious holidays.[10]

However, as we shall see, there are also many cases of majority precedence that do not prohibit or disadvantage minorities in pursuing their cultural practices, and where liberty or adversely impacting the free exercise of culture is not at issue. Rather, the privileging of the majority culture is deemed to raise questions about equal citizenship and fairness and broader issues of inclusiveness and belonging. It is this institutional and symbolical privileging of the established majority culture that I propose to examine.

Liberal nationalists and Quebecan interculturalists tend to grant considerable latitude to the majority to institutionalize and endorse the historic identity as it sees fit. For example, Charles Taylor argues that states with 'historic nations' are at liberty to secure and publicly express their identities so long as the fundamental liberties of minorities are respected.[11] Michael Walzer agrees that there 'doesn't seem to be any requirement of equal provision or equal protection for minority cultures, so long as basic rights are respected'.[12] And David Miller argues that while the fundamental rights of democratic citizens are unassailable, their equal treatment and respect might justifiably be overridden where historic nations and national identity are at stake.[13]

The Bouchard-Taylor report stipulates a more nuanced and demanding standard.[14] In addition to respecting the fundamental liberties of all citizens, two other conditions obtain. First, any majority precedence must be left *de facto* and not be 'converted into the precedence of law, i.e. into a hierarchy'. Second, public bodies of governance that have a regulatory role should be free of the symbolic intervention of the majority culture. So, for example, while the cross on the Quebec flag or giving prominence to Christian religions in the school course on ethics and religious cultures is each deemed legitimate, a cross on the wall of the National Assembly or in public courtrooms or prayers at municipal meetings is considered illegitimate.

I think more needs to be said in relation to assessing the legitimacy of symbolic majority cultural precedence in general and majority cultural symbols in particular. Let me then turn to one controversial case of such precedence in order better to tease out some points for thinking about this issue.

Learning from *Lautsi*

The well-known *Lautsi* case involving a challenge to the display of the crucifix in an Italian school classroom provides a useful vehicle for exploring symbolic majority precedence.[15] I want to draw on the case and some of the commentary on it to develop four points for thinking about the acceptability or otherwise of symbolic majority precedence in general. These points relate to a symbol's public resonance, the issue of state neutrality, the impact on minorities, and the impact on the majority. Let me consider each in turn.

Resonances of the symbol

The Italian government went to great lengths to defend the crucifix in its classrooms as being something other than a religious symbol. While it might be a religious sign in religious contexts, in a state classroom its resonance was much enlarged. It stood for the foundational values of democracy and Western civilization, the government maintained. Italy's Supreme Administrative Court was more specific in its judgment, contending that the crucifix symbolized 'liberty, equality, human dignity and religious toleration, and accordingly also of the secular nature of the State – principles which underpin our Constitution'.[16] At first blush, these claims seem like arguing black is white: the crucifix as symbol of secular values and the secular state! Yet, historically, Christianity is one of the fountainheads of these values and it did sanction the division between political and religious authority that was eventually inscribed as a cardinal liberal principle. In Italy, there is a very real sense in which the crucifix could be said to encapsulate a sizeable chunk of Western history, both the harmony and the discord. In Protestant countries the crucifix is likely to have a narrower reference, the cross rather than the crucifix conveying the comparable Christian contribution to Western civilization. Context is everything.

In any event, the Grand Chamber of the European Court of Human Rights rejected this line of defense, preferring to see the crucifix as a symbol that could not shake its predominantly religious connotations. That contention strikes me as a mistake if only because it contradicts the Grand Chamber's own caution that the domestic courts disagreed about the meaning of the crucifix and it was not for their Court to determine such things. The European Court of Human Rights is, of course, bound by and tasked with interpreting the provisions of the Convention for the Protection of Human Rights and Fundamental Freedoms. Fortunately, political theorists are not so restricted. Rather than test the classroom crucifix against the provisions of the Convention, I propose a 'political' assessment. The enlarged connotations of the crucifix claimed by the Italian authorities could easily enough be tested empirically. One need only survey Italian citizens to establish how the crucifix is popularly viewed. Should it turn out that Italians generally do not see this symbol in the manner portrayed by their government and the Administrative Court this need not count against retaining crucifixes in the class room. Instead, it might be taken to suggest only that the government has some work still to do in educating its citizens in appreciating the broader significance of this Church symbol as they outlined it in Strasbourg.

Of course, this assumes that the Italian government was sincere in its claims about the enlarged connotations of the crucifix and that they were not designed simply to rebut a legal challenge. Thus, another test of the seriousness of the claims about the crucifix would be whether the Italian authorities were willing to teach these enlarged, civic meanings to the school children who daily face the crucifix on their wall. To be sure, such a course carries its own risks since potent symbols usually have fierce defenders of particular interpretations. Sometimes it is better to let the meaning of symbols go unheralded, as it were, allowing individuals and groups to attach their own significance to them. However, whether diplomatic or strategic ambiguity remains a realistic option once a symbol has been serially challenged and forensically examined in the courts, *à la Lautsi*, is an open question.

The Italian government also advanced the argument that the crucifix was a 'passive' symbol, which could not be likened to the active influence on pupils of didactic teaching or particular class activities. On this point, the Grand Chamber concurred. In fact, passivity or activity has little to do with it. Try telling Jews, for example, that a swastika is only passively sitting on a wall. A crucifix has seemingly been sitting passively above the president's chair in Quebec's National Assembly since 1936. What is important is the current force and signification of the symbol and not its motionlessness. The Bouchard-Taylor report is right at least on that score in questioning the appropriateness of the crucifix in the National Assembly (even if it is vague in explaining why exactly the symbol is inappropriate in this context).[17]

There is another sense though in which symbols might be passive or, perhaps better, clinically dead. That occurs when they have lost their historical or original resonance. For example, the official coat of arms of the state of New South Wales in Australia is the Red Cross of Saint George. Accordingly, my university, the University of New South Wales, uses the same emblem. Thus, if Jewish and Muslim students at UNSW wish to don a university sweatshirt, it means emblazoning a large cross on their non-Christian chests. Is this domination or alienation at work? I don't think so and apparently nor do most minority students at UNSW. The cross in the university's emblem (and even in the state's coat of arms) has lost its historical resonance for current generations. Jewish and Muslim staff and students do not quake in their boots reminded of the Crusades or expecting the Spanish Inquisition. On the contrary, most students graduate without ever noticing the emblem's historical connection to Christendom. The cross, in this context, is little more than a design logo, part of the furniture of the university. The historical potency of a symbol can drain away leaving it to be filled with new and sometimes mundane resonance.

In this respect, I disagree slightly with prominent legal academic Joseph Weiler, who represented eight third-party governments before the Grand Chamber hearing. In his oral submission, Weiler insisted that the crucifix could be understood neither as only a national symbol nor as having only religious significance, since '[i]t is both' given European history.[18] In support, he cited the example of the photograph of the Queen of England hanging in England's classrooms:

> Like the Cross, that picture has a double meaning. It is a photo of the Head of State. It is, too, a photo of the Titular head of the Church of England . . . Would it be acceptable for someone to demand that the picture of the Queen may not hang in the school since it is incompatible with their religious conviction or their right to education since – they are Catholics, or Jews, or Muslims? Or with their philosophical conviction – they are atheists?

The example is, indeed, instructive but for the opposite reason to Weiler's purpose. Queen Elizabeth may well be 'Defender of the Faith and Supreme Governor of the Church of England', but it is doubtful that many of her subjects in England have this ceremonial and arcane aspect of her job description in mind when they see her or her image. For the vast majority, it is her other ceremonial role that defines her – head of state or Sovereign of the Realm. Moreover, among those who do make the connection to the Church of England, the symbolism of it – like the establishment of the Church itself – has a hollow ring. In this day and age most non-Anglicans in Britain are likely to consider their ineligibility to assume the monarchy, for example, to be a relief rather than a disability.

This is not to say that the crucifix has undergone the same secular transformation on the continent as has the British monarchy. It is to say that the signification of symbols is dynamic and contingent.

The dynamism and contingency can also work in the opposite direction. Longstanding symbols and signs that passed unnoticed for generations may suddenly acquire resonance and excite controversy. Although not an example of majority cultural precedence, this kind of symbolic retrieval occurred recently at Harvard Law School. The august institution replaced its official shield after students protested its incorporation of a family crest of an eighteenth-century slaveholder.[19] Similar controversies have erupted over the Wills Memorial Building at the University of Bristol in the UK (tobacco and slave trade profiteering), and a mathematics building at the University of Melbourne named after an anatomy professor and leading eugenicist at the institution in the early twentieth century.

The point remains: the interpretation of symbols for the purposes of assessing majority cultural precedence should not be anchored to some form of 'originalism' but rather to how they are predominantly perceived in our own time.

The chimera of value neutrality

Another key argument put by the Italian government and especially the third-party interveners in *Lautsi* was that secularism in the sense of removing vestiges of religion from public space is not neutral. As Weiler puts it, black is not the absence of color; black *is* a color.[20] Opting for a blank wall over a crucifixed wall is to take sides *against* the public presentation of religious symbols and *for* a particular laic view of secularism.[21]

Some might counter that white being the presence of all color, a blank white wall is the most multicultural solution to the question of adorning classroom walls, and the fact that this solution coincides with a certain secularist or laic position is purely coincidental. Similarly, Rawlsian political liberals are fond of arguing that government policies will always differentially impact sections of the population and is unavoidable; what is key is that a policy is not justified with the aim of privileging particular groups or conceptions of the good life. The argument is a bit of a dodge. While some such distinction between 'neutrality of aim' and 'neutrality of effect', as Rawls otherwise puts it, seems warranted in the case of effects that may incidentally follow upon the enactment of principles, matters are more complicated when serious consequences are invariably or commonly related to the embraced aims. Then, the fact that the effects were unintended is morally qualified by the knowledge that one's actions are likely to give rise to them. Collateral damage is not a morally neutral category for which no one is responsible. In this context, Rawls's counsel that '[w]e must accept the facts of commonsense political sociology' might well be taken to heart as a refutation of the claim to neutrality, rather than, as he intends it, simply a frank recognition that one's acclaimed neutral principles have unfortunate consequences.[22]

Of course, one may still have good reason to prefer one non-neutral position to another non-neutral position – the blank wall over the crucifix, say. But Weiler's point – which I endorse – is that this is exactly where the debate needs to be conducted, as contending positions and visions regarding what this or that society is and should be or should not be. What is unacceptable is the notion that one position automatically trumps others because it alone does not take a position. On this account, the Grand Chamber's judgment deferring to the 'margin of appreciation' available to member states and effectively returning the conundrum to Italy to sort out was entirely appropriate. What the crucifix represents in twenty-first-century Italy and whether it should remain hanging in Italian classrooms and, if so, why – are properly questions for the Italians to answer.

There are, however, two further points that they, like others faced with such issues, should consider.

Impact on minorities

Bouchard argues that majority cultural precedence should *benefit* minorities and not only the majority culture for it to be legitimate.[23] This mirrors the more common suggestion of those defending minority cultural rights that such measures should benefit also the broader community.[24] The problem is that a notable benefit to minorities is unrealistic across many of Bouchard's own cases of legitimate majority precedence. Maintaining the official burial of heads of state in Catholic churches, for example, does not benefit minorities in any obvious way.

If a distinct benefit is asking too much, it is surely reasonable to expect that majority cultural symbols will at least not representationally *target, threaten or marginalize* minorities. Each of these amounts to an assault on the standing of minority members as equal members of the society. I have already mentioned the provocation of the swastika. Erecting a Confederate flag in the south of the United States is also interpreted as a hostile gesture by many folks there. The provocative symbol continues to arouse controversy or worse. Many commentators, for example, suggested that the prominence of the flag and its symbol influenced the white gunman who killed nine people in a black church in Charleston South Carolina in June 2015.[25]

It is nevertheless the case that societies built on conflict and conquest will likely trade in some symbols and commemorations that prompt painful memories for groups still among them. Those of us from settler societies know how our celebratory national days can represent tragedy and sadness for indigenous people. For example, Australia Day falls on January 26, and celebrates the arrival of European settlement, and is, for that reason, a day of sorrow for many Aboriginal Australians. In these circumstances, there is a special obligation to ensure that the history being honored is complete and honest, as against triumphalist and jingoistic, and that the commemorative events are otherwise sensitive and inclusive. This point leads directly to the next.

Impact on the majority

Although requiring a benefit to minorities from majority cultural symbols may not be realistic, Bouchard's intuition that attention needs to be paid to the *overall* impact of symbolic majority precedence remains valid. Thus, rather than only focusing on the impact on minorities as a test of legitimacy, it is important also to consider the impact of majority precedence on the cultural *majority* itself. Arguably, it is how the symbolic precedence impacts on the cultural majority – and *vice versa* – that has the most repercussions for minorities.

Specifically, does the cultural majority leverage or exploit the symbolic precedence in question in order to proclaim its exclusive ownership of the public culture? If so, the majority precedence is not simply honoring an historical and institutional cultural legacy and identity open to appreciation by all citizens, but is using it to exclude and demean the standing of other members of the political community. In a democracy, symbolic majority precedence cannot be legitimate if used in that manner.

Majority precedence and multiculturalism

Liberal multiculturalism is both remedial and prophylactic. It is an attempt to correct or redress the longstanding denial of rights, equality, and inclusion of cultural minorities and their members and also an attempt to protect against their further and future abridgement. Multiculturalists who adhere to the third school of thought outlined at the beginning of this essay accept the legitimacy of some majority precedence (beyond language and public holidays) while also being committed to ensuring the rights, equality, and inclusion of cultural minorities. The argument, as we have seen, is mainly about which kinds of majority precedence are and are not defensible and on what grounds. Above, I critically considered some decision criteria and suggested some points of my own. I assumed that the main challenge is an overbearing dominant culture or cultural majority and that the solution is a certain kind of multiculturalism, one that seeks to check that dominance rather than eliminate all traces of it.

Both sides to this assumption have been assailed. Christian Joppke, for example, argues that the influence of liberal universalistic values has been such that Western states now define even their particularity in terms of them. The main tension in Western states concerning the incorporation of minorities, he contends, is thus not between liberalism and a particular national culture or identity. Rather, it is between two vying tendencies within liberalism: 'one that prescribes autonomy and reason over heteronomy and faith, and thus an ideal way of life; and a second that limits itself to procedural toleration and noninterference in a society marked by an irredeemable multiplicity of ways of life'.[26] Joppke frames the first mode as an illiberal form of liberalism since its strictures on how one should live are so demanding, especially for many traditional and religious minorities. In contrast, he contends that the advent of multiculturalism in Western societies was premised on the toleration mode of liberalism. It ran into strife only when confronted with illiberal minorities, as happened in Europe with high levels of Muslim immigration, for then the dilemma became tolerating intolerance.

Joppke's main focus is on the conditions Western states set for acquiring citizenship. The distinction between an autonomy-cum-identity liberalism and a toleration-cum-procedural liberalism fits this context both normatively and, to a large extent, descriptively, as most Western states now ask newcomers only to abide by liberal democratic institutions and law and not to be true believers in liberalism. Joppke's discussion nevertheless ranges widely over the social and political landscape of contemporary liberal democracies, suggesting a wider canvass for his argument. At this more general societal level, however, the distinction between an autonomy and toleration mode of liberalism and the latter's connection to multiculturalism is, I think, misleading.

Joppke himself acknowledges that the toleration mode of liberalism does not render the value of autonomy obsolete.[27] Indeed, the picture of how autonomy operates as governing values is complex and far from a uniform one. First, all liberal states today inculcate liberal identities in their born citizens through a mandatory school education. Such an education need not be dismissive of faith or tradition, but it is autonomy orientated in teaching young people how to reason and think critically about issues for themselves. Second, liberal states also valorize autonomy under the force of law in myriad respects, everything from the protection of fundamental rights and liberties to mandatory informed consent protocols before taking out a mortgage or undergoing a medical operation. But third, and crucially, liberal states *for the most part* simply presume the autonomous agency of their citizens; that is, they ascribe autonomy to them. They do not – indeed, could not without ceasing to be liberal states – routinely monitor their citizens' decisions for their reasonableness or reflectiveness or whether they have been unduly influenced by inadequate information or other subverting factors. Liberal citizens are thus at liberty to engage in non-autonomous behavior in much of their everyday lives.[28] A considerable realm of autonomy-based liberalism effectively coincides with a tolerant liberal state.

At the same time, Joppke's toleration mode of liberalism is not so unwaveringly tolerant. There *are* unequivocal accounts of toleration-based liberalism, Chandran Kukathas's vision of a liberal archipelago in which groups are free to conduct their affairs pretty much as they please being perhaps the most extreme contemporary statement.[29] But this vision is not Joppke's toleration mode of liberalism. As noted, Joppke contends that 'toleration' liberalism was challenged only when confronted with the intolerance of illiberal minorities. In Western societies, minorities tend to be deemed illiberal and intolerant where their traditions or practices (especially, regarding women and free speech) breach respect for the autonomy of their own members or others. Multiculturalism as a public policy arose first in Canada and Australia, both places that tied the policy expressly to individual rights and liberties and whose public cultures place premium value on the autonomy of the individual.

The contrast Joppke appears to be reaching for, then, is not between an autonomy mode and a toleration mode of liberalism but rather between two autonomy modes of liberalism. Following Rawls, this contrast is conventionally framed in terms of 'comprehensive', 'perfectionist' or ethical liberalism, on the one hand, and 'political' liberalism, on the other.[30] The former prescribes autonomous agency as the ideal way people should live in every aspect of their lives, while the latter is content with limiting the role of autonomy to establishing only individuals' political rights and responsibilities. However, even this contrast is misleading since political liberalism also generates rights and obligations at odds with various religious and traditional practices.[31] And as Rawls himself acknowledges, invoking autonomy for a specifically 'political' purpose can well shape the identities of citizens more generally, as indeed it typically does.[32] So we are left with a contrast between an autonomy liberalism that is comprehensive, perfectionist, and relentless and an autonomy liberalism that is something less than those things and in this sense is, yes, more tolerant and accommodating of diversity.

I do not deny that this last contrast is an important cleavage in contemporary politics. But it is not the only or always even the main axis of contention regarding the place of cultural minorities. The force of historical, established, or national cultures continues to be felt in many Western societies, notwithstanding the reluctance of their respective political classes formally to define their national distinctiveness in such terms. Joppke grants one way in which this influence is manifest when he notes how particular national contexts variously emphasize liberal democratic values.[33] So, for example, France emphasizes republican solidarity and a certain equality at the expense of liberty, the United States champions liberty at the expense of equality, and Australia stresses liberty and egalitarianism while pretty much ignoring solidarity. Each is a product and expression of its own history and cultural formation.

National cultures also intrude through the cultural inflection of liberal values. To dominant cultural majorities, freedom, for example, just appears to be how *they* have conventionally lived, dressed, and played. These deep assumptions are not a problem when societies are culturally homogeneous. They are exposed as assumptions and become a problem with the arrival of diversity, when some folk wish to do these things a bit differently – for example, by wearing a hijab instead of, say, a scarf. Suddenly, what liberty and equality might sanction no longer seems to be in accord with the established cultural norms and patterns of behavior. Unless the minority practice involves some harm or violation of basic rights, the sentiment 'This is what we do here' lends itself to muscular legislation rather than to 'muscular liberalism', to use David Cameron's phrase.[34] There is nothing liberal or tolerant about prohibiting clothing where it involves no harm.

Perhaps the most enduring intrusions of national culture and identity occur directly rather than through the cultural inflection or particular assignment of liberal values. A society's practices and institutions are commonly established by and to serve the interests of a founding cultural group or groups, who were present at the creation, as it were. The crudest cases of this legacy involve the direct and invidious discrimination against minority members. There are also indirect and more subtle forms of discriminations that arise with the established institutions – the organization of the working and school week, the scheduling of public events like elections, dress codes and burial practices, what sounds may emanate from houses of worship, to name just a few examples. The reports of the expert commissions considered in this book are replete with these sorts of cases. Then there are the cases that may or may not impinge on liberal values, such as what may be done to the human body (piercing, tattooing, breast implants, vaginoplasty, and gender reassignment – yes; scarification and female genital cutting or nicking – no; infant male circumcision – maybe).

If this account of the situation of minorities is correct, might not one conclude that multiculturalism is redundant since the remedy is essentially a fairer and more rigorous adherence to liberal values against the encroachments of a dominant culture? Liberal multiculturalism, after all, relies on just these values.[35] To conclude thus would be a grave mistake however. Even if the appropriate treatment of cultural minorities is implicit in core liberal values, liberal institutions have become more responsive to the plight of cultural minorities in no small part due to the emergence of multiculturalism as an idea, movement, and policy regime (in places) since the 1970s. Multiculturalism resembles feminism in this regard. Though the rights of women are implicit in liberal principles, it took a feminist movement before liberal institutions began seriously to be reformed in line with women's interests. Feminism changed the liberal landscape. Even folk opposed to feminism were changed by it in the sense that what they oppose today looks decidedly different from and more accepting than previously. The center of gravity of the debate has been shifted.

Much the same is true of multiculturalism.[36] Even as it was expressly rejecting multiculturalism, the Council of Europe, for example, was drawing lessons from it on how cultural diversity might better be accommodated under the badge of interculturalism.[37] And the same Angela Merkel who in 2010 pilloried the idea of a multicultural society for Germany,[38] became in 2012 the leading defender of Muslim's and Jews' right to continue the practice of infant male circumcision in the aftermath of the Cologne court decision outlawing the practice for violating the German Constitution, and in 2015 Europe's most outspoken defender of accepting and incorporating unprecedented numbers of refugees and Muslims from Syria.

Multiculturalism remains vital as a check on the majority culture from overreaching its proclaimed commitment to liberal values or from adhering to unjustifiably narrow interpretations or applications of them. Not only must the voices of reaction and xenophobia be combatted, so must also the tendency of dominant majorities to extend their domination. In this respect, multiculturalism may be likened to a wedding ring. Even if one removes the ring (or drops the word 'multiculturalism') there is still the matter of one's vows. As a policy rubric, multiculturalism is a sign to others and a reminder to ourselves of the commitments we have made.

Which brings me to my final point. Liberal democratic values – liberty, equality, and even solidarity or inclusion – will only take us so far. Particular cases often reveal one of these values to be pitched against the other or others, while each value admits of many interpretations. Appealing to the free exercise of religion in this or that legal code is typically the beginning of a long and complicated interpretative exercise. In the end, how we interpret liberal values and the legal codes that seek to enshrine them often appeal to other considerations and commitments. Among these should be considered answers to a few pragmatic questions: what are we trying to achieve here beyond the particulars of this case? What kind of society are we seeking to create? How best to make a culturally diverse liberal democracy work? Since these questions will impose themselves eventually, we should collectively learn to ask them sooner rather than later.

Notes

* I thank Katayoun Alidadi, Veit Bader, and Simon Thompson for their helpful comments on versions of this essay, and Christian Joppke and Jocelyn Maclure for their challenging responses at the Max Planck Institute conference.

1 G Bouchard and C Taylor, *Building the Future: A Time for Reconciliation. Report. Commission de consultation sur les practiques d'accommodement reliees aux differences culturelles* (Gouvernement du Québec 2008).

2 B Parekh, *The Future of Multi-Ethnic Britain: The Parekh Report* (Profile Books 2000).

3 B Stasi, *Laïcité et République. Rapport de la Commission de réflexion sur l'application du principe de laïcité dans la République*, Rapport de commission, Paris, République française, 2003.

4 M-C Foblets and C Kulakowski, *Round Tables on Interculturalism*, Brussels, Ministry for Employment and Equal Opportunities, in charge of Migration and Asylum Policy, 2010; and M-C Foblets, Chapter 11, this volume.

5 See, for example, J Rawls, *A Theory of Justice* (Harvard University Press 1971); C Larmore, *Patterns of Moral Complexity* (Cambridge University Press 1987); J Rawls, *Political Liberalism* (Columbia University Press 1993); and B Barry, *Culture and Equality* (Polity Press 2001).

6 See, for example, IM Young, *Justice and the Politics of Difference* (Princeton University Press 1990); V Bader, *Democracy or Secularism? Associational Governance of Religious Diversity* (University of Amsterdam Press 2007); and A Patten, *Equal Recognition: The Moral Foundations of Minority Rights* (Princeton University Press 2014).

7 See, for example, Y Tamir, *Liberal Nationalism* (Princeton University Press 1993); W Kymlicka, *Multicultural Citizenship: A Liberal Theory of Minority Rights* (Oxford University Press 1995); D Miller, *On Nationality* (Oxford: University Press 1995); and M Canovan, *Nationhood and Political Theory* (Edward Elgar 1996).

8 G Bouchard, 'What Is Interculturalism?' (2011) 56 *McGill Law Journal* 435; C Taylor, 'Interculturalism or Multiculturalism?' (2012) 38 *Philosophy and Social Criticism* 413.

9 See, for example, B Parekh, *Rethinking Multiculturalism: Cultural Diversity and Political Theory* (Macmillan 2000); and T Modood, *Multiculturalism: A Civic Idea*. 2nd edition (Polity Press 2013).

10 The 2009 Swiss ban on minarets also seems to pit a minority's free exercise claim against majority precedence, in this case, the majority's appreciation of how public space should symbolically appear. However, much more besides was going on in that case. Cf., for example, D Miller, 'Majorities and Minarets: Religious Freedom and

Public Space' (2016) 46 *British Journal of Political Science* 437; and C Laborde, 'Miller's Minarets: Religion, Culture, Domination' in SJ Fine, D Butt, and Z Stemplowska (eds), *Political Philosophy, Here and Now: Essays in Honour of David Miller* (Oxford University Press, forthcoming).

11 C Taylor, 'The Politics of Recognition' in A Gutmann (ed.), *Multiculturalism and 'The Politics of Recognition'* (Princeton University Press 1992).

12 M Walzer, 'Comment' in Gutmann (n 11), 101.

13 Miller (n 10); and D Miller, *Justice for Earthlings: Essays in Political Philosophy* (Oxford University Press, 2013), 98.

14 Bouchard and Taylor (n 1).

15 The case covers the judgments of the Veneto Administrative Court (2005) and Supreme Administrative Court (2006) and the Lower Chamber (2009) and Grand Chamber (2011) decisions of the European Court of Human Rights. I will concentrate on the latter two cases. See, respectively, ECtHR November 2, 2009, Case No. 30814/06, *Lautsi v. Italy*; and ECtHR March 11, 2011, Case No. 30814/06, *Lautsi and Others v. Italy*.

16 *Lautsi*, 2011 Eur. Ct. H.R. §§ 15–16.

17 Bouchard and Taylor (n 1), 152.

18 'Oral Submission by Professor Weiler on Behalf of Armenia, Bulgaria, Cyprus, Greece, Lithuania, Malta, Russia, and San Marino – States Who Intervene as Third Parties in the Lautsi Case before the Grand Chamber of the European Court of Human Rights', July 1, 2010: available at www.ilsussidiario.net/News/Politics-Society/2010/7/1/EXCLUSIVE-Oral-Submission-by-Professor-Joseph-Weiler-before-the-Grand-Chamber-of-the-European-Court-of-Human-Rights/96909/.

19 'Harvard Law School Is Ditching Its Controversial Crest Linked to a Slaveholder', *Reuters*, March 15, 2016.

20 'Oral Submission' (n 18).

21 J Weiler, 'EJIL Editorial – Lautsi: Crucifix in the Classroom Redux' (2010) 21 *European Journal of International Law*: www.ejiltalk.org/new-issue-of-ejil-vol-211-out-in-this-issue/.

22 Rawls, *Political Liberalism* (n 5), 193. On the need of political liberals to take responsibility for the broader cultural impact of their political principles, see J Tomasi, *Liberalism beyond Justice: Citizens, Society, and the Boundaries of Political Theory* (Princeton University Press 2001).

23 Bouchard (n 8), 455; and G Bouchard, 'Quebec Interculturalism and Canadian Multiculturalism', in N Meer, T Modood, and R Zappata-Barrero (eds), *Multiculturalism and Interculturalism: Debating the Dividing Lines* (Edinburgh University Press 2016).

24 A situation that Patrick Charlier and Nathalie Denies capture with the concept of 'highest common denominator' in discussing the Belgian case in Chapter 12, this volume.

25 K Rogers, 'Charleston Shooting Reignites Debate about Confederate Flag', *The New York Times*, June 19, 2015.

26 C Joppke, *Immigration and Citizenship* (Polity Press 2010), 137.

27 Joppke (n 26), 138.

28 For further elaboration of these points, see GB Levey, 'Confronting Autonomy in Liberal Practice' in M-C Foblets, AD Renteln and M Graziadei (eds), *The Paradoxes of Personal Autonomy* (Routledge, 2017).

29 C Kukathas, *The Liberal Archipelago: A Theory of Diversity and Freedom* (Oxford University Press 2003).

30 Rawls, *Political Liberalism* (n 5).

31 W Kymlicka, *Contemporary Political Philosophy: An Introduction.* 2nd edition. (Oxford University Press 2002), 237.

32 Rawls, *Political Liberalism* (n 5), 199–200.

33 Joppke (n 26), 123–137.

34 *BBC News*, 'State Multiculturalism Has Failed, Says David Cameron', February 5, 2011: www.bbc.co.uk/news/uk-politics-12371994.

35 There are, of course, accounts of multiculturalism that view liberal values as just as much part of the problem as national cultures. It goes without saying that these accounts are considered less acceptable in Western societies than are liberal versions of multiculturalism.

36 W Kymlicka, *Multicultural Odysseys: Navigating the New International Politics of Diversity* (Oxford University Press 2007).

37 Council of Europe Committee of Ministers, 118th Sess, *Living Together as Equals in Dignity. White Paper on Intercultural Dialogue* (Council of Europe 2008).

38 *BBC News*, 'Merkel Says German Multicultural Society Has Failed', October 17, 2010: www.bbc.co.uk/news/world-europe-11559451.

Bibliography

Bader, Veit. *Democracy or Secularism? Associational Governance of Religious Diversity*. Amsterdam: University of Amsterdam Press, 2007.

Barry, Brian. *Culture and Equality*. Cambridge: Polity Press, 2001.

BBC News, 'Merkel Says German Multicultural Society Has Failed', October 17, 2010: www.bbc.co.uk/news/world-europe-11559451.

BBC News, 'State Multiculturalism Has Failed, Says David Cameron', February 5, 2011: www.bbc.co.uk/news/uk-politics-12371994.

Bouchard, Gérard. 'Quebec Interculturalism and Canadian Multiculturalism'. In *Multiculturalism and Interculturalism: Debating the Dividing Lines*, ed. Nasar Meer, Tariq Modood, and Ricardo Zappata-Barrero, 77–103. Edinburgh: Edinburgh University Press, 2016.

Bouchard, Gérard. 'What Is Interculturalism?' *McGill Law Journal* 56 (2011): 435–468.

Bouchard, Gérard, and Charles Taylor. *Building the Future: A Time for Reconciliation. Report. Commission de consultation sur les practiques d'accommodement reliees aux differences culturelles*. Montréal: Gouvernement du Québec, 2008.

Canovan, Margaret. *Nationhood and Political Theory*. Cheltenham, UK: Edward Elgar, 1996.

Council of Europe Committee of Ministers, 118th Sess. *Living Together as Equals in Dignity. White Paper on Intercultural Dialogue*. Strasbourg: Council of Europe, 2008.

Foblets, Marie-Claire, and C Kulakowski. *Round Tables on Interculturalism*. Brussels: Ministry for Employment and Equal Opportunities, in charge of Migration and Asylum Policy, 2010.

Joppke, Christian. *Immigration and Citizenship*. Cambridge: Polity Press, 2010.

Kukathas, Chandran. *The Liberal Archipelago: A Theory of Diversity and Freedom*. Oxford: Oxford University Press, 2003.

Kymlicka, Will. *Multicultural Citizenship: A Liberal Theory of Minority Rights*. Oxford: Oxford University Press, 1995.

Kymlicka, Will. *Contemporary Political Philosophy: An Introduction*. 2nd edition. Oxford: Oxford University Press, 2002.

Kymlicka, Will. *Multicultural Odysseys: Navigating the New International Politics of Diversity*. Oxford: Oxford University Press, 2007.

Laborde, Cécile. 'Miller's Minarets: Religion, Culture, Domination'. In *Political Philosophy, Here and Now: Essays in Honour of David Miller*, ed. Sarah Jane Fine, Daniel Butt, and Zofia Stemplowska. Oxford: Oxford University Press, forthcoming.

Larmore, Charles. *Patterns of Moral Complexity*. Cambridge: Cambridge University Press, 1987.

Levey, Geoffrey Brahm. 'Confronting Autonomy in Liberal Practice'. In *Personal Autonomy in Plural Societies: A Principle and Its Paradoxes*, ed. Marie-Claire Foblets, Alison Dundes Renteln, and Michele Graziadei, 38–50. London: Routledge, 2017.

Miller, David. *On Nationality*. Oxford: Oxford University Press, 1995.

Miller, David. *Justice for Earthlings: Essays in Political Philosophy*. Oxford: Oxford University Press, 2013.

Miller, David. 'Majorities and Minarets: Religious Freedom and Public Space', *British Journal of Political Science* 46 (2016): 437–456.

Modood, Tariq. *Multiculturalism: A Civic Idea*. 2nd edition. Cambridge: Polity Press, 2013.

Parekh, Bhikhu. *Rethinking Multiculturalism: Cultural Diversity and Political Theory*. Basingstoke: Macmillan, 2000.

Parekh, Bhikhu. *The Future of Multi-Ethnic Britain: The Parekh Report*. London: Profile Books, 2000.

Patten, Alan. *Equal Recognition: The Moral Foundations of Minority Rights*. Princeton, NJ: Princeton University Press, 2014.

Rawls, John. *A Theory of Justice*. Cambridge, MA: Harvard University Press, 1971.

Rawls, John. *Political Liberalism*. New York: Columbia University Press, 1993.

Reuters. 2016. 'Harvard Law School Is Ditching Its Controversial Crest Linked to a Slaveholder', *Reuters*, March 15.

Rogers, Katie. 'Charleston Shooting Reignites Debate about Confederate Flag', *The New York Times*, June 19, 2015.

Stasi, Bernard. *Laïcité et République. Rapport de la Commission de réflexion sur l'application du principe de laïcité dans la République*, Rapport de commission. Paris: République française, 2003.

Tamir, Yael. *Liberal Nationalism*. Princeton, NJ: Princeton University Press, 1993.

Taylor, Charles. 'The Politics of Recognition'. In *Multiculturalism and 'The Politics of Recognition'*, ed. Amy Gutmann, 25–73. Princeton, NJ: Princeton University Press, 1992.

Taylor, Charles. 'Interculturalism or Multiculturalism?' *Philosophy and Social Criticism* 38 (2012): 413–423.

Tomasi, John. *Liberalism beyond Justice: Citizens, Society, and the Boundaries of Political Theory*. Princeton, NJ: Princeton University Press, 2001.

Walzer, Michael. 'Comment'. In *Multiculturalism and 'The Politics of Recognition'*, ed. Amy Gutmann, 99–104. Princeton, NJ: Princeton University Press, 1992.

Weiler, Joseph. 'EJIL Editorial – Lautsi: Crucifix in the Classroom Redux', *European Journal of International Law* 21 (2010): www.ejiltalk.org/new-issue-of-ejil-vol-211-out-in-this-issue/.

Weiler, Joseph. 'Oral Submission by Professor Weiler on Behalf of Armenia, Bulgaria, Cyprus, Greece, Lithuania, Malta, Russia, and San Marino – States Who Intervene as Third Parties in the Lautsi Case before the Grand Chamber of the European Court of Human Rights', July 1, 2010: www.ilsussidiario.net/News/Politics-Society/2010/7/1/EXCLUSIVE-Oral-Submission-by-Professor-Joseph-Weiler-before-the-Grand-Chamber-of-the-European-Court-of-Human-Rights/96909/.

Young, Iris Marion. *Justice and the Politics of Difference*. Princeton, NJ: Princeton University Press, 1990.

16 The European Court of Human Rights

Fundamental assumptions that have a chilling effect on the protection of religious diversity

Eugenia Relaño Pastor[*]

Introduction

The mandates of the four expert commissions being studied in this volume were based on the notion that every society needs to be cohesive, but also respectful of diversity.[1] They also faced the challenge of finding ways to protect diversity while fostering a common sense of belonging among the members of society. As a result, the recommendations released by the four commissions to the public authorities relied on an unquestionable reality: pluralism is not just a topic in the policymakers' agenda but, to cite the European Court of Human Rights' (ECtHR) jurisprudence, it is also 'indissociable from a democratic society'.[2] We all live in multi-ethnic, multi-faith, multicultural and multi-community societies, as the Parekh Report states,[3] and handling the differences between different communities and their seemingly incompatible values is nowadays integral to democracy.[4] The reports analysed in this volume show that states may find it challenging to uphold such values as tolerance, respect and pluralism, and this essay argues that these values are not neutral in the ECtHR case law.

As enshrined in Article 9, freedom of thought, conscience and religion form the foundations of a 'democratic society' within the meaning of the European Convention on Human Rights (ECHR), and one of the concepts more often used to interpret the scope of Article 9 is 'pluralism'. In the majority opinion in *Kokkinakis v. Greece*,[5] the Court held that pluralism is, in its religious dimension, one of the most vital elements that make up the identity of believers and their conception of life, but that it is 'also a precious asset for atheists, agnostics, sceptics and the unconcerned'.[6] Pluralism, in that sense, is both the means and the end of fostering genuine religious freedom and a basic good for a healthy democracy.[7] It is, moreover, not only a good in and of itself, but also one that nourishes other liberal goods such as respect for diversity.[8]

Diversity has played a fundamental role in a number of opinions (especially those involving Jehovah's Witnesses) that seemingly advance religious pluralism. It has also been a key element in a series of prominent cases in which the ECtHR has failed to protect religious pluralism, particularly in cases concerning Islam in public life. In these cases, the scope of religious pluralism was restricted, paradoxically, for the sake of preserving democratic values, with the result that the freedom to manifest religious identity and to live with dignity was circumvented.

This essay shows how religious diversity has been disregarded in the Strasbourg decisions, which also proves the unwillingness of the ECtHR to contextualise the position of the rights holders, who have been understood in the Court along the majoritarian lines as a monolithic dissident cultural community.

Although in *Kokkinakis* the Court held that pluralism was an essential democratic value and reiterated a value-pluralist tradition based on the recognition of diversity and accommodation of competing claims, the so-called Islamic cases have marked a further turn from *Kokkinakis*. The Court no longer focuses on the minimal conditions under which pluralism could flourish, but instead seeks to establish the limits of pluralism. The Court holds assumptions about religion, freedom of religion, secularism and public order that are used as justifications for restricting religious pluralism, while regarding religious freedom and democracy as potentially antagonistic. Unfortunately, the ECtHR decisions analysed in this essay represent neither a prudent balancing of interests nor an incidental 'free electron' in Strasbourg's legal reasoning. Rather, they reflect the Court's secular logic that has shadowed the interpretation of Article 9 by undermining the scope of religious pluralism.

Believers' voices beyond secular norms

The Strasbourg Court's legal discourse has a strong impact not only on the parties involved in the specific cases but also on the human rights adjudications throughout Europe. The Court enjoys a high degree of prestige in establishing standards for the protection of freedom of religion, and its impact extends beyond the European borders. By examining various Court judgments, particularly those concerning the right to wear religious garments, it is possible to conclude that notions of religious pluralism that take into account other normative orders have been neglected by the Court, which has adopted a constitutional legal approach to the problem of conflicting rights concerning religious identity and law. As a result, the voices of the rights holders and duty bearers, as well as their cultural norms, have been ignored. In fact, a controversy ensued in 2004 in Canada when concerns about Islamic courts escalated in response to an independent report (Boyd Report) that recommended that religious family arbitration continue to be allowed and that a series of measures be adopted to protect vulnerable individuals.[9]

Some decisions, such as *Dahlab*,[10] *Şahin*,[11] *Sefika Köse and 95 Others*,[12] *Dogru*[13] and *S.A.S.*,[14] reflect Europe's struggle with public Islam and how believers – particularly women – are easily made to disappear behind general interpretations of the Canadian constitutional principles. Drawing on stereotypical images of the religious dress practices, ECtHR presented a negative image of religious symbols as incongruous with the democratic principles in Europe, while the believers were deemed 'undemocratic'. The *Dahlab* case involved a primary school teacher who had converted to Islam and was banned from teaching for wearing a headscarf. The Court held that the teacher's right under Article 9 to put her religious beliefs on display was outweighed by the state's interest in protecting the rights and freedoms of others and preserving public order and safety: 'It therefore appears difficult to reconcile the wearing of an Islamic headscarf with the message of tolerance, respect for others'.[15]

As Gunn has suggested, such forms of 'otherisation' were caused by the fear of secularists that some public manifestations of Islamic practices interfered with the image of a secular Europe.[16] That perception of Islam is set against the backdrop of the constitutional principle of neutrality and the wide margin of appreciation that states have. As a result, the identity and the autonomy of the individuals are left out of consideration. The judicial (mis)perception of Islam as monolithic and institutional, when juxtaposed with the principles of secularism and democratic values, effectively silences the rights holders.[17] In *Şahin* the Court ruled:

> [I]t is the principle of secularism, as elucidated by the Constitutional Court (see paragraph 39 above), which is the paramount consideration underlying the ban on the wearing of religious symbols in universities.[18]

The European Court adheres to a rather light and narrow conception of democracy, thereby endangering the participation of cultural minorities in the name of preserving the secular nature of the state. The Strasbourg reasoning is just the opposite to the one used by all four expert commission reports, where the attention was on the measures that could enhance the participation of minority groups commonly subject to religious and cultural discrimination.[19]

Similarly, in *Köse*, the Court opposed the exercise of religious freedom in order to prevent chaos. This case concerned the prohibition for students of religiously oriented public secondary schools to wear the Islamic headscarf within the limits of the school. Relying on Article 9 of the Convention, the applicants complained, in particular, that the ban on wearing the Islamic headscarf in the schools in question constituted an unjustified infringement of their right to freedom of religion and in particular their right to manifest their religion. The Strasbourg Court upheld a uniformity criterion against the recognition and respect for cultural diversity:

> It should be noted that school rules of this kind are general rules that apply to all pupils independently of their religious beliefs and serve among other things the legitimate aim of preserving the neutral character of secondary education, which is intended to protect adolescents when they are at an impressionable age.[20]

Sometimes, the uniformity standard adopted by the Court is justified on the basis of the 'national traditions' of the particular state. The reason the Court has commonly used to restrict diversity in the Islamic cases is the presumed incompatibility of Islam with the prevalent national tradition, which, paradoxically, refers to a homogeneous society. However, the protection of national traditions is not intended to undermine the rights of the minorities per se. As the Bouchard-Taylor Report states, if a historic identity is integral to a society's past, present and future, it should be taken into account and protected as long as the fundamental liberties of minorities are also respected.[21]

In *Dogru*, the applicant (age 11) went to her physical education and sports classes wearing her headscarf and refused to take it off despite repeated requests to do so by her teacher, who explained that the wearing of a headscarf was incompatible with the dress code for the physical education classes. The Caen Administrative Court found that the applicant's attitude had created an atmosphere of tension within the school and that her expulsion from the school had been justified on the basis of the relevant factors, disregarding the proposal she had made at the end of January to wear a hat instead of her headscarf. In its determination as to whether such restriction was 'prescribed by law' if it was directed towards one or more of the legitimate aims set out in that paragraph, and whether it was 'necessary in a democratic society' to achieve the aims concerned, the ECtHR concluded:

> Where questions concerning the relationship between State and religions are at stake, on which opinion in a democratic society may reasonably differ widely, the role of the national decision-making body must be given special importance This will notably be the case when it comes to regulating the wearing of religious symbols in educational institutions, in respect of which the approaches taken in Europe are diverse.[22]

It is also remarkable in *Dogru* that the Court showed its concern for 'the rights and freedoms of others' even while ignoring the voice of Ms Begin Dogru and rejecting her offer to wear a hat instead of her headscarf as a way to comply with the school's internal rules.

> The Court considers, having regard to the foregoing, that the penalty of expulsion does not appear disproportionate, and notes that the applicant was able to continue her schooling by correspondence classes. It can be seen that the applicant's religious convictions were fully taken into account in relation to the requirements of protecting the rights and freedoms of others and public order.[23]

In these cases, the legal reasoning focuses on the state, on one hand, and on religion, on the other. The Court was mainly concerned with the preservation of the state's constitutional principles, namely democracy, tolerance, secularism, public order and the rights and freedoms of others. The applicant's religion is made synonymous with a 'non-democratic religion', and the applicant is set against the constitutional notions of democracy and the Court's definition of secularism. In finding a balance between 'democratic secularism' (State) and the scope of freedom of religion (Religion), the arguments of the applicant regarding their own religious nomos are overtly overlooked.[24]

Moreover, in asking whether the wearing of the hijab conformed to democratic values such as gender equality and tolerance, there is already a denial of the autonomy of the woman. In her dissenting opinion in *Şahin*, Judge Tulkens stressed:

> The ban on wearing the headscarf is therefore seen as promoting equality between men and women. However, what, in fact, is the connection between the ban and sexual equality? The judgment does not say. Indeed, what is the signification of wearing the headscarf? . . . What is lacking in this debate is the opinion of women, both those who wear the headscarf and those who choose not to.[25]

Many existing empirical studies show that the vast majority of face-veiled women claimed that wearing the face veil was a personal and autonomous choice,[26] without any corresponding family or social pressure, and sometimes against the will of their relatives. The issue of autonomy, consent, choice and pressure is a tricky one. But what three of the reports show is the relevance of the opinion, and the experience of each individual concerned in the particular case. The Bouchard-Taylor Commission held hearings in 15 regions that lasted a total of 31 days. During the 328 hearings, the commission heard the testimonies of 241 individuals.[27] The Parekh Report explicitly demonstrated what it meant 'to be black in Britain' or 'to live with racism'.[28]

The European Court has not paid attention to any empirical studies on the veil, or even to the complexity of the issue in the daily life of veiled women. In *S.A.S. v. France*, the Human Rights Centre of Ghent University and the Open Society Foundation presented empirical studies as third-party interveners, showing that the use of the veil was context-specific.[29] Nevertheless, the European Court ignored it. Consequently, the Strasbourg Court has failed to contextualise the position of the rights holders whose existence is swallowed up in the image of what the majority perceive as a cultural monolithic community, and it has failed to do justice to such individuals. In fact, the question in the cases mentioned above was not whether France, Turkey or Switzerland had infringed upon the applicant's freedom of religion, but whether the applicant had infringed upon relevant constitutional principles, with which the ECtHR itself could not legitimately interfere. Applicants are, instead, pushed and pulled in various directions, due to their divergent 'normative commitments' (personal/religious as well as those imposed by the state), whereas the ECtHR should provide space within which persons are allowed to interpret how the commitments make sense to each of them. That requires that judges take into account the specific context of the people involved and the coexistence of diverse normative and religious obligations in which the applicants could be caught.

In spite of a number of legal-moral orders that could be applied to a particular case (national order, international human rights law, religious norms and the person's own perception of religion), the Strasbourg Court has tended to be more concerned with the legitimacy of the state than with the recognition of other different legal-moral regimes. In *Refah Partisi and Others v. Turkey*, the Court rejected a multiplicity of different legal orders within the same state:

> A difference in treatment between individuals in all fields of public and private law according to their religion or beliefs manifestly . . . cannot maintain a fair balance between, on the one hand, the claims of certain religious groups

who wish to be governed by their own rules and on the other the interest of society as a whole, which must be based on peace and on tolerance between the various religions and beliefs.[30]

The Strasbourg Court has not been sensitive enough to admit interpretations of the freedom of religion from different perspectives. However, the existence of legal pluralism on many levels is a reality now in Europe, where, currently, there are at least three mutually influencing orders: the constitutional traditions of each of the European Member States, the EU's fundamental rights protection system and the European Convention on Human Rights. Since the ECtHR has sometimes alluded to comparative law for a better understanding of the right at issue,[31] it seems unjustified that the internal legal pluralism could be discouraged, particularly in the context of wearing religious garments.[32] It appears to be more legitimate to contextualise rights against the backdrop of the rights holders' communities than to rely on interpretations deriving from extraterritorial jurisdictions.

On the other hand, a legal subject is an active agent who should not have to struggle between the national law and his or her own traditions. In order to contextualise the exercise of any right, the state law should enable equal recognition of the personal and/or religious culture and respect for state law. In that respect, contextualisation does not mean to widen the 'margin of appreciation', as Prof. Weiler suggested in *Lautsi v. Italy*:

> The European Court of Human Rights is not an Oracle. It is a dialogical partner with the Member States Parties to the Convention . . . When there is a diverse constitutional practice among Convention States – and certainly there is in this area – the Court needs to listen, not only preach, and to be seen to be listening.[33]

That kind of contextualisation takes into consideration only one side of inter-normativity and not the position of the rights holders, who could be influenced by different heteronomies. In *S.A.S. v. France*, the applicant defined herself as 'a devout Muslim' who wore the burqa and the niqab in accordance with her religious faith, her culture and her personal convictions. According to her explanation, 'neither her husband nor any other member of her family put pressure on her to dress in this manner'.[34] The applicant added that 'she wore the niqab in public and in private, but not systematically: she might not wear it, for example, when she visited the doctor, when meeting friends in a public place, or when she wanted to socialise in public. She was thus content not to wear the niqab in public places at all times but wished to be able to wear it when she chose to do so, depending in particular on her spiritual feelings. There were certain times (for example, during religious events such as Ramadan) when she believed that she ought to wear it in public in order to express her religious, personal and cultural faith. Her aim was not to annoy others but to feel at inner peace with herself'.[35]

As illustrated by the above cases, ECtHR has not properly addressed practices affecting identity and autonomy. The Court should approach a conflict of rights by starting with the perspective of the rights holders and only subsequently moving on to human rights and constitutional perspectives. The question is not only about the impact of possible normative overlapping (religious obligations, state normative orders and human rights order) on affected individuals, but also about the extent to which the individual's own understanding of autonomy and freedom has been incorporated into the analyses of the judges. If it is the intention of the ECtHR to do justice to the scope of lived diversities, it is necessary to empower rights holders when facing normative overlapping conflicts as an essential part of any legal pluralistic balancing of rights.

Discouraging religious diversity: assumptions and misconceptions

Assumptions about religion

Any attempt to define the scope and the content of the right of religious freedom involves assumptions about the underlying nature of religion itself.[36] Such assumptions often rest on unarticulated premises concerning cultural aspects of main and theistic religions.[37] In the context of liberal theories of religious freedom, religion is a matter of the private belief of an autonomous individual[38] based on his or her own conscience and belief, whereas, in the Islamic tradition, many Muslims regard themselves as belonging to a religious community they have not chosen. These two different approaches to the meaning of religion have implications for the manifestation of practices in the public sphere and the scope of neutrality. The background conception of *religion as conscience* defines the basic shape of Article 9 jurisprudence, and it is also a foundational principle in the Stasi Report.[39]

In *Kokkinakis*, the Court outlined the general principles of Article 9, which have been reiterated subsequently in the following judgments. The Court has declared that 'the freedom of thought, conscience and religion' is not only the first general principle but also one of the foundations of a 'democratic society' within the meaning of the Convention. In its religious dimension, it is one of the most vital elements that make up the identity of believers and their conception of life; it is also a precious asset for atheists, agnostics, sceptics and the unconcerned. Pluralism, indissociable from the notion of a democratic society, which has been dearly won over the centuries, depends on it.[40]

The Court has also endorsed a secularist move to relegate religion to the privacy of conscience, as was shown in cases such as *Şahin*[41] and *Sinan Isik v. Turkey*.[42] This kind of secularism places religion in the conscience of the individual. Additionally, the Court not only tends to portray religion as conscience, but also tends to favour the cerebral and the internal over the active and symbolic dimensions of religion or belief, assuming the lesser importance of living out one's religion or belief,[43] as the Court concluded in *Dahlab*:

> [E]ven if it is particularly important to the appellant and does not merely represent an expression of a particular religious belief but complies with an imperative requirement of that belief, the wearing of a headscarf and loose-fitting clothes remains an outward manifestation which, as such, is not part of the inviolable core of freedom of religion.[44]

Assumptions about freedom of religion

The Strasbourg jurisprudence has analysed freedom of religion in terms of a binary opposition between belief and practice, between *forum internum* and the *forum externum*, privileging the former over the latter.[45] The *forum internum* is defined as the internal sphere of personal thought, conscience or belief, and the protection of this sphere is absolute.[46] However, the two sides of the binary are interconnected. One of the main concerns for the Strasbourg Court is to determine whether a certain act 'counts' as religious practice for the purposes of legal protection. The Court has long been criticised for minimising the scope of Article 9(1) of the ECHR by adopting a restrictive approach to determine what counts as a manifestation of religion or belief.[47] Nonetheless, the Court has gradually started to 'count' practices that, though not necessarily required by a religion, are still motivated or inspired by it.[48] However, the Court has tended to favour those who conform to what is prescribed by the religious groups, while disempowering those who may disagree within the group.[49] In doing so, ECtHR has only relied on an objective test. An objective test is akin to the Court's scrutiny, which takes into account the practices that the religious tenets recognise as mandatory, whereas a subjective approach to freedom of religion is based 'not on what others view the claimant's religious obligations as being, but rather on what the claimant views these personal religious "obligations" to be'.[50] The Court's preference for an objective approach has resulted in an essentialisation and a simplification of the complex relationships between believer, belief and orthodox practice, and thus in the marginalisation of minority voices.

For example, *Gatis Kovalkovs v. Latvia* concerns the practice of religion in prison, wherein the applicant alleged that he was prevented from performing the rituals of Vaishnavism as a result of the confiscation of his incense stick. The Court accepted that burning incense sticks could be regarded as motivated or inspired by religion, but the Latvian government presented information provided by members of the Riga Vaishnavist congregation that informed that the obligation to observe the basic rituals of Vaishnavism, including the burning of incense sticks, was 'conditional'. The Court had depended exclusively on the information provided by the Riga Vaishnavist congregation and eventually concluded:

> [R]estricting the list of items permitted for storage in prison cells by excluding items (such as incense sticks) which are not essential for manifesting a prisoner's religion is a proportionate response to the necessity to protect the rights and freedoms of others.[51]

There is also a second set of cases in which the Court regards the applicants' practices as a 'manifestation' of their religion largely based on essentialist understandings of group identity; in other words, stereotyping. The Court has relied on this strategy in dealing with religious groups with which we are not familiar, to essentialise them in orthodox ways. This tendency has been employed in cases involving Sikh and Muslim applicants (*Mann Singh v. France*, *Dahlab*, *Şahin*). In these cases the Court has tended to neglect the applicant's voice and to focus on their practices or symbols as if they had a separate existence. The Strasbourg judges invoke symbols without regard to the applicants' contexts. In the *Mann Singh* case, which concerned a Sikh man who had been denied the renewal of his driver's licence for refusing to take off his turban for the picture, the Court assessed whether Mann Singh's wearing of his turban fell within the scope of Article 9 of the ECHR. The Court concluded:

> [A]ccording to the applicant, the Sikh faith compels its members to wear the turban in all circumstances. It is not only considered at the heart of their religion, but also at the heart of their identity. Therefore, the Court notes that this is an act motivated or inspired by a religion or belief.[52]

The Court privileged the religious symbol over the intent of the applicant, because the Sikh religion was what had compelled him to wear the turban, even if the applicant may have exercised his autonomy in believing in the Sikh religion. This suggests that those who do not follow the 'core' practice of wearing a turban may be regarded less as members than others.

In addition, the cases concerning the prohibiting of Muslim women from wearing the headscarf is more emblematic of how the individual believers are being 'objectivised' because of their religious/cultural practices. As Peroni points out: 'The Court leaves out the applicant's subjectivity and objectivizes his practice by the using of the passive voice instead of the active voice (passivization). The Court separates the object from its wearer and objectivizes his religious practice'.[53] In *Dahlab* the Court concluded: 'It therefore appears difficult to reconcile the *wearing of an Islamic headscarf* with the message of tolerance, respect for others and, above all, equality and non-discrimination that all teachers in a democratic society must convey to their pupils'.[54] Or in *Şahin*, the majority of the Strasbourg judges held:

> [W]hen examining the question of *the Islamic headscarf* in the Turkish context, it must be borne in mind the impact which wearing such a symbol, which is presented or perceived as a compulsory religious duty, may have on those who choose not to wear it . . . [so that] this religious symbol has taken on political significance in Turkey in recent years.[55]

Likewise, in *S.A.S.*, the Court's assessment of the case starts thus:

> The ban on *wearing clothing* designed to conceal the face, in public places, raises questions in terms of the right to respect for private life (Article 8 of the Convention) of women who wish to wear the full-face veil for reasons related to their beliefs, and in terms of their freedom to manifest those beliefs (Article 9 of the Convention).[56]

Not the individual but solely the object (the headscarf) is at the centre of analysis. The headscarf acquires a life of its own, pushing aside the lives and intentions of the applicants.

Another way of stereotyping religious minorities is by disseminating negative stereotypes about some specific religious garments. In all of the cases about Muslims and religious garments, the ECtHR constructed a negative image of the headscarf as a symbol that is non-compliant with the values of European democracy. For example, in *Dahlab*, the Grand Chamber referred to the wearing of the headscarf as a 'powerful external symbol' which 'appeared to be imposed on women by a religious precept that was hard to reconcile with the principle of gender equality'; in *Refah Partisi*, the Court stated that measures taken at universities to prevent certain fundamentalist religious movements from exerting pressure on students who did not practise that religion or on those who belonged to another religion could be justified under Article 9.2 of the Convention; and in his dissent against the majority in *Şahin*, Judge Tulkens concluded:

> While everyone agrees on the need to prevent radical Islamism, a serious objection may nevertheless be made to such reasoning. Merely wearing the headscarf cannot be associated with fundamentalism and it is vital to distinguish between those who wear the headscarf and 'extremists' who seek to impose the headscarf as they do other religious symbols. Not all women who wear the headscarf are fundamentalists and there is nothing to suggest that the applicant held fundamentalist views.[57]

Like Judge Tulkens, the Bouchard-Taylor Report noted:

> [M]any citizens believe that the wearing of the headscarf attests to the inferior status that women appear to be granted in Islam . . . [but] the crucial question is to ascertain who has the right to decide on the meaning of an individual's expressive acts?[58]

Surprisingly, in *S.A.S.*, the Court was aware of the negative stereotype of Muslim women as oppressed and in need of protection, as it had been at the heart of the debates surrounding the ban on full-face veils in Europe. While the French government made gender equality (the veil as a symbol of patriarchal authority) one of its main arguments in favour of the ban, the Strasbourg Court refuted this negative stereotype. It did, however, find that this argument was ultimately immaterial:

> The Court would first emphasise that the argument put forward by the applicant to the effect that the ban introduced by sections 1 to 3 of the Law of 11 October 2010 was based on the erroneous supposition that the women concerned wore the full-face veil under duress, is not pertinent.[59]

Another ECtHR assumption that undermines religious diversity is the premise about what counts as a 'proper' manifestation of religion and whether the manifestation is supposed to be in line with practices that are shared by the group as a whole. There are cases (mostly concerning Jehovah's Witnesses) in which the Court has adopted a sociological approach to determining the scope of freedom of religion by highlighting what other co-religionists believe. In *Thlimmenos v.*

Greece, a case concerning access to employment as an accountant being denied to a member of Jehovah's Witnesses due to a past conviction for refusing to serve in the military for religious reasons, the Court noted that the applicant was a member of the Jehovah's Witnesses, a religious group committed to pacifism.[60] The Court offered similar justification in *Bayatyan v. Armenia*, concerning the conviction of a Jehovah's Witnesses' applicant for draft evasion, stressing that the applicant was 'a member of Jehovah's Witnesses, a religious group whose beliefs include the conviction that service, even unarmed, within the military is to be opposed'.[61]

In other cases, such as *Kosteski v. the former Yugoslav Republic of Macedonia*, the Court held that the applicant was supposed to 'be or to act as a Muslim believer'. This case concerned a Muslim who was fined for absence from work on a Muslim holiday. The applicant had complained that this constituted an interference with the inner sphere of belief because he was required to prove his faith. The Court observed:

> [T]he courts' decisions on the applicant's appeal against the disciplinary punishment imposed on him made findings effectively that the applicant had not substantiated the genuineness of his claim to be a Muslim and that his conduct on the contrary cast doubt on that claim in that there were no outward signs of his practising the Muslim faith or joining collective Muslim worship.[62]

In other cases such as *Jakóbski v. Poland* and *Eweida*, however, the Court adopted a more open approach to diversity and accepted the applicants' practices as protected by Article 9 of the ECHR, even when their religiously mandated or prescribed character was actually contested by the governments involved in the cases. In *Jakóbski*, the applicant alleged under Article 9 of the Convention that he had been refused a meat-free diet in prison, contrary to the requirements of his religion, Buddhism, and the Court declared:

> [T]he applicant's decision to adhere to a vegetarian diet can be regarded as motivated or inspired by a religion and was not unreasonable. Consequently, the refusal of the prison authorities to provide him with a vegetarian diet falls within the scope of Article 9 of the Convention.[63]

In *Eweida and Others v. the United Kingdom*, concerning four Christian applicants wishing to manifest their religion at work – two of them by visibly wearing a cross around their necks – the Court stated:

> Ms Eweida's insistence on wearing a cross visibly at work was motivated by her desire to bear witness to her Christian faith. Applying the principles set out above, the Court considers that Ms Eweida's behaviour was a manifestation of her religious belief, in the form of worship, practice and observance, and as such attracted the protection of Article 9.[64]

It is worth underlining the Court's shift towards offering broader protection for individualistic manifestations of religion. Since *Eweida* and *S.A.S.*, the applicant no longer needs to prove adherence to a set of prescribed precepts in order for a religious claim to fall within the scope of protection of Article 9 ECHR. It is hard to determine the extent to which the views of the applicants or the religious precepts of their religious groups matter, rendering uncertain the question of how much protection the Strasbourg Court was seeking to provide to ensure internal diversity within a religious group.

Assumptions about secularism and neutrality

The meaning of the concept of secularism is linked to the meaning of public sphere and to the scope of the right to religious freedom. As the four expert commission reports show, the concept of secularism is bound to the peaceful ways in which societies reconcile cohesion, equality and diversity and their own conceptions of citizenship.[65]

The national identities of the state parties to the ECHR are characteristically secular identities. Nonetheless, how they interpret and apply secularism and the idea of religious neutrality can vary significantly. One of the common starting points is the liberal idea about the separation of the public and private spheres. Religion is seen as being distinct from the state, and to that extent 'disestablished' and 'privatised'. Religion is also viewed as belonging to the private sphere, and not a part of a 'neutral' public sphere that seeks to maintain its neutrality through rigorous commitment to individual rights. This approach is also common among practitioners of human rights who have applied the human rights framework to freedom of religion from a biased point of departure, namely by considering religious belief essentially private and individualist rather than communitarian in organisational orientation.[66]

On the surface at least, states may not have a cultural or religious goal, and they are required to remain 'neutral' regarding all religions and beliefs and between religion and non-religion, by both protecting the neutrality of their public spheres and not interfering in the autonomous spheres of conscience and belief. Many courts in the European states stress the role of the state as the neutral and impartial organiser of religious life. But, like it or not, religious difference is seen as a threat to public order. Therefore, it is difficult to explain how states, on the one hand, are bound to the endeavour of managing religious pluralism and tolerance through a policy of strict neutrality while, on the other hand, they claim that it is legitimate for the state to prohibit public religious manifestations that the state believes could undermine its essential political foundations, especially when many consider those foundations to *be religious* (*Lautsi*).[67] As the Bouchard-Taylor Report states: '[S]ecularism is more or less integral and rigid or flexible and open, depending on the way in which the dilemmas that arise when principles of secularism come into conflict are resolved'.[68]

For example, in the mid-1990s, the Strasbourg Court upheld two state bans on films deemed offensive to majoritarian Christian sensibilities: *Otto Preminger-Institute v. Austria*[69] and *Wingrove v. United Kingdom*.[70] In both cases, well-established religions within a state, either because they are an official religion or have a large number of adherents, had their core doctrines recognised as manifestations of religious belief. The Court endorsed a public sphere grounded in a kind of secular morality, which was supported by the Christian majority of the population. As a result, the Court provided evidence of a different treatment of the claims of majority and minority religious groups when conflicts arose. Although states may find it difficult to deal with values such as tolerance, respect and pluralism, the cases mentioned above show that these values are not neutral but are, rather, vehicles to legitimise certain assumptions about the proper reach of religion in the public sphere.

In addition, the concept of 'secularism' or 'neutrality' has been challenged by two assumptions coming from different angles: from the Christian national majorities and from Muslim minorities. The former challenge is advanced by Christians who question the neutrality of what they regard as a form of unjust 'secular establishment'. Since liberal democracies were born in Western Christianity, it would be difficult for Christians to move away from a form of liberal secularism to a form of 'fundamentalist' secularism that excluded Christian collective claims.[71] In his concurring opinion in *Lautsi*, Judge Bonello argued that removing the crucifix from Italian schools would not be neutral but an aggressive espousal of secularism:

> A court of human rights cannot allow itself to suffer from historical Alzheimer's. It has no right to disregard the cultural continuum of a nation's flow through time, nor to ignore what, over the centuries, has served to mould and define the profile of a people.[72] . . . A European court should not be called upon to bankrupt centuries of European tradition. No court, certainly not this Court, should rob the Italians of part of their cultural personality.[73] . . . Freedom of religion is not secularism. Freedom of religion is not the separation of Church and State. Freedom of religion is not religious equidistance – all seductive notions, but of which no one has so far appointed this Court to be the custodian. In Europe, secularism is optional, freedom of religion is not.[74]

Judge Power's concurring opinion reinforces that of Judge Bonello, that neutrality does not require a secularist approach, but Judge Power introduces an important nuance about the duty of the state to respect the religious freedom of all persons and groups.[75] In his separate opinion, Judge Bonello pointed to the fact that within the European tradition, education was rooted, historically, within the Christian tradition. To prohibit public schools, regardless of the wishes of the body politic, from displaying a symbol of the Christian – or indeed any other religious – tradition, and to require of the state that it pursue not a pluralist but a secularist agenda, risks venturing into the territory of intolerance – a concept that is contrary to the values of the Convention:

> Neutrality requires a pluralist approach on the part of the State, not a secularist one. It encourages respect for all world views rather than a preference for one . . . Within such a pluralist and religiously tolerant context, a Christian symbol on a classroom wall presents yet another and a different world view. A truly pluralist education involves exposure to a variety of different ideas including those which are different from one's own . . . Education would be diminished if children were not exposed to different perspectives on life and, in being so exposed, provided with the opportunity to learn the importance of respect for diversity.[76]

Judges Power and Bonello concurred in favour of the visibility of the religion of the majority in the public square: Why should Christians not have the right to live their lives in a public culture that recognises Christian mores and values? From this perspective, neutrality should be presented as an equal right to public religious manifestations and it should be associated with a form of democratic pluralism that allows religious diversity (mainly majority Christians) to freely develop.

The second challenge to the notion of secularism is advanced by religious minorities, particularly Muslims, whose objection is not to the presence of Christian values, symbols or practices in the public sphere, but to the exclusion or restriction of Islamic manifestations of religion on the basis of what is seen as interpretations of secularist neutrality according to the Strasbourg Court. As I have pointed out above, the differential treatment of Islam and Christianity in ECtHR jurisprudence is justified by portraying Islam as a threat to democracy, and Christianity as compatible with the values and rights in the Convention.[77] There is a range of ECtHR cases that deals directly with Islam, mainly on the question of wearing the Islamic headscarf (with the exception of *Refah*). The Court's major decisions that illustrate the relationship between Islam and secular public spheres are documented in *Şahin*, *Refah*, *Dogru* and *S.A.S.* In *Şahin*, the Court accepted that the domestic regulations at issue interfered with Şahin's right to manifest her religion under Article 9.1 but held that they were a valid limitation under Article 9.2 because, as they pursued the legitimate aims of protecting the 'rights and freedoms of others' and 'public order', they could be regarded as 'necessary in a democratic society'. It was held that the 'necessity' of the interference had to be based on two principles: secularism and equality.

In *Refah*, the Grand Chamber upheld a ban on the largest political party in Turkey, the Refah (Welfare) Party, for violating the constitutional principles of secularism and democracy through its activities. Even in the absence of such threats, the Court found that both the sharia and 'plural religiously based legal systems' were (even if democratically adopted) inherently incompatible with the ECHR and its concomitant notions of democracy and the rule of law. This case was a missed opportunity to rethink the notion of legal pluralism in the context of a secular society. As Judge Kovler stressed in his concurring opinion:

> I also regret that the Court, in reproducing the Chamber's conclusions (paragraph 119 of the judgment), missed the opportunity to analyse in more detail the concept of a plurality of legal systems, which is linked to that of legal pluralism and is well-established in ancient and modern legal theory and practice. Not only legal anthropology but also modern constitutional law accepts that under certain conditions members of minorities of all kinds may have more than one type of personal status. Admittedly, this pluralism, which impinges mainly on an individual's private and family life, is limited by the requirements of the general interest.[78]

While in the *Refah* decision, the Court's concern was about the limits of liberal pluralism, in *Şahin*, the Court was seemingly concerned with competing conceptions of secularism in the public sphere.[79] The Court's legal reasoning was the same in *Dogru* as in *Şahin*, but something had been added. In the Court's conception of *laïcité*, the state was neutral, blind and indifferent to religious diversity in order to honour the underlying non-discrimination principle. The Court agreed with the Stasi Commission's conception of *laïcité*.[80] The Court also observed that the purpose of that restriction on manifesting a religious conviction was to adhere to the requirements of secularism in state schools, as interpreted by the Conseil d'Etat:

> The Court also notes that in France, as in Turkey or Switzerland, secularism is a constitutional principle, and a founding principle of the Republic, to which the entire population adheres and the protection of which appears to be of prime importance, in particular in schools.[81]
>
> In France, the troubles have given rise to various forms of collective mobilisation regarding the question of the place of Islam in Republican society. It is in this context that, on 1 July 2003, the President of the Republic instructed a commission to study the application of the principle of secularism in the Republic. That commission, known as the 'Stasi commission', after the name of its chairman, produced a report for the President of the Republic on 11 December 2003. The picture it presented of the threat to secularism bordered on the alarming . . . In this context it is natural that many of our fellow citizens demand the restoration of Republican authority and especially in schools. It is with these threats in mind and in the light of the values of our Republic that we have formulated the proposals set out in this report.[82]

Islam, symbolised in these cases by the headscarf, posed a threat to the secular character of the Republic. If the restoration of the Republican authority implies reinforcing the neutral, collective identity of the majority by excluding the collective identity of particular religious minorities, French neutrality is not so neutral after all. Neutrality becomes an ethical and cultural claim (French Republicanism) and a way to preserve the tenet of one particular social group by excluding the 'others'.

Assumptions about public order and the rights of others

The idea that religious freedom may be limited to protect either public order or the rights and freedoms of others has been a key question in many of the cases discussed above. There is a clash of values when teachers and students in Switzerland, France or Turkey choose to wear headscarves and the state, acting in the name of the rights and freedoms of others, seeks to prevent them from doing so. In cases involving Islam, such as in *Refah*, *Şahin* or *Dogru*, the Court has advanced a wider conception of public order by using substantive notions of secularism and democracy that, however, excluded all forms of consensus of competing claims.

The Court's case law is not clear as to what may constitute 'the rights and freedoms of others' outside the scope of rights protected by the Convention. However, the majority in *S.A.S.* had upheld a legitimate aim of ensuring the possibility of 'living together' through 'the observance of the minimum requirements of life in society', which is understood to be one facet of the 'rights and freedoms of others' within the meaning of Articles 8.2 and 9.2 of the Convention:

> The Court takes into account the respondent State's point that the face plays an important role in social interaction. It can understand the view that individuals who are present in places open to all may not wish to see practices or attitudes developing there which would fundamentally call into question the possibility of open interpersonal relationships, which, by virtue of an established consensus, forms an indispensable element of community life within the society in question. The Court is therefore able to accept that the barrier raised against others by a veil concealing the face is perceived by the respondent State as breaching the right of others to live in a space of socialisation which makes living together easier.[83]
>
> It indeed falls within the powers of the State to secure the conditions whereby individuals can live together in their diversity. Moreover, the Court is able to accept that a State may find it essential to give particular weight in this connection to the interaction between individuals and may consider this to be adversely affected by the fact that some conceal their faces in public places.[84]
>
> Consequently, the Court finds that the impugned ban can be regarded as justified in its principle solely in so far as it seeks to guarantee the conditions of 'living together'.[85]

As was pointed out by the dissenting opinion of Judges Nussberger and Jäderblom, 'the very general concept of "living together" does not fall directly under any of the rights and freedoms guaranteed within the Convention'.[86] More than a decade ago, the Stasi Commission reinforced one single meaning for the expression 'living together'.[87] The reasons issued by the French government and subsequently reiterated by the Strasbourg Court in explanation of not tolerating the full-face veil reveal its biased interpretations of its symbolic meaning. In *S.A.S.*, the full-face veil appeared to be linked to the self-confinement of an individual that was cut off from others even while living among them. The French government and the ECtHR presume the existence of a right to see the face of others in a public space, while this is certainly not a legal right.[88] The question is, what if a woman wanted or decided to 'cut' herself off from social life? Should the states adopt a paternalistic attitude that renders such women voiceless and subordinated?[89] Muslim women are presented in cases like *S.A.S.* as women incapable of being autonomous, and as a consequence, as Anne Phillips has pointed out, it is a 'patronizing denial of human agency and responsibility to represent individuals from minority ethnic groups as compelled by the culture to behave in certain ways'.[90]

All these interpretations were called into question by the applicant, who claimed to wear the full-face veil only owing to her spiritual feelings and did not consider it an insurmountable barrier to communication or integration. But even assuming that such interpretations of the full-face veil were correct, it has to be stressed that there is no specific constitutional right to be protected from seemingly shocking or provocative models of cultural or religious expressions of identity, even if some are foreign to the traditional French and European lifestyles and sensibilities:

> [I]t can hardly be argued that an individual has a right to enter into contact with other people, in public places, against their will. Otherwise such a right would have to be accompanied by a corresponding obligation. This would be incompatible with the spirit of the Convention. While communication is admittedly essential for life in society, the right to respect for private life also comprises the right not to communicate.[91]

Although the Court has clearly elaborated in its jurisprudence the state's duty to ensure mutual tolerance between opposing groups and has stated that 'the role of the authorities is not to remove the cause of tension by eliminating pluralism, but to ensure that the competing groups tolerate each other',[92] in *S.A.S.*, the Court endorsed criminalising the wearing of the full-face veil, thereby clearly undermining pluralism and tolerance. The Burqa cases are not primarily about empowering women and securing gender equality but they are about competing worldviews and political, institutional, cultural and religious powers.[93]

In *Eweida*, the Court ruled that the right to publicly manifest one's religion was a fundamental right because a healthy democratic society had to tolerate and sustain pluralism and diversity, but also because individuals who have made religion a central part of their lives have the right to communicate their conviction to others.[94] Unfortunately, in *S.A.S.*, the Strasbourg Court overstepped its own precedents in *Eweida* and sought to justify its general ban on religious dress and symbols in the interest of preserving social cohesion or of upholding the conditions for 'living together'. If we accept this argument, a new meta-legal concept called 'living together' emerges, which, in effect, represents a moral imposition of a kind of public dignity at the hands of the social majority, which may justify restrictions on the full exercise of individual liberty.[95] This represents a dangerous totalitarian turn, as well as an illegitimate intrusion of the state into the terrain of defining human dignity, a strictly personal and non-transferrable act that is determined exclusively by each individual. From a legal standpoint, dignity may not be externalised and turned into a public ethics of 'living together' in order to protect individuals from their own dignity and from the exercise of their own rights and fundamental freedoms. This has no legal basis, and clearly risks becoming an instrument of oppression.

Conclusion

The four expert commissions under analysis in this volume agreed that, in their corresponding societies, people had to inhabit a shared space and future despite many differences. Each member of society has a role in a collective project, and that role is not exempt from conflicts. Conflicts of values are an intrinsic part of human life, and conflicts between lifestyles suggest that there is no single right answer in choosing between them. In order to manage these kinds of conflicts, legal actors and courts should start by recognising the plurality of the ethos inhabiting the shared space and the centrality of the autonomous person to choose their own *nomos*. The point of departure for the mandates of the four commissions was to remember that the liberal state could not remain neutral towards certain core values, such as the moral equality of persons according to one's own potential. The state had to be able, in principle, to protect each citizen's decision, which it would be unable to enforce if it favoured a specific conception of the world and of good. This commitment needs creative forms and it must not be linked to a particular political party. It is time for Europe to think on a long-term basis. We have the human rights instruments and our tradition of respecting diversity.

The Strasbourg Court is home to a value-pluralist tradition, which rests on the recognition of diversity and accommodation of competing claims. The Strasbourg Court puts forth meaningful slogans such as: 'Pluralism, tolerance and broad-mindedness are hallmarks of a "democratic society"'; 'Democracy does not simply mean that the views of a majority must always prevail: a balance must be achieved which ensures the fair and proper treatment of people from minorities and avoids any abuse of a dominant position'; 'The Court must seek to reconcile universality and diversity and that it is not its role to express an opinion on any religious model whatsoever'. All these precepts seem no longer valid in the cases described in this essay.

The analysis of the cases presented here indicates that plurality has been wiped off the map of the European Court of Human Rights. The Strasbourg Court has missed many opportunities to rule in favour of pluralism and diversity within groups and individual justice. But it also has missed the opportunity to incorporate the insights of the scholars of social-legal pluralism and multicultural studies. The Court has failed to (1) ensure the proper treatment of people from minority groups and to commit to oversight over any abuse of a dominant position; (2) recognise a personal autonomous choice in exercising the right to freedom of religion, which is being sacrificed to abstract principles and to essentialist stereotyping approaches about 'otherness'; (3) harmonise the principles of secularism, equality and liberty and not to weigh one against the other; (4) emphasise the notion of human dignity as the central value to protect the individual and religious minorities from certain forms of group libel; and (5) to define public order in terms of equal dignity of all members of society and not to act in a way that the public order undermines another's dignity by imposing an 'external public dignity' based on the abstract principle of *living together*.

The Strasbourg Court should re-inscribe 'pluralism' into its rulings as the central question informing human rights adjudication. It is through pluralisation that a reconstituted idea of human rights might emerge and it is through a contextualised legal analysis of the individual's own perception of religious normativity that human dignity can be better protected. Since the European Court of Human Rights is an epiphenomenon of the Member States of the Council of Europe, it may be time to reimagine our liberal societies with updated and pluralist conceptions of human rights and human beings.

Notes

* Senior Research Fellow in the Law & Anthropology Department at the Max Planck Institute of Social Anthropology in Halle (Germany) and Legal Adviser in the Migration and Equal Treatment Area at the Spanish Ombudsman.

1 B Parekh, *The Future of Multi-Ethnic Britain: The Parekh Report* (Profile Books 2000) 40–57; M-C Foblets and C Kulakowski, *Round Tables on Interculturalism*, Brussels (Ministry for Employment and Equal Opportunities, in charge of Migration and Asylum Policy 2010), 5; G Bouchard and C Taylor, *Building the Future: A Time for Reconciliation. Report. Commission de consultation sur les pratiques d'accommodement reliées aux différences culturelles* (Gouvernement du Québec 2008) 183–196.

2 *Kokkinakis v. Greece* App no. 14307/88 (ECtHR, 25 May 1993) para 31.

3 B Parekh, *The Future of Multi-Ethnic Britain*, 2.

4 G Bouchard and C Taylor, *Building the Future: A Time for Reconciliation*, 18.

5 *Kokkinakis v. Greece*.

6 *Kokkinakis v. Greece* para 31.

7 'There can be no democracy without pluralism' in *United Communist Party of Turkey v. Turkey, United Communist Party of Turk. v. Turkey* App no. 133/1996/752/951 (ECtHR, 31 July 1998) para 43.

8 ZR Calo, 'Pluralism, Secularism and the European Court of Human Rights' (2010–2011) 26 *Journal of Law and Religion* 262.

9 G Bouchard and C Taylor, *Building the Future: A Time for Reconciliation*, 51.

10 *Dahlab v. Switzerland* Admissibility Decision App no. 42393/98 (ECtHR, 15 February 2001).

11 *Leyla Şahin v. Turkey* App no. 44774/98 (ECtHR, 10 November 2005).

12 *Köse v. Turkey* Admissibility Decision App no. 26625/02 (ECtHR, 24 January 2006).

13 *Dogru v. France* App no. 27058/05 (ECtHR, 4 March 2009).

14 *S.A.S. v. France* App no. 43835/11 (ECtHR, 1 July 2014).

15 *Dahlab v. Switzerland* para 1.

16 Jeremy Gunn, 'Fearful Symbols: The Islamic Headscarf and the European Court of Human Rights in Sahin v. Turkey' (2008–2009) 3 *Annuaire droit et religions* 339; Javier Martínez-Torrón, 'The Islamic Cases in the Court of Strasbourg' in A. Yildiz and M. Tumay (eds), *Islam, Human Rights and Secular Values* (Liberte 2012) 343–391 and Carolyn Evans, 'The "Islamic Scarf" in the European Court of Human Rights' (2006) 7(1) *Melbourne Journal of International Law* 52.

17 AR Jackson and DA Gozdecka, 'Caught between Different Legal Pluralisms: Women Who Wear Islamic Dress as the Religious 'Other' in European Rights Discourse' (2011) 64 *Journal of Legal Pluralism* 91, 95.

18 *Şahin v. Turkey* para 116.
19 For the concept of 'interculturalism' see G Bouchard and C Taylor, *Building the Future: A Time for Reconciliation*, 257–258; B Parekh, *The Future of Multi-Ethnic Britain*, 76–90; M-C Foblets and C Kulakowski, *Round Tables on Interculturalism*, 46.
20 *Köse v. Turkey* para 12.
21 G Bouchard and C Taylor, *Building the Future: A Time for Reconciliation*, 214–216.
22 *Dogru v. France* para 63.
23 *Dogru v. France* para 76.
24 AR Jackson and DA Gozdecka, 'Caught between Different Legal Pluralisms', 98.
25 *Şahin v. Turkey*, Dissenting opinion of Judge Tulkens, para 11.
26 Mainly the one conducted by the University of Ghent, on a sample of 27 women wearing or having worn the full-face veil in Belgium (out of approximately 200 to 300 on the Belgian territory), see E Brems, Y Janssens, K Lecoyer, S Ouald Chaib and V Vandersteen, *Wearing the Face Veil in Belgium: Views and Experiences of 27 Women Living in Belgium Concerning the Islamic Full Face Veil and the Belgian Ban on Face Covering* (University of Ghent 2012), together with research carried out in France by the Open Society Foundations, involving 32 women, who wore the full-face veil, entitled *Unveiling the Truth: Why 32 Muslim Women Wear the Full-Face Veil in France* (Open Society Foundations 2011) and published in April 2011. It indicated that the women interviewed were not coerced into wearing the veil, that many had decided to wear it despite opposition from their families, that one-third did not wear it as a permanent and daily practice, and that the majority maintained active social lives. The report also revealed that the ban had contributed to discontent among these women and had reduced their autonomy, and that the public discourse accompanying it had encouraged verbal abuse and physical attacks against them by members of the public.
27 G Bouchard and C Taylor, *Building the Future: A Time for Reconciliation*, 17.
28 B Parekh, *The Future of Multi-Ethnic Britain*, 8 and 58.
29 *S.A.S. v. France* paras 102–105.
30 *Refah Partisi (The Welfare Party and others v. Turkey)* App nos. 41340/98, 41342/98, 41343/98 and 41344/98 (ECtHR, 31 July 2001) para 70.
31 *Eweida and Others v. UK* App nos. 48420/10, 59842/10, 51671/10 and 36516/10 (ECtHR, 15 January 2013). Relevant Comparative Law, Third countries paras 1–2.
32 AR Jackson and DA Gozdecka, 'Caught between Different Legal Pluralisms', 107.
33 Professor Joseph Weiler's testimony before the Grand Chamber of the European Court of Human Rights, on behalf of the nations that intervened in favour of Italy against Ms Soile Lautsi, regarding the exposition of crucifixes in public schools in Italy, 30 June 2010 (video of Weiler's testimony is available at <https://dotsub.com/view/65bc5332-aa10-4b8c-bc50-d051e8f4fcc7> accessed 13 April 2018).
34 *S.A.S. v. France* para 11.
35 *S.A.S. v. France* para 12.
36 K Greenawalt, *Religion and the Constitution: Free Exercise and Fairness*, vol. 1 (Princeton University Press 2006) 123–156; G Stopler, 'The Challenge of Strong Religion in the Liberal State' (2014) 32 *Boston University International Law Journal* 429 and MC Nussbaum, *Liberty of Conscience: In Defense of America's Tradition of Religious Equality* (Basic Books 2008).
37 PG Danchin, 'Islam in the Secular Nomos of the European Court of Human Rights' (2011) 32(4) *Michigan Journal of International Law* 663 and 676.
38 RW Lovin, 'Religion and Political Pluralism' (2007–2008) 27 *Mississippi College Law Review* 97.
39 'L'Etat laïque, garant de la liberté de conscience, outre la liberté de culte ou d'expression, protége l'individu; il permet librement à tous de choisir, ou non, une option spirituelle et religieuse, d'en changer ou d'y renoncer . . . La défense de la liberté de conscience individuelle contre tout prosélytisme vient aujourd'hui compléter les notions de séparation et de neutralité centrales dans la loi de 1905', Commision de Reflexion sur l'Application du príncipe de laïcité dans la Republique, Rapport au President de la Republique, Remis le 11 décembre 2003 (Stasi Report), 14.

40 *Kokkinakis v. Greece* para 31.
41 'In a secular regime, religion is shielded from a political role. It is not a tool of the authorities and remains in its respectable place, to be determined by the conscience of each and everyone', *Şahin* para 39.
42 'The fact of asking for no information to be shown on identity cards is closely linked to the individual's most deeply held beliefs. Accordingly, the Court considers that the issue of disclosure of one of an individual's most intimate aspects still arises', *Sinan Isik v. Turkey* App no. 21924/05 (ECtHR, 2 May 2010) para 51.
43 Carolyn Evans, 'Individual and Group Religious Freedom in the European Court of Human Rights: Cracks in the Intellectual Architecture' (2010–2011) 26 *Journal of Law and Religion* 321.
44 *Dahlab v. Switzerland* para 3.
45 'Religious freedom is primarily a matter of individual thought and conscience. This aspect of the right set out in the first paragraph of Article 9, to hold any religious belief and to change religion or belief, is absolute and unqualified', *Eweida and Others* para 80. See L Peroni, 'Deconstructing "Legal" Religion in Strasbourg' (2014) 3(2) *Oxford Journal of Law and Religion* 242.
46 P Petkoff, 'Forum Internum and Forum Externum in Canon Law and Public International Law with a Particular Reference to the Jurisprudence of the European Court of Human Rights' (2012) 7(3) *Religion and Human Rights* 183–214.
47 L Peroni, 'Deconstructing "Legal" Religion in Strasbourg', 245 and J Martinez-Torrón, 'The (Un)protection of Individual Religious Identity in the Strasbourg Case Law' (2012) 1(2) *Oxford Journal of Law and Religion* 381.
48 'Even where the belief in question attains the required level of cogency and importance, it cannot be said that every act which is in some way inspired, motivated or influenced by it constitutes a "manifestation" of the belief . . . the existence of a sufficiently close and direct nexus between the act and the underlying belief must be determined on the facts of each case', *Eweida* para 82.
49 L Peroni, 'The European Court of Human Rights and the Intragroup Religious Diversity: A Critical Review' (2014) 89 *Chicago-Kent Law Review* 673.
50 L Peroni, 'The European Court of Human Rights', 675.
51 *Gatis Kovalkovs v. Latvia* App no. 35021/05 (Third Section Decision, 31 January 2012) para 68.
52 *Mann Singh v. France* App no. 24479/07 (ECtHR, 13 November 2008) para 3.
53 L Peroni, 'Religion and Culture in the Discourse of the European Court of Human Rights: The Risks of Stereotyping and Naturalising' (2014) 10(2) *International Journal of Law in Context* 201–203.
54 *Dahlab v. Switzerland* para 13.
55 *Şahin v. Turkey* para 115.
56 *S.A.S. v. France* para 106.
57 *Şahin v. Turkey*, Dissenting opinion of Judge Tulkens, para 10.
58 G Bouchard and C Taylor, *Building the Future: A Time for Reconciliation*, 148.
59 *S.A.S. v. France* para 137.
60 *Thlimmenos v. Greece* App no. 34369/97 (ECtHR, 6 April 2000) para 42.
61 *Bayatyan v. Armenia* App no. 23459/03 (ECtHR, 7 July 2011) para 111.
62 *Kosteski v. the former Yugoslav Republic of Macedonia* App no. 55170/00 (ECtHR, 13 July 2007) para 39.
63 *Jakóbski v. Poland* App no. 18429/06 (ECtHR, 7 March 2011) para 45.
64 *Eweida and Others v. UK* para 89.
65 M-C Foblets and C Kulakowski, *Round Tables on Interculturalism*, 59; B Parekh, *The Future of Multi-Ethnic Britain*, 42; B Stasi, *Laïcité et République. Rapport de la Commission de réflexion sur l'application du principe de laïcité dans la République*, 17 and G Bouchard and C Taylor, *Building the Future: A Time for Reconciliation*, 131–150.

66 K-H Ladeur, 'The Myth of the Neutral State and the Individualization of Religion: The Relationship between State and Religion in the Face of Fundamentalism' (2009) 30(6) *Cardozo Law Review* 2448–2451.
67 M Evans, 'Freedom of Religion and the European Convention of Human Rights: Approaches, Trends and Tensions' in P Cane, C Evans and Z Robinson, *Law and Religion in Theoretical and Historical Context* (Cambridge University Press 2009) 312–313.
68 G Bouchard and C Taylor, *Building the Future: A Time for Reconciliation*, 157.
69 *Otto Preminger-Institute v. Austria* App no. 13470/87 (ECtHR, 20 September 1994).
70 *Wingrove v. United Kingdom* App no. 17419/90 (ECtHR, 25 November 1996).
71 P Danchin, 'Islam in the Secular Nomos of the European Court of Human Rights', 683.
72 *Lautsi v. Italy*, Concurring opinion of Judge Bonello para 1.1.
73 *Lautsi v. Italy*, Concurring opinion of Judge Bonello para 1.2.
74 *Lautsi v. Italy*, Concurring opinion of Judge Bonello para 2.5.
75 See 'The Public Sphere, the Private Sphere and the Demands of Secularism' in Bouchard-Taylor Report, 142–143.
76 *Lautsi v. Italy*, Concurring opinion of Judge Power.
77 'Robertson lists "the western Christian Church" as one of the six core principles underpinning the Council of Europe' quoted in M Evans, 'Freedom of Religion and the European Convention of Human Rights', 314.
78 *Refah Partisi (the Welfare Party and Others v. Turkey)*, Concurring opinion of Judge Kovler.
79 P Danchin, 'Islam in the Secular Nomos of the European Court of Human Rights', 701–702.
80 'Quelles sont aujourd'hui les mesures susceptibles de favoriser une mise en œuvre apaisée du principe de laïcité? De quelles évolutions cette clarification devrait-elle s'accompagner pour promouvoir une laïcité garante de la cohésion nationale et du respect des différences de chacun?', B Stasi, *Laïcité et République*, 3.
81 *Dogru v. France* para 72.
82 *Dogru v. France* para 21.
83 *S.A.S. v. France* para 122.
84 *S.A.S. v. France* para 141.
85 *S.A.S. v. France* para 142.
86 *S.A.S. v. France*, Dissenting opinions, para 5.
87 'Pour que chaque citoyen puisse se reconnaître dans la République, elle soustrait le pouvoir politique à l'influence dominante de toute option spirituelle ou religieuse, afin de pouvoir vivre ensemble', B Stasi, *Laïcité et République*, 9.
88 E Brems, 'Face Veil Bans in the European Court of Human Rights: The Importance of Empirical Findings' (2013–2014) 22 *Journal of Law and Policy* 536. After S.A.S, the ECtHR delivered two judgments, both involving the Belgian burqa ban, using similar legal reasoning: Affaire Belcacemi et Oussar c. Belgique, App. no 37798/13 (ECtHR, 11 July 2017), Dakir v. Belgium, App. no 11/12/17 (ECtHR, 11 July 2017).
89 EY Krivenko, 'The Islamic Veil and Its Discontents: How Do They Undermine Gender Equality' (2012) 7 *Religion and Human Rights* 18.
90 A Phillips, *Multiculturalism without Culture* (Princeton University Press 2007) 14.
91 *S.A.S. v. France*, Dissenting opinions, paras 6–8.
92 *Serif v. Greece* App no. 38178/97 (ECtHR, 14 December 1999) para 53.
93 N Gal-Or, 'Is the Law Empowering or Patronizing Women? The Dilemma in the French Burqa Decision as a Tip of the Secular Law Iceberg' (2011) 6 *Religion and Human Rights* 331.
94 *Eweida and Others v. UK* para 94.
95 E Relaño, 'Las dos caras del doctor Jekyll: las prohibiciones del velo integral a debate' (April 2015) 5 *Debates jurídicos* 29, www.rightsinternationalspain.org/.

17 Charting perspectives, positions and recommendations in four commission reports

Reasonable accommodation for religion or belief as barometer

*Katayoun Alidadi**

Introduction

Taking reasonable accommodation for religion or belief as the point of reference, this essay aims to offer a close comparison of the official reports of the four diversity commissions which form the subject matter of this edited volume, i.e. the Parekh Commission (Commission for Multi-Ethnic Britain, report issued 2000), the Stasi Commission (France, report issued 2003), the Bouchard-Taylor Commission (Quebec, report issued 2008) and the Foblets-Kulakowski Commission (Belgian Round Tables on Interculturalism,[1] report issued 2010).

While the legal concept of reasonable accommodation has its roots in employment discrimination law in the United States,[2] the idea behind accommodating diverse needs and customs is a more general one, which implies a level of openness to minority practices. In a sense, reasonable accommodation constitutes a play on what the Round Tables Report referred to as 'the basic principle of reciprocity' in pluralist societies.[3] Under an assimilative approach, 'burdens' of inclusion would fall solely on the shoulders of the 'Other', who is expected to change old habits and practices and to adopt new ways of living. As a corrective, under a reciprocal 'interculturalist' approach, the receiving/majority society may call upon its people or its institutions to also share in the burdens, to institute specific changes to remove obstacles, or to address discrimination in a spirit of solidarity with the newcomers, in order to enable their full and equal participation in a pluralistic society, 'provided this does not impose unreasonable or disproportionate demands'.[4]

Despite the fact that not all reports were concerned primarily with this topic, or even predominantly concerned with *religious* diversity challenges, all four reports include vital reflections and particular approaches towards the accommodation of religious minorities. Moreover, the reasonable accommodation debate reveals an undeniable linkage across the commission reports, embodying a cross-national conversation that testifies to the porous and interdependent nature of context-driven reflections, understandings and positions on managing diversities. Perhaps unsurprisingly, this conversation has not been immune to misunderstandings and (un)conscious misreadings. Far from benign, such misconceptions on reasonable accommodation have had a distorting effect on academic and policy reflections, analyses and recommendations. As ideas, concepts or proposed suggested solutions travel across borders and timeframes, this risk of miscomprehension or distortion is one which should be carefully considered, in particular by future expert commissions when they engage in conscious or unconscious borrowing from earlier reports. In the conclusion, I offer some notes on such comparative cross-national exercises in the management of religious, ethnic and cultural diversity.

Expert commissions and context-driven reflections on managing ethnic, cultural and religious diversity

The Parekh Commission – short for the Commission on the Future of Multi-Ethnic Britain – was established by the Runnymede Trust in January 1998, 'to consider the political and cultural implications of the changing diversity of British people'. In contrast to the other three, it was commissioned not by a government agency but, rather, by a private organization.[5] Its extensive report reviews political philosophy on ethnic minorities, assesses the challenges in multicultural Britain, and makes numerous policy recommendations.[6]

Written at the dawn of the new millennium, with an overall concern for 'race and diversity', the Parekh Report readily acknowledged the relevance of 'culturally-based' racism, of which religious intolerance was deemed an expression.[7] Befitting a report that understands difference-recognition as key to ensuring equality,[8] the Parekh Report came out in favor of legally enshrining reasonable accommodation for religion or belief, despite or perhaps due to the precariously poised protections against religious discrimination under British law at that time. Indeed, even at a time when discrimination based on religion or belief had not yet been explicitly prohibited in the UK (this would change with the adoption of the 2003 Employment Equality Regulation), the Parekh Report recommended a legally enforceable obligation to commit to reasonable accommodation in areas of employment, education and other areas. Additional research and advocacy for good practices were also called for that complemented (but did not substitute[9]) this recommendation.

The French Stasi Commission – the *Commission de reflexion sur l'application du principe de laïcité dans la Republique* – was installed during the presidency of Jacques Chirac, on July 3, 2003. It is the only commission to have been presided by a politician[10] and one with the shortest 'turnover period'.[11] The commission held public consultations, deliberated on relevant examples from abroad, and issued a report focused on the concept and understandings of the key normative concept of *laïcité*; the 20 members were unanimous in declaring the importance of *laïcité*, a concept which is open to divergent interpretations. In stark contrast to the multicultural approach adopted by the Parekh Commission, which saw Britain as a 'community of communities',[12] the Stasi Report explicitly rejected such an viewpoint and employed the concept of *laïcité* to anchor a French republican, universalist embedded ideology in its report.[13]

This focus on *laïcité*, rather than on the actual social challenges at hand makes the Stasi Report the most abstract, monothematic and top-down of the four reports, notwithstanding the extensive public consultations of the commission and some broader-ranging recommendations. As the Stasi Report notes, the legal corpus regarding the concept of *laïcité* is quite limited.[14] Still, since the French Constitution of 1946, the principle of *laïcité* has gained constitutional status[15] and the current Constitution of 1958 declares France to be 'une république indivisible, laïque, démocratique et sociale'. The French Constitutional Court's jurisprudence on *laïcité* remains less extended than on the freedom of conscience and opinion. At the international and supranational level, it is the freedom of (and restrictions on) religion and belief rather than *laïcité* or related concepts of secularism that are often at the core of inquiry at supranational courts and international monitoring bodies.[16]

To be sure, the commission's report contains much more than what it became known for – namely, recommending a ban on the wearing of ostentatious and conspicuous religious dress and symbols in French primary and secondary public schools.[17] It includes a number of progressive proposals such as the accommodation of minority religious holidays.[18] Indeed, when it comes to religious accommodations, the 'Stasi approach' ranges from restrictive (religious dress) and pragmatic (religious diets) to accommodating (religious holidays).

Some four years following the Stasi Report, the Bouchard-Taylor Commission – the Consultation Commission on Accommodation Practices Related to Cultural Differences – was set up on February 8, 2007 by Quebec Premier Jean Charest in response to a media storm and public discontent concerning reasonable accommodation.[19] Fierce public critique was particularly triggered by the *Multani* decision of the Canadian Supreme Court.[20] After one year of research and wide public consultation, a 310-page report was issued in May 2008. While the three other commissions were made up of 20 members or more, the Bouchard-Taylor Commission only had two members. The report is co-signed by the two chairs, historian Gérard Bouchard and philosopher Charles Taylor, and not the 15-member advisory committee.

Engendered by a 'crisis of accommodation', the commission's reconciliatory and therapeutic function had been laid out; the commission was to resolve or at least dissipate the immediate tensions and polarization surrounding religious accommodation. The commission decided to interpret its mandate broadly,[21] so that 'the debate on reasonable accommodation [was perceived] as the symptom of a more basic problem concerning the sociocultural integration model established in Québec since the 1970s'. The report sought to reassure the public that 'the foundations of collective life in Québec are not in a critical situation' but that at most there was a crisis of perception. It then proceeded to advocate a normative approach under 'open secularism' as an extension of the existing Canadian approach.

The orientation towards interculturalism and open secularism allows for a contrast to be drawn with the Stasi Report.[22] The Bouchard-Taylor Commission clearly rejected a 'very rigid system of secularism', which would not 'acknowledge any obligation to accommodate the religious beliefs and practices of minority groups', and recommended, much like the Parekh Commission, a substantive notion of equality in which reasonable accommodation would hold a key place.[23]

Finally, the Belgian Round Tables on Interculturalism[24] or the 'Foblets-Kulakowski Commission' issued its report in 2010. The Belgian Round Tables was the successor to the Belgian *Commission du dialogue interculturel/ Commissie voor Interculturele Dialoog*, which issued its report in 2005.[25] The Round Tables Report, in pursuance of the governmental mandate, focuses 'on the difficulties experienced, in particular, by the persons and communities to which we refer in this report as "ethnic, cultural and religious minorities"'.[26] But the situation of Muslims in Belgium was their core interest. Specific references to the Islamic headscarf expressed the recognition that that question had 'become a catalyst' for the discussions on cultural and religious diversity in Belgium. Recent developments in France and Belgium, in particular, the terrorist attacks in Paris (November 13, 2015) and Brussels (March 22, 2016), drew further and global attention to Belgium's 'poorly integrated immigrant population of mainly Moroccan and Turkish descent'.[27]

Despite the commendable compromises it advanced on numerous topics, the Round Tables Report was arguably the result of the most divided of the four commissions under review, and one which experienced various hiccups along the way.[28] This appears in stark contrast to the Stasi Report, which claimed a broad consensus and support for the idea of *laïcité* as a unifying factor among the Commissioners. Because of the divergent points of view and attempts to formulate compromise solutions, the overall normative approach of the Round Tables appears less clear cut than in the case of the other three commissions. Some proposals advanced something close to an exclusive understanding of *laïcité*, while other proposals took the need for a more substantive equality in Belgian society to heart; this makes for an interesting juxtaposition of progressive and more cautious recommendations in the same document. Nadia Fadil, in Chapter 13 of this volume, offers some crucial insights into the negotiations among the members on the Round Tables.

The 'recommendation light' approach with regard to the particularly challenging topic of reasonable accommodation may not be representative of the entire report, which indeed made some rather brave policy proposals. As noted, unlike the Parekh Report, which advised adopting a legal obligation to institute reasonable accommodation *and* conducting additional research, the Belgian Round Tables effectively deferred the issue of reasonable accommodation and pointed to the need for more empirical research on the Belgian situation.[29] The report also strongly advocated good accommodation practices and the 'citizen route'; however, unlike the Bouchard-Taylor Report from which it borrowed this reference, the 'citizen route' in Belgium remains a voluntary one, as it does not stand in the shadow of a legal duty to reasonable accommodation as is the case in Quebec.

Charting the cross-national conversation on reasonable accommodation

The four commissions reflected on and addressed reasonable accommodation for religion or belief to differing degrees.[30] As said, the Parekh Commission was forthright in recommending a legal duty of reasonable accommodation on the grounds of religion or belief in employment and beyond. The French Stasi Commission addressed reasonable accommodation for religion or belief including in employment (a domain which it sees 'not saved' from the same religious identity claims as in other areas)[31] even if that was not the main concern.

In contrast, reasonable accommodation was front and center in Quebec, albeit more so in the area of health and public schools than of employment. A core difference was that the Quebec report departed from a very different socio-legal context, with solid legal mandates of reasonable accommodation in place under Canadian law, so that its recommendations still need to be read in light of that socio-legal background. It should come as no surprise that textually identical recommendations could signify very different things in different national or temporal contexts.[32] This becomes clear when the 'citizen route' is advocated to the Belgian case, where reasonable accommodation in employment is not a legal right and exclusive interpretations of neutrality have tended to restrict the participation of (Muslim) minorities.

This section discusses some important lessons drawn from comparing understandings and recommendations directly related to reasonable accommodation in the four expert commission reports, in particular, the changes in policy advice in the British case, the French Stasi Commission views on reasonable accommodation in conversation with other reports, and the transposition of the 'citizen route' from one legal context (Québec) to a very different one (Belgium).

Reasonable accommodation recommendation in the UK: ancient history?

At the time the Parekh Commission convened, legislation on religion or belief discrimination was controversial. On the one hand, many public bodies had formally declared they would refrain from discrimination on religious grounds in recruitment or in the workplace, but on the other, it was argued, among others, by the British government, that existing legislations protecting against racial discrimination could not be extended to cover religion or belief. The Parekh Commission took stock of the practical and theoretical concerns surrounding the introduction of laws protecting against discrimination on grounds of religion or belief but provided solid counter-arguments. For instance, it dismissed concerns related to the definition of religion or belief.[33] Subsequent European developments bolstered these arguments, in particular, since Article 13 of the Treaty of Amsterdam (now Article 19 TFEU) provided the EU with a legal basis for action against religion or belief discrimination.

When it came to the concept of reasonable accommodation, the Parekh Commission recommended to extend this right to religion or belief, viewing accommodation in various areas of social life like business, schools and public institutions as an inherent aspect of the protection against religion or belief discrimination.[34] The legal developments in the protection against religion or belief discrimination in the UK have been rich.[35] And yet, the principle of reasonable accommodation has not been adopted in legislation. Moreover, since the Parekh Commission subsequent commissions focusing its work on religion or belief in the British public sphere have failed to recommend a legal duty of reasonable accommodation, preferring to avoid thorny and contentious discussions since recent cases have revealed not only tensions but also conflicts between religious accommodation and claims of sexual and other minorities.[36] Indeed, with debates on religious and other components of diversity intensified since, the prospect on reasonable accommodation seems much more restrained today in the UK, and the CORB Report (2015) recommends 'no immediate change in the law'.[37] If many recommendations included in the Parekh Report, despite the initial criticism, have in the meanwhile been realized,[38] then the reasonable accommodation recommendation is an outlier.

The Stasi Commission's misunderstanding of reasonable accommodation

As a commission set up to reflect on the application of the *laïcité* principle, a significant portion of the Stasi Commission Report was dedicated to a discussion of this 'foundational value of the French republic'. The commission operated under the remarkable assumption that a consensus garnered among its 20 members on the concept of *laïcité* also reflected the public opinion.[39] The views of those challenging the legitimacy of the notion were essentially discredited and considered easily remedied, by corresponding efforts to step up the fight against discrimination. The commission did not engage with the proverbial elephant in the room, namely that certain understandings of *laïcité* were at the core of, or at least contributed to the perpetuation of stigmatization and discrimination of religious minorities in France. Indeed, the report – under the motto '*affirmer une laïcité ferme qui rassemble*' – seems less about finding and delineating ways to manage diversity than about defending and renewing the potency of *laïcité* under conditions that seem to contradict its established status.[40]

The general approach adopted by the Parekh Report and the Stasi Report is strikingly different. If the Parekh Report took a view of the state as 'a community of communities', the Stasi Report aspired to the possibility and desirability of assimilation as a set of expectations towards newcomers. The Stasi Commission adopted a top-down, principled approach, rejecting the segregating realities of '*communautarism*' (communalism) elsewhere which were said to be avoided in France,[41] while the Parekh Commission adopted a more bottom-up approach (i.e. taking the reality of diversities as its point of departure). However, it should be clear that the Parekh Report adopted an equally principled approach by celebrating multiculturalism not only in its descriptive form but also normatively. For the Stasi Commission, *laïcité* represented the 'middle ground', which rejects '*communautarism*' but does not fail to account for the realities of 'diversity or pluralism' in French society.[42] But these new realities are framed in problematic terms: diversity is a problem, even a threat to the French model. At its core, the Stasi Report did not regard diversity as a positive development, but as one that needed to be under watch to prevent escalation.[43] What's more, the Stasi Report is pro assimilation as a policy, as long as it serves the purpose of maintaining the historically conceived French republican universalist ideas since 'all who live in French territory must abide by *laïcité*', and this concept operates as the prime vehicle for restoring the social peace. The *status quo* requires assimilation of newcomers. This then, as Talal Asad has suggested, amounts to a *de facto cuius regio eius religio* principle.[44]

In contrast to the Parekh Report, which enumerates the various ways in which Muslims are disadvantaged in the workplace and beyond, the Stasi Report largely discredits the 'felt sentiments' of those who consider themselves disadvantaged in French society. For instance, it refers to '*le sentiment éprouvé* of being subject to discrimination or being outside the national community by many of those who inhabit our territory' as contributing to the lack of minority integration and the appeal of extremist groups.[45]

In contrast to the Belgian Round Tables Report, which mandated the 'principle of reciprocity', and thereby also a 'certain openness' to diversity, in the French context, the notions of 'reciprocity' and *laïcité* functioned as 'border guards' that demanded deference to the principle of separation of church and state and shared values, such as the equality of men and women, as a part of the 'republican pact'.[46] If anything, reciprocity in the French context refers to the notion that such demands on minorities are justified merely because of the state's respect of the principle of freedom of religion and non-interference in religious affairs, or so it would seem.[47]

This normative approach is reflected and played out in the recommendations, including with regard to reasonable accommodation. With regard to private sector employment, the French report makes a distinction between religious dress, requests for time off and the issue of halal food. The Stasi Commission studied examples from abroad and incorporated the 'Quebec notion' of reasonable accommodation in its report. However, at least in one important instance, the meaning of reasonable accommodation is misappropriated to serve the purpose of the commission's understanding of *laïcité*. In a peculiar twist, the term 'reasonable accommodation' was used to denote its polar opposite, a situation in which the entire burden of compliance fell on the shoulders of the 'citizen'.[48]

Even more remarkable, *laïcité* not only requires newcomers or minorities ('simply a citizen') to comply with the majoritarian code – i.e. the peculiar understanding of 'reasonable accommodation' noted – but, in fact, also necessitates a change in religious dogma altogether since the report states that 'laïcité requires an interpretative effort to reconcile religious dogma with the laws that govern society, so that "la vie ensemble" is made possible'.[49]

Despite its claims, this articulated penchant for 'moderate' strands of religion – religion 'worn lightly'[50] – makes of *laïcité* anything but a 'neutral' concept but in fact a very dogmatic one meant to serve the '*vivre ensemble*'.[51]

One of the core requirements of *laïcité* is the neutrality of the state and its civil servants. The requirement of 'strict neutrality' (implying a prohibition on religious dress) for French civil servants – irrespective of whether they came in direct contact with the public – had already been established by Conseil d'Etat jurisprudence since the 1950s.[52] The Stasi Commission not only advised enacting these jurisprudential rules regarding public servants into law[53] but it also recommended that the principle of 'strict neutrality' be extended to public service contracts. This technique, commonly applied to strengthen the enforcement of anti-discrimination law (e.g. the U.S. Civil Rights Act of 1964), would considerably expand the domain of *laïcité* into the private sector labor market.[54]

In the private sector, the Stasi Report recognizes that requests for accommodations have been a fixture in the French workplace since the 1960s, when large corporations employing (male) immigrant Muslim workers implemented – at times extensive – workplace accommodations, such as changing canteen menus, adapting break time in consideration of Ramadan, allowing workers time off during Islamic holidays, and providing prayer space. The commission viewed these accommodations as good practices, apt for the time and the circumstances under which they were introduced. But they had a different opinion of the present-day conditions, especially of 'the fundamentally different' new generation of Muslim workers, in so far as legitimate 'needs' (*besoins*) had been purportedly transformed into 'claims' (*revendications*) that challenged the principle of '*vivre-ensemble*'. Examples included the headscarf[55] and the refusal to shake hands with male colleagues, issues that involved second-generation female workers, who were mostly born and raised in France. Another example concerns the refusal to accept the authority of a female boss.[56] The Stasi Commission recommended rejected accommodation in such cases, in particular, if it affected the principle of equality between men and women and of 'la mixité' in public services, and this had to be enforced by law.[57] Similarly, the Belgian Round Tables Report also rejected any accommodation when a threat to gender equality was perceived.

The Stasi Commission squarely opposed a right to reasonable workplace accommodation when it recommended legal action that would enable 'companies who are facing difficulties' to restrict, with the consent of the social partners, religious dress and symbols for reasons of 'security, customer contacts or the social peace in the company'.[58] Thus, employers could ban headscarves and other religious dress for employees irrespective of the level of customer contact the job entails and needing to link restrictions to particular job functions. This amounted to a significant departure from French labor and anti-discrimination law principles calling for a contractual case-by-case approach.[59] The proposal further differs from this approach since it allows an employer to alter earlier oral or even written dress accommodations in the interest of, *inter alia*, the 'social peace'. Taken together with the extension of the public-sector duty of strict neutrality to government service contracts, these provisions would have severely restricted employment opportunities for Muslim women and others who habitually 'dress religiously' and thereby – in the view of the Stasi Commission – engage in 'auto-discrimination'.

However, there are a few areas where the approach of the Stasi Commission towards religious practices is more flexible. For instance, the commission considers the taking into account of dietary restrictions (e.g. halal food) in the workplace as a practical matter that had to remain 'compatible with the well-functioning of the service, according to the principle that the Québécois call "reasonable accommodations"'.[60] With regard to time off during religious holidays, the approach of the Stasi Commission followed existing French administrative practices, which – compared to other European countries – is very progressive.[61] The recommendation on 'credit days' was also advanced in the Round Tables Report.[62] But unlike the Belgian Round Tables Report, which adopted an audacious proposal to adjust the public holiday calendar to reflect changes in the religious demographics, the Stasi Commission did not see 'any reason to call into question a calendar conceived principally with Catholic holidays . . . in mind'.[63] It recommended the recognition of the most sacred days of the 'two largest monotheistic religions in France' (i.e. Yom Kippur and Aïd-el-Kebir) as additional, paid holidays (see Chapter 6, by Patrick Weil).

Bouchard-Taylor framework: favoring the 'citizen route' and separating it from the 'legal route'

As said, the Bouchard-Taylor Commission interpreted its mandate to take stock of the 'accommodation crisis' in a broad fashion, which led to an extensive review of interculturalism, immigration, secularism, and of Quebec identity. With $5 million at its disposal,[64] the commission mandated various research projects and engaged in extensive public consultations.[65] While sketching the recent eventful history involving accommodation cases in Quebec and Canada, the commission sought to reassure the public that there had not been any real 'accommodation crisis', but at best a 'crisis of perceptions', in part due to 'sudden media enthusiasm and rumors'. What the commission did see emerging was 'an identity counter-reaction movement that has expressed itself through the rejection of harmonization practices. Among some Québecois this counter-reaction targets immigrants, who have become, to some extent, scapegoats'.[66] In comparative perspective, the commission saw the situation in Quebec as enviable, claiming that 'fears warranted elsewhere are not necessarily justified here'.[67]

The Bouchard-Taylor Report's appeasing ethics is already apparent in the title *Building the Future: A Time for Reconciliation*, but this ethical stance may also explain the conceptual notions it proposed, as well as the newfangled definition it gave to the key term of reasonable accommodation. Considering that Canada and Quebec were considered exemplary in the area of diversity management on the international stage, the high regard the co-chairs enjoyed globally, and the budgetary and mandate freedom granted to the commission, the approach characterizing the commission report was bound to become influential beyond Quebec. That various terms introduced in the Bouchard-Taylor Report – such as 'harmonization practices' and the 'citizen route' – were subsequently adopted and advocated within the framework of the Belgian Round Tables is therefore not surprising; ideas – whether good or bad – travel, in particular across close language communities. However, while the literal terms may have traveled unscathed, the same cannot be said of their meanings, which were 'lost in transposition'. Whether due to misunderstanding or (un)conscious misreading, the result of the transplanted Bouchard-Taylor framework has become a justification for much less progressive standards in the Belgian case.

A closer look at what I will call 'the Bouchard-Taylor framework' is thus warranted. Although this is not emphasized in the report, I argue that much of this framework derives from, and relies on, an unwarranted redefinition of reasonable accommodation. In an effort to move away from 'reasonable accommodation' (*sensu strictu*, in the Bouchard-Taylor sense) and the so-called 'legal route', the key concept of reasonable accommodation became the underdog, since the Bouchard-Taylor framework gives center stage to newly coined terms, such as 'harmonization practices', 'concerted adjustments' and 'optional agreements'. In the proposed analytical framework, the 'legal route' and the 'citizen route' are artificially but starkly contrasted:

> Under the legal route, requests must conform to formal, codified procedures that the parties bring against each other and that ultimately decree a winner and a loser. Indeed, the courts impose decisions most of the time. The legal route is that of reasonable accommodation. Requests follow a much different route under the second path, which is less formal and relies on negotiation and the search for a compromise. Its objective is to find a solution that satisfies both parties and it corresponds to concerted adjustment. *Generally speaking, we strongly favour recourse to the citizen route and concerted adjustment, for several reasons: a) it is good for citizens to learn to manage their differences and disagreements; b) this path avoids congesting the courts; and c) the values underlying the citizen route (exchanges, negotiation reciprocity, and so on) are the same ones that underpin the Québec integration model. In quantitative terms, we have noted, moreover, that most requests follow the citizen route and only a small number rely on the courts.*[68]

The adherence to this new typology is already operative in the proposed definition of reasonable accommodation, rendering the concept of reasonable accommodation an easily refutable straw man: in the glossary of terms, reasonable accommodation is defined as

> [a]n arrangement that falls under the legal sphere, more specifically case law, aimed at relaxing the application of a norm or a statute in favour of an individual or a group of people threatened with discrimination for one of the reasons specified in the Charter.[69]

This is in contrast to the common understanding of the legal concept of reasonable accommodation, which makes no distinction between the 'citizen route' or the 'legal route'.[70] As any law or legal concept, it is applicable in the social context of every-day life as well as when claims are presented in legal forums. To borrow from John Griffiths, (effective) laws have general effects (in everyday life) as well as specific effects (when legally enforced).[71] There is no fundamental difference in the requirements of reasonable accommodation, whether applied on the ground or mandated by courts. If judges held people to different standards than they were expected to adhere to in their everyday dealings under applicable laws, the result would be uncertainty, unfairness, maybe even a bit of chaos. However, under the Bouchard-Taylor framework, things work differently; 'concerted adjustments' (the citizen route) need not follow the same rules as courts would in case of 'reasonable accommodation', even if the accommodations/adjustments would fall under the legal mandates. In particular, the legal rules, 'undue hardship' standards and guidelines form mere 'inspiration'[72] when it comes to the 'hybrid notion' of concerted adjustment, i.e. which sometimes stems from a legal obligation and sometimes does not have a clear legal underpinning.[73] Since the road to litigation is not closed off, just discouraged and frowned upon, it is not clear if 'more flexible understandings' are then to replace the standards developed in legislation and under case law, and what the effects will be of a divergence between the two – artificially separated – spheres of the law and of society. Indeed, the framework fails to acknowledge that all negotiations (i.e. the 'citizen route') to some extent are conducted 'in the shadow of the law' and that disputes go to courts when no satisfactory negotiation is achieved.[74] Solutions, whether 'imposed' or negotiated, may be very similar in the end. While it is certainly true that the citizen route 'in strictly quantitative terms . . . carries much more weight than the legal route', it would be a mistake to discredit the value of such legal route as both the ever-present 'shadow' as well as the *ultimum remedium*, in particular when the law mediates power differentials between parties. The 'conflict mode', which the commission links with the courts, is not due to the forum (the court or the public sphere), but rather reflects the nature and stage of the relationship of the parties involves.[75]

To be sure, the anti-conflict/litigation stance may come as a surprise if one reads the entire report. After all, in various places, it emphasizes the point that legal cases involving reasonable accommodation have been limited in Canada. It seems internally incoherent to argue that the 'citizen route' must be favored to avoid congesting the courts and, in the same breath, to recognize the paucity of actual legal decisions on reasonable accommodation in the court system. The call for the 'citizen route' and the dejudicialization[76] and decentralization of the accommodation process in the Bouchard-Taylor Report thus can only be understood as a counter-reaction to the (public) *perception* of an overzealous recourse to the court system for issues of multicultural nature and an attempt to temper polarization and public resentment through a more reconciliatory effort.[77] In this sense, Bouchard and Taylor's assessment of the situation in Quebec and their proposed remedy are not aligned, and this, above all, reflects an internal inconsistency.

Further, and more critically, is the veneration of the 'harmony law model' (in the words of Laura Nader) in the Bouchard-Taylor framework. The promotion of the 'citizen route' with concerted adjustments as a *replacement* for reasonable accommodation misrecognizes the legacy of social justice litigation, in particular in the area of human rights, equal treatment and diversity, where

Canada's model has been hailed as highly progressive. This may be seen as an external inconsistency in the framework.

The role of law in the struggle for social and economic justice for religious minorities has been immense in the Canadian context, leading to a legal and social change, perhaps unparalleled in the Western world. It is not clear why that legacy is being discredited so easily, in the name of reconciliation, while there is no talk of a real crisis.

Despite the 'common sense' understanding that individuals should try to work out conflicts among themselves, it must be recognized that the availability and effective use of the law in democratic societies remains crucial (and should be defended as such). Legal anthropologist Laura Nader has compellingly argued in *The Life of the Law* that plaintiffs are 'the life of the law' and social justice litigation is central to the democratic enterprise.[78] Nader linked the rise of the alternative dispute resolution movement in the U.S. since the 1960s with the erosion of the plaintiff's power in a rigged system which too often favors the powerful interests of large corporations. In the case of ethno-religious minority litigation, power imbalances are also often prevalent. Indeed, the progression in the legal protections against discrimination (implying a right to reasonable accommodation in the Canadian case) and a sophistication of the understanding of equality itself would not have been possible without some (unpopular) litigation bringing to light practices and realities of obstacles to inclusion and participation.[79] If disputes remained underground, negotiated behind closed doors, silently contained instead of 'kicking up a fuss', what would have justified the adoption of laws and policies to address urgent societal tensions? Thus, not only is the underlying meaning 'lost in transposition', the Bouchard-Taylor framework can be reproached for various errors of its own making. It does disservice to those who are unjustifiably scapegoated according to the same report.

Nader took issue with similar arguments, seeking to relieve courts of consumer disputes, which has silenced the voices of many individuals and trampled upon their faith in justice and 'the system', by closing the regular legal routes. Certainly, the Bouchard-Taylor Commission is not going as far as to close the gateway to legal recourse, but its framework is easily manipulated to serve such discourse.

This may be the more relevant considering the two areas – education and health[80] – on which the Bouchard-Taylor Commission focused. Particularly in these areas, with structural and real power differentials between respective parties, it is important not to give bureaucrats the last word and allow plaintiffs the remedy of justice through the public court system if needed.

Subsequent developments in Canada in the failed attempt to establish a 'Secular Charter' for Quebec public institutions show that reasonable accommodation was no transitory political issue in Quebec and will continue to drive debates, divisions and tensions.[81] Indeed, 'accommodations has emerged as a new dimension in the politics of collective identity in Québec, adding questions about secular democracy to longstanding concerns about language policy'.[82] If, notably, companies and employers are not considered to be struggling with accommodations,[83] it is because the business sector largely stands behind reasonable accommodation and madc that clear in response to the proposals for a Secular Charter in Quebec. The business community opposed the Charter, seeing no need for restrictions *à la française* on religious distinctions in Quebec institutions.[84]

Belgian Round Tables: borrowing concepts, tabling hard issues

The recommendations of the Belgian Round Tables Report are a mixed bag.[85] Some recommendations range from what can be considered courageous or ambitious[86] in the socio-political context, opening the door to a 'greater openness to diversity', and the critical reception of these aspects of the report bear this out. Other recommendations are characterized by conservatism and validate the *status quo*.[87] Similar to the Stasi Commission, the Round Tables Commission adopted the stance that certain achievements or values – 'basic principles' – were off-limits. For instance, 'no cultural, religious and/or philosophical arguments may be invoked to avoid compulsory courses or to withdraw from the system of co-education'.[88] With regard to education, which was arguably the most urgent areas tackled in the Round Tables, the degree of accommodation depends on the particular issue involved with three quite distinct proposals being made in the areas of religious/philosophical dress, religious time and the issue of diets.

First, when it comes to religious dress code in public schools, the Steering Committee adopted a proposal which sought to 'split the difference'. The proposal consisted of a 'general freedom for students[89] to wear religious or philosophical symbols in the last three years of secondary school and a complete ban during the first three years of secondary school'.[90] Second, with regard to the school calendar, it advanced a reform in line with the proposed modification of the civil calendar, where holidays of minority religious traditions would supplement 'secular' and Catholic holidays. Finally, with regard to dietary restrictions, its proposals fed into the same kind of pragmatism also observed in the preceding three reports, with the difference that it recognized the limitations of this approach.[91]

The Round Tables Report took note of the problem of high unemployment among ethnic and ethno-religious minorities and the alarming ethno-stratification of the labor force in Belgium as (potential) sources of social conflict, calling for 'a more proactive economic policy to include persons belonging to ethnic, cultural or religious minorities in the public space and on the job market'.[92] Making a distinction between the private and the public sector, the report also addressed the mushrooming of 'neutrality policies' in private companies: 'Whatever the position taken in the debate on the neutrality required in the public services, neutrality is not a legal obligation for private companies – notably those whose activities are purely commercial in nature – and we must therefore avoid it being arbitrarily used to oppose' the freedom of religion or belief under Article 9 ECHR.[93]

While not front and center as in the Quebec case, the Belgian government had expressed considerable interest in the topic of reasonable accommodation, taking cues from Quebec's experience. The report frames the 'profound gap' between members of its Steering Committee – between those in favor of accommodations following the Quebec example, and their opponents[94] – as an indication of a larger societal polarization on the issue. The Steering Committee had hoped to rely on a newly commissioned sociological study on Belgian practices of reasonable accommodation in employment, but this study's results only became publicly available after the Round Tables Report was issued.[95] Surely the findings[96] could have usefully fed into the Round Tables debate, but it is very doubtful that data on actual practices would have made proponents and opponents – with normatively anchored stances – bury their hatchets.

The resulting 'non-recommendation' on reasonable accommodation in the Round Tables Report amounts to a missed opportunity.[97]

> The Steering Committee invites people to give further scrutiny to any benefits or disadvantages that may result from extending the concept of reasonable accommodation, as defined in the Law of 10 May 2007, in order that such accommodations no longer be limited solely to persons with disabilities, but may also be applied to other situations, notably those relating to religious or philosophical convictions. . . . As regards reasonable accommodation, the Steering Committee wishes to expand the 'citizen route' by creating a series of useful tools for handling legitimate demands more effectively, and thus *avoiding the need to have recourse to the legal route.*[98]

If the Round Tables was '[a] collaborative endeavour, in search of viable solutions', then one can safely assume that no viable compromise was possible; no 'common denominator' was in sight and the highly relevant issue of reasonable accommodation basically remained tabled. The literal transplantation[99] of the Bouchard-Taylor framework and the embrace of the same predilection for the 'citizen route' in a crucially different legal and social context, as illustrative of a cross-national conversation on reasonable accommodation between expert commissions, seemed to have exacerbated the polarization.[100] The only thing 'the majority of the Steering Committee members' – i.e. not even all members – agreed on was that 'reflection must be continued'. The only suggestion made prior to this reflection was to develop tools 'to satisfy in a pragmatic way the requests for accommodation made for cultural or religious reasons' but apparently wholly within a 'citizen route'. This raises so many questions. For instance, which authority can set standards in the citizen route?[101] What recourse is available to the requesting parties in case of a rejection? If none, how beneficial would this be for a potentially vulnerable party? In any event, it should be clear that proceeding 'in the shadows', without a legal framework, may potentially victimize the vulnerable instead of empowering them.

In a 'dissenting statement' attached to the final report, Edouard Delruelle, one of the chairs of the 2005 Commission for Intercultural Dialogue and the then co-director of Belgium's equality body (the predecessor of the current Unia), expressed his personal disagreement with the recommendation on reasonable accommodation. Fiercely opposing an amendment of the Anti-Discrimination Law of 2007, and arguing that any 'analogy with the situation of a person with a disability and freedom of belief . . . [was] a fundamental error in law', Delruelle warned against what he called the 'instrumentalisation' of anti-discrimination law, 'to resolve a different problem, namely that of peaceful coexistence in a culturally pluralistic society . . . [which] can only be solved by a *concerted adjustment* by social actors'.

This amounts to cherry picking from the Canadian legal framework and from Bouchard-Taylor's reliance on concerted adjustments. In the Canadian context, the role of law is hardly rejected in disputes involving the 'peaceful coexistence in a culturally pluralistic society'. In fact, duties of reasonable accommodation are part and parcel of the substantive equality approach advocated by the Canadian Supreme Court. If anything, the call for concerted adjustments, a concept conceived by a mandated commission after decades of legal progress in this area, must be interpreted *in light of* the socio-legal Canadian situation. If the main report points to misunderstandings and the retorted effects of the Bouchard-Taylor framework, this dissenting opinion, by a member who no doubt carried some weight in the deliberation, screams it out. Taken out of context, the 'common sense' predilection for the 'citizen route' in the Bouchard-Taylor framework effectively blocks progress in other countries towards the Canadian model by being 'instrumentalized' to entrench the *status quo*, which is far less progressive in Belgium than in Quebec and the rest of Canada. Interestingly, in such context, the unintended but tangible effect of the 'citizen route' approach is that it rejects, rather than embraces, open secularism.

Conclusion

This essay offers a comparative analysis of the four expert commission reports under review in this volume from the angle of reasonable accommodation. The different timing, context, composition, working methods and rationale for each commission can explain, to some extent, the level of attention given to religious diversity and reasonable accommodation for religion and, to a much less extent, belief. Yet despite the difference in focus, all four reports address the topic of reasonable accommodation in a meaningful way, and this – largely – fits into an at least implicitly adopted conception of equality.

On the basis of the preceding discussion, one can argue that reasonable accommodation, by itself, is a 'bad traveler', an elusive concept that is vulnerable to misunderstanding and distortion when it travels from one context to another. The French Stasi Report employed 'reasonable accommodation' to denote an approach under which all burden of accommodation fell on the newcomer 'in return for' the respect for the freedom of conscience in France. The report and framework of Bouchard-Taylor added to the confusion, as apparent in the Belgian Round Tables Report. Concepts and propositions, removed from their direct legal and social background, are exposed to an inherent risk of misunderstanding and misconstruction, a pitfall commissions are not immune to. The call in the Bouchard-Taylor Report to refocus on the 'citizen route' was arguably misguided in the specific Canadian context. But that the commission's preferred 'concerted adjustments' still take place in the 'shadow of the law' and with extensive legally enshrined duties of reasonable accommodation the effects of privileging the 'citizen route' would remain modest at best.[102] In that context, perhaps the applicable law does not determine the outcome of every negotiation on the ground fully, but it will not fail to impact and guide the avenues taken. In contrast, the same call for 'concerted adjustments' and rejection of the 'legal route' with reasonable accommodation carries a very different tone when transported to a context where legal duties of reasonable accommodation are hotly contested. In this way, a progressive example inadvertently becomes the pretext for blocking progressive policies elsewhere. The discussion above shows that this is a process can be mediated by expert commission activities. With the borders for national debates highly porous, it becomes important for such expert commissions to appreciate that policies advocated and suggested, whether or not followed by the government in question, can affect discussions in other countries. One may accordingly question whether this 'distorted' influence is an unavoidable side-effect of the cross-national interest in such commission reports. How, then, can ideas and concepts such as reasonable accommodation, with contextual origins but seemingly universal application, go from being 'bad travelers' to 'global citizens'? To what extent can the inevitable jetlag be contained when venturing overseas?

Perhaps one answer is to engage more and more deeply – rather than less or superficially – with comparative, cross-national exercises in the management of religious, ethnic and cultural diversity, including in the context of expert commissions. Of the four reports under review, none manifestly demonstrated this comparative, cross-national engagement as a part of their core exercise. Certainly, the reports contain references to experiences in other countries and some even engage with approaches borrowed from other expert commission reports,[103] but all in all these remain rather ancillary considerations. To some extent, this may be due to the mandates, which are national-derived and which regard the triggering challenges as specific to a given context. While factors may be specific to a national context, this volume bears witness that the challenges of cultural and religious diversity *an sich* are not and neither is the technique of engaging experts to reflect on what does or does not work when it comes to the management of religious diversity. In that sense, this amounts to a plea for a solid integration of a thoroughly contextualized and cross-national comparative analysis, to which this very volume, in a general fashion, seeks to contribute.

Notes

* Assistant Professor of Legal Studies at Bryant University in Smithfield, Rhode Island, and Research Associate at the Max Planck Institute for Social Anthropology in Halle, Germany. An earlier version of this essay was presented at the 'Religious and Cultural Diversity in Four National Contexts' conference, organized at the Max Planck Institute for Social Anthropology (Department of Law and Anthropology) on June 26–27, 2015. Many thanks to the participants for their questions and helpful suggestions for improvement.

1 'Rondetafels van de Interculturaliteit' could also have been translated as Round Tables on Interculturality, but this follows the 'official' translation to interculturalism.

2 See Title VII of the U.S. Civil Rights Act of 1964 (42 U.S. Code Chapter 21), amended in 1972 to explicitly include a legal mandate for employers to accommodate the religious practices and observances of their employees or prospective employees, unless to do so would create an 'undue hardship' on the employer. The concept was thus first introduced in the U.S. in the context of religious discrimination, and later expanded to persons with disabilities. The same was the case in Canada, although it was a judicial decision and not a legislative amendment that paved the way.

3 Belgian Round Tables Report, 10.

4 Ibid.

5 The Runnymede Trust is an independent British think-tank promoting racial equality, see www.runnymedetrust.org/.

6 Commission on the Future of Multi-Ethnic Britain, *The Future of Multi-Ethnic Britain* (London: Profile Books, 2000), 434 et seq.

7 One could say that in recognizing religious discrimination in Britain, and the need for corresponding policies, the Parekh Commission avoided the pitfall that the Stephen Lawrence Inquiry Commission in Britain faced, which by all accounts had ignored strands of anti-Muslim racism in its report. See ibid., 61–62. See Khalida Khan, 'Where's the Muslim in Macpherson's Black and White Britain?' *Q News*, March 1999.

8 'When equality ignores relevant differences and insists on uniformity of treatment, it leads to injustice and inequality'. Parekh Report, Preface, ix.

9 This is in contrast to the Belgian Round Tables Report, in which the call for research effectively substitutes any proposition for legal enshrining of a reasonable accommodation duty in the workplace, see *infra*.

10 The late Bernard Stasi had, among others, served as the ombudsman of the (French) Republic (*médiateur de la République*).

11 Less than six months. The report, with 78 pages, is also the shortest of the four.

12 See, for instance, Parekh Report, Preface, ix.

13 The core function of the commission may have been to reverse the jurisprudence of the 1989 advice of the Conseil d'Etat on the wearing of religious garbs in public schools. L'assemblée générale du Conseil d'Etat, avis le 27 novembre 1989. For further discussion on this jurisprudence, see p. 30 et seq. Following this case, various school regulations which prohibited the wearing of religious symbols were found to be invalid. See, to that effect, Conseil d'Etat, 2 novembre 1992, *Kherouaa*.

14 'la grande loi est celle du 9 décembre 1905 complétée par celle du 2 janvier 1907 sur l'exercice public des cultes'. Stasi Report, p. 19.

15 For historical reasons, the Law of 1905 and the requirements of *laïcité* do not apply to the region of Alsace-Moselle (which in 1905 was part of Germany), where the system of funding recognized religions persists. On February 21, 2013, the Constitutional Council of France upheld the Napoleonic Concordat of 1801, after it was appealed that it violated the secular nature of the French Republic by the 'Association pour la promotion et l'expansion de la laïcité'. See Guillaume Perrault, 'L'Alsace-Moselle garde le concordat', *Le Figaro*, February 22, 2013.

16 The Stasi Commission reviewed the jurisprudence of the European Court of Human Rights, to bolster its position that freedom of religion or belief is not absolute but subject to restrictions. The case of *Cha'are Shalom ve Tsedek v. France* App no. 27417/95 (ECtHR, June 27, 2000) is utilized to illustrate the large margin of appreciation for states as well as the disputable position that 'religious diversity thus finds its limits in the confrontation with the requirements of *laïcité*'. Thus, peculiarly, *laïcité* is seen as the guarantor of freedom of religion as well as the justification for restrictions on religious manifestations at the same time.

17 The law 2004-228 of March 15, 2004 encadrant, en application du principe de laïcité, le port de signes ou de tenues manifestant une appartenance religieuse dans les écoles, collèges et lycées publics was signed into law by President Jacques Chirac on March 15, 2004 and came into effect at the beginning of the 2004–2005 school year.

18 For more on this proposal, see Chapter 6, by Patrick Weil, in this volume.
19 Lori Beaman, 'Introduction: Exploring Reasonable Accommodations', in *Reasonable Accommodation: Managing Religious Diversity*, ed. Lori Beaman (Vancouver: UBC Press, 2012), 3 et seq.
20 Canadian Supreme Court, *Multani v Commission scolaire Marguerite-Bourgeoys*, [2006] 1 S.C.R. 256, 2006 SCC 6.
21 Thus rejecting the 'narrow' approach to strictly consider the legal dimension of reasonable accommodation.
22 The Bouchard-Taylor Commission explicitly rejected the 'strict limits on freedom of religious expression' that it saw in the French case of legislation that restricted the wearing of religious signs in public schools. It gives three reasons for considering this inappropriate for Canada. See Bouchard-Taylor Report, 20.
23 Ibid., 159–160.
24 'Interculturalism' had been favored over 'multiculturalism' by the Belgian government for the Round Tables mandate.
25 However, the approaches were very different and where the Round Tables on Interculturalism makes certain policy choices, the Commissie voor Interculturele Dialoog merely set out the possible policy choices and published an anthology of public reactions.
26 Belgian Round Tables Report, 13.
27 Roger Cohen, 'The Islamic State of Molenbeek', *The New York Times*, April 11, 2016.
28 The Round Tables 'Steering Committee' started with 29 members but only 22 remaining members signed the report, with two adding dissenting statements. For more background, see Chapter 13 by Nadia Fadil and Chapter 11 by Marie-Claire Foblets in this volume.
29 For the commissioned research study on practices of reasonable accommodation (only available after the issuance of the final report), see Ilke Adam and Andrea Rea, 'Culturele diversiteit op de werkvloer: praktijken van redelijke aanpassing in België.' Brussels: Research commissioned by the Centre for Equal Opportunities and Opposition to Racism, 2010.
30 One measure could be the number of times the term 'reasonable accommodation' figures in the reports. Unsurprisingly, the term is most frequently used in the Bouchard-Taylor Report (use of the term 'reasonable accommodation' 67 times, 356 times of 'accommodation'; The Parekh Report invokes 'reasonable accommodation' only in four instances, reasonable and accommodation occur independently 13 times; 'accommodement raisonable' is used four times in the Stasi Commission Report; in the Round Tables Report there are 19 instances of the use of 'reasonable accommodation'). But the same idea can either be conveyed with different notions – e.g. the concept of 'concerted adjustment' coined in that report – or be discussed more generally under the rubric of equality/non-discrimination or human rights (freedom of religion). It is also relevant that the reports are of very different lengths.
31 The section title – 'Un monde du travail qui n'est plus épargné' – is itself revealing for the view on the place of religion in the workplace.
32 For instance, recommending ways to temper damage awards in certain kinds of litigation takes on a different meaning and significance depending on the legal context: Does the country have solid laws and a (perceived) problem of extremely high damage awards or is this a legal environment where, instead, access to justice and diffident laws and legal knowledge make certain kinds of litigation extremely scarce? In one instance, it can be seen as a corrective towards a workable system, in another it solidifies obstacles and limits progress towards justice.
33 Parekh Report, 240.
34 '*We recommend that legislation be introduced in Britain prohibiting direct and indirect discrimination on grounds of religion or belief.* Employers, schools and other institutions should be under a duty to make reasonable adjustments to accommodate a person's religious observance or practice, provided that this can be achieved

without undue adverse effects on the employer's business or on the general conduct of the school or other institution. Since many employers and schools do already make reasonable adjustments, the good practice they have developed should be evaluated and shared more widely. *We recommend that a statement of general principles be drawn up on reasonable accommodation in relation to religious and cultural diversity in the workplace and in schools, and that case-study examples of good practice be provided'.* Parekh Report, 240–241 (emphasis added).

35 See, for instance, Bob Hepple, *Equality: The New Legal Framework* (Oxford: Hart, 2011).

36 In the UK, the *Ladele* case placed the tensions between religious accommodation and increasing marriage equality for same-sex couples on display. *Ladele v Islington LBC* [2009] EWCA Civ 1357; [2010] 1 W.L.R. 955 (CA (Civ Div)), see Russell Sandberg, 'The Right to Discriminate', *Ecclesiastical Law Journal* 13, no. 2 (2011): 157–181.

37 Commission on Religion or Belief in British Public Life, *Living with Difference: Community, Diversity and the Common Good* (Cambridge: The Woolf Institute, 2015), 72; 77 ('We do not recommend any immediate changes in the law with regards to freedom of religion or belief or to discrimination on grounds of religion or belief'.) For more on CORB, see Chapter 2 by Tariq Modood and Chapter 5 by Ralph Grillo in this volume.

38 As Tariq Modood argues in Chapter 2 of this volume.

39 However, this may be an overly optimistic viewpoint, and the report itself recognizes that newcomers are not always familiar with the concept or find it legitimate. For those, it is necessary to promote an understanding of *laïcité* by fighting discrimination and pursuing equal opportunity policies. Stasi Report, 50 (all translations of the Stasi Report are my own).

40 '[T]he emergence of new religious practices requires a renewed application of the principle of *laïcité*'. Ibid., 50.

41 The Stasi Commission contrasts France with the Netherlands, with pillarization, housing segregation and the emergence of 'black schools' since the 1960s. Ibid., 34 et seq. In contrast, it is said the French situation had improved thanks to the rejection of pillarization and a strong French cultural identity, which promotes integration.

42 Ibid., 16.

43 Ibid., 7, para 3; 15, para 2.

44 Talal Asad, 'French Secularism and the "Islamic Veil Affair"', *The Hedgehog Review*, Spring and Summer (2006), 94.

45 Stasi Report, 7.

46 Ibid., 15.

47 See Ibid., 15. 'la protection des religions non-majoritaires permettent d'exiger en retour un effort'.

48 Ibid., 16–17. Elsewhere in the report, 'reasonable accommodation' is used in a more general sense, outside the context of employment. On p. 23, the report notes that France knows 'many reasonable accommodations' which 'temper' the requirements of neutrality and the separation of church and the state. It lists examples, such as the funding of chaplains in the army, hospitals and prisons, taking into account Jewish and Muslim burial practices at secular cemeteries, the institution of free Sundays so that religious education could be organized, and the law of 1987 foreseeing a beneficial tax regime for religious associations.

49 Ibid., 16.

50 Stanley Fish, 'Our Faith in Letting It All Hang Out (op-ed)', *New York Times*, 2006.

51 This notion would serve an important role in the burqa debate in France, which followed almost a decade after the report was published. In ECtHR, *SAS v France*, July 1, 2014, France invoked three grounds for limiting the right to wear a burqa in public spaces under Article 9.2 ECHR: respect for the equality between men and women, for human dignity and for the minimum requirements of life in society. The court did not accept the French government's position that the ban was valid due to gender

equality or human dignity concerns, but nonetheless accepted the French government's claim that a ban was necessary for 'living together' harmoniously, within France's (wide) margin of appreciation. A minority opinion (issued by judges Nussberger of Germany and Jäderblom of Sweden) saw the majority decision 'sacrific[ing] concrete individual rights guaranteed by the Convention to abstract principles'. In contrast to the majority, they found a consensus *against* the necessity of banning full-face veils in European countries and put weight on the fact that human rights institutions and organizations opposed such a ban. Also, they argued that even if 'living together' was a legitimate aim within the meaning of paragraph 2 of Articles 8 and 9, the ban was still a disproportionate measure.

52 French Council of State, May 3, 1950, *Demoiselle Jamet* and advice of May 3, 2000, *Melle Marteaux*. Stasi Report, 22.

53 Stasi Report, 55.

54 On *laïcité* in private sector employment, see the debate surrounding the Baby-Loup day care dispute: Cour de cassation (social chamber), March 19, 2013, n. 11-286845, *Bull.* 2013, V, n. 75; Cour de Cassation (grand chamber), *Mme A. c. Association Babyloup*, June 25, 2014; Anne Jouan, 'Baby-Loup: La Cour de cassation confirme le licenciement de la salariée voilée', *Le Figaro*, June 24, 2014.

55 The 'insistence' of Muslim women to wear a headscarf on the job is qualified as 'auto-discrimination': by insisting on the wearing of a religious symbol, young women are thus seen as depriving themselves of 'any possibility' of employment or, if they are already employed, of promotion and advancement.

56 Stasi Report, 44–45.

57 Ibid., 54. The same approach is taken in the health care context, ibid. 61.

58 Ibid., 62.

59 See a discussion of some case law, including the 2002 case *Tahri contre Téléperformance France* in ibid., 26–27.

60 Ibid., 64.

61 Ibid., 65.

62 Ibid.

63 Ibid., 65.

64 The other commission reports do not mention budgets.

65 Bouchard-Taylor Report, 17.

66 Ibid., 18.

67 Ibid.

68 Ibid., 19 (emphasis added).

69 See ibid., 290. This concept finds its antithesis of sorts in the notion of 'concerted adjustment', i.e. 'all adjustment requests that arise and are handled in the citizen sphere'. A third, less utilized concept is that of 'optional agreements', i.e. 'concerted agreements that are granted by virtue of a non-legal obligation'.

70 This is admittedly ever so cursory in the commission report: 'The notion of accommodation should, strictly speaking, cover the legal obligation overall and thus extend to the citizen sphere. However, *for convenience and in keeping with the choice that we made*, we usually restrict it to the legal sphere'. Bouchard-Taylor Report, 63 (emphasis added).

71 John Griffiths, 'The Social Working of Anti-discrimination Law', in *Non-Discrimination Law: Comparative Perspectives*, ed. M.L.P. Loenen and Peter R. Rodrigues (The Hague: Kluwer, 1999), 313–330; John Griffiths, 'The Social Working of Legal Rules', *Journal of Legal Pluralism* 48, no. 1 (2003): 1–84.

72 '[U]nder the citizen route, managers are encouraged at the outset to handle all requests on an equal footing without wondering about the legal, ethical or other nature of the obligation'.

73 Bouchard-Taylor Report, 65. It is not clear who decides whether this is the case or not. It need not surprise that in practice, this may have significant importance in negotiations on the ground.

74 While this is the case in general, at times a more principled public decision is sought for strategic reasons. See, for example, the background of the seminal *Plessy v. Ferguson* case (1896) in the U.S., which was 'staged' by the New Orleans Comité des Citoyens to challenge racial segregation. See Mark Elliott, *Color-Blind Justice: Albion Tourgée and the Quest for Racial Equality from the Civil War to Plessy v. Ferguson* (New York: Oxford University Press, 2006),

75 On p. 167 the Bouchard-Taylor Report argues that '[j]udicialization *naturally* tends to polarize positions and fuel antagonism between the parties'. I would respectfully disagree; the polarization of parties is not necessarily the result of the judicial process itself, but rather precedes it, especially in the case of the parties who fail to reach agreement and look to legal avenues for a solution. Furthermore, not all legal disputes are social battlefields.

76 Despite the fact that on numerous occasions the report considers religious and cultural accommodations cases to be 'too rare'.

77 Obviously, the fact that multicultural issues find their way into courts does not necessarily point to 'abuse'. First, there are also signs that many other techniques of resolution (or even plainly 'lumping it', 'avoidance' or 'exit') are at play. Second, this could be seen as a natural development in the context of a changing society that is searching for solutions, and allows for the identification of problematic issues. On the other hand, this does not mean adequate access to justice is thereby achieved.

78 Laura Nader, *The Life of the Law: Anthropological Projects* (Berkeley, CA: University of California Press, 2002).

79 See the Bouchard-Taylor Report, 160, noting the changes in Canadian law over the last 25 years towards a conception of equality that 'pays closer attention to differences'. The commission notes that 'this change has taken concrete form, in particular, in a legal tool or provision called reasonable accommodation, which is basically dictated by the general principle of equality and fairness'.

80 Reasonable accommodation for religion in Canada has its origin in employment discrimination, starting with the 1985 *O'Malley* case. However, what triggered the Quebec accommodation crisis (and the focus of the Bouchard-Taylor Report) was not employment, but rather education, health and public services. Reasonable employment accommodation cases, instrumental in developing the legal doctrine, were thus largely insignificant in the public debates. See Sujit Choudhry, 'Rights Adjudication in a Plurinational State: The Supreme Court of Canada, Freedom of Religion, and the Politics of Reasonable Accommodation', *Osgoode Hall Law Journal* 50, no. 3 (2013), 587.

81 In November 2013, the Quebec government introduced a bill to the Quebec National Assembly which aimed at stressing secularism and religious neutrality of the state and limiting religious accommodations within the Quebec Charter of human rights and freedoms (Bill no. 60: 'Charter affirming the values of State secularism and religious neutrality and of equality between women and men, and providing a framework for accommodation requests').

82 Choudhry, 'Rights Adjudication in a Plurinational State', 587.

83 Bouchard-Taylor Report, 39.

84 Rhéal Séguin, 'Secular Charter Toxic to Quebec's Economy, Business Group Warns PQ', *The Globe and Mail*, January 30, 2014 ('Major businesses in Quebec are calling on the Parti Québécois government to withdraw its controversial secular charter bill, saying it is creating a climate of intolerance and exclusion toward immigrants . . . the province's influential business lobby group Conseil du patronat argued that 97 per cent of the major employers it surveyed said they experienced no problems with respect to religious accommodations. The debate over religious symbols was a non-issue, the lobby group said').

85 Perhaps one can say this about the preceding commission reports, but it certainly holds in the case of the Belgian report, which emphasizes that it is the result of a tension-fraught *compromis* exercise.

86 See, for instance, the proposal to adapt the calendar of legal holidays, both the civil calendar and the school calendar, where the Steering Committee proposes that the calendar of legal holidays be reformed as follows: five existing holidays are retained (January 1, May 1, July 21, November 11 and December 25), each person can freely choose two holidays 'depending on his or her culture or religion', and three new non-religious legal holidays are created (some suggestions provided are the international days celebrating equality, diversity and the fight against discrimination).

87 The proposal regarding religious dress in public schools is a compromise, but clearly shows signs of conservatism. The reasonable accommodation 'recommendation' is also an example, see *infra*.

88 Round Tables Report, 20.

89 The parents' religious and parental freedom is apparently left out of the equation.

90 Round Tables Report, 24.

91 'As regards food in the school canteen, the Steering Committee recommends being pragmatic. For example, providing vegetarian menus could be a welcome alternative – albeit not an entirely satisfactory one – for some students'. Ibid., 25.

92 Ibid., 28.

93 Ibid., 29. See, on the legitimacy of neutrality policies under EU anti-discrimination law: CJEU, Case C-157/15, *Achbita and Centrum voor Gelijkheid van kansen en voor racismebestrijding v G4S Secure Solutions NV*, March 14, 2017.

94 Round Tables Report, 33.

95 Adam and Rea, 'Culturele diversiteit op de werkvloer: praktijken van redelijke aanpassing in België'.

96 Discussed in Katayoun Alidadi, 'Studie over redelijke aanpassingen voor religie op Belgische werkvloer', *De Juristenkrant* no. 221 (2011).

97 Surprisingly then, the report identifies this recommendation as one of the seven 'whose symbolic and practical impact is undoubtedly the strongest'. See Round Tables Report, 59–60.

98 Ibid., 34–35.

99 Ibid., 34. ('Accommodations in these areas can take several forms. The first, which is preferred, is what Québecers call the "citizen route" ("*la voie citoyenne*"), and which looks for "concerted adjustment": among people of good will, it should be possible to find solutions that are mutually satisfactory. The second is the "legal route" ("*la voie juridique*"): in the absence of agreement, the request is put before a judge, who rules either to dismiss the request on the grounds that it is "unreasonable", or to impose an accommodation on the other party').

100 Ibid. Some of the 'struggles' of the Committee members reveal their lack of engagement with the Canadian model, and other models, where there is a right to reasonable accommodation, as well as general human rights jurisprudence. E.g. questions such as 'how to evaluate the sincerity of such a belief?' seemed to have been the stopping point for some reflections.

101 To be sure, the Round Tables offer some suggestions: e.g. intercultural mediators could facilitate negotiations on accommodation requests, Belgium could set up a 'perfectly neutral and expert body that can issue quasi-judicial opinions' etc., but these themselves raise many of the same questions. On which non-legal standards would this body rely? Why would someone go to or follow these opinions if no sanctions are available in the eventual case accommodations are arbitrarily but legally rejected?

102 See Robert Mnookin and Lewis Kornhauser, 'Bargaining in the Shadow of the Law: The Case of Divorce', *Yale Law Journal* 88 (1979); Robert H. Mnookin, R. Cooter, and S. Marks, 'Bargaining in the Shadow of the Law: A Testable Model of Strategic Behavior', *Journal of Legal Studies* 11 (1982).

103 See, for instance, the Stasi Report reference to the Netherlands, *supra* note 41.

Bibliography

Adam, Ilke , and Andrea Rea. 'Culturele diversiteit op de werkvloer: praktijken van redelijke aanpassing in België'. Brussels: Research commissioned by the Centre for Equal Opportunities and Opposition to Racism, 2010.

Alidadi, Katayoun. 'Studie over Redelijke Aanpassingen Voor Religie Op Belgische Werkvloer'. [In Dutch]. *De Juristenkrant* 221 (2011): 6–7.

Asad, Talal. 'French Secularism and the "Islamic Veil Affair"'. *The Hedgehog Review* Spring and Summer (2006): 93–106.

Beaman, Lori. 'Introduction: Exploring Reasonable Accommodations'. In *Reasonable Accommodation: Managing Religious Diversity*, edited by Lori Beaman. Vancouver: UBC Press, 2012.

Choudhry, Sujit. 'Rights Adjudication in a Plurinational State: The Supreme Court of Canada, Freedom of Religion, and the Politics of Reasonable Accommodation'. *Osgoode Hall Law Journal* 50, no. 3 (2013): 575–608.

Commission on Religion or Belief in British Public Life. *Living with Difference: Community, Diversity and the Common Good.* Cambridge: The Woolf Institute, 2015.

Commission on the Future of Multi-Ethnic Britain. *The Future of Multi-Ethnic Britain.* London: Profile Books, 2000.

Elliott, Mark. *Color-Blind Justice: Albion Tourgée and the Quest for Racial Equality from the Civil War to Plessy v. Ferguson.* New York: Oxford University Press, 2006.

Fish, Stanley. 'Our Faith in Letting It All Hang Out (Op-Ed)'. *New York Times*, 2006.

Griffiths, John. 'The Social Working of Anti-Discrimination Law'. In *Non-Discrimination Law: Comparative Perspectives*, edited by M.L.P. Loenen and Peter R. Rodrigues. The Hague: Kluwer, 1999.

Griffiths, John. 'The Social Working of Legal Rules'. *Journal of Legal Pluralism* 48, no. 1 (2003): 1–84.

Grim, Brian J. 'Religion, Law and Social Conflict in the 21st Century: Findings from Sociological Research'. *Oxford Journal of Law and Religion* 1, no. 1 (April 1, 2012): 249–271.

Hepple, Bob. *Equality: The New Legal Framework.* Oxford: Hart, 2011.

Jouan, Anne. 'Baby-Loup: La Cour de cassation confirme le licenciement de la salariée voilée'. *Le Figaro*, June 24, 2014.

Khalida, Khan. 'Where's the Muslim in Macpherson's Black and White Britain?' *Q News*, March 1999, 26–27.

Mnookin, Robert H., R. Cooter, and S. Marks. 'Bargaining in the Shadow of the Law: A Testable Model of Strategic Behavior'. *Journal of Legal Studies* 11 (1982): 225–251.

Mnookin, Robert, and Lewis Kornhauser. 'Bargaining in the Shadow of the Law: The Case of Divorce'. *Yale Law Journal* 88 (1979): 950–997.

Nader, Laura. *The Life of the Law: Anthropological Projects.* Berkeley, CA: University of California Press, 2002.

Perrault, Guillaume. 'L'Alsace-Moselle garde le concordat'. *Le Figaro*, February 22, 2013.

Séguin, Rhéal. 'Secular Charter Toxic to Quebec's Economy, Business Group Warns PQ'. *The Globe and Mail*, January 30, 2014.

Index